Analysis of Structural Composite Materials

MONOGRAPHS AND TEXTBOOKS IN MATERIAL SCIENCE

Edited by

R. F. Bunshah
UNIVERSITY OF CALIFORNIA
LOS ANGELES, CALIFORNIA

A. Sosin
UNIVERSITY OF UTAH
SALT LAKE CITY, UTAH

1. METAL DEFORMATION PROCESSES: FRICTION AND LUBRICATION
   Edited by John A. Schey
2. CONCEPTS OF FIBER-RESIN COMPOSITES
   R. L. McCullough
3. OPTICAL MICROSCOPY FOR THE MATERIALS SCIENCES
   James H. Richardson
4. LIQUID METALS: CHEMISTRY AND PHYSICS
   Edited by Sylvan Z. Beer
5. ATLAS OF BINARY ALLOYS: A PERIODIC INDEX
   Karl P. Staudhammer and Lawrence E. Murr
6. ANALYSIS OF STRUCTURAL COMPOSITE MATERIALS
   Sabodh K. Garg, Vytas Svalbonas, and Gerald A. Gurtman
7. STRESS WAVE PROPAGATION IN SOLIDS: AN INTRODUCTION
   Richard J. Wasley

Additional volumes in preparation

Analysis of Structural Composite Materials

Sabodh K. Garg
Systems, Science and Software
La Jolla, California

Vytas Svalbonas
Grumman Aerospace Corp.
Bethpage, New York

Gerald A. Gurtman
Systems, Science and Software
La Jolla, California

MARCEL DEKKER, INC. New York 1973

MARCEL DEKKER, INC.
95 Madison Avenue, New York, New York 10016

LIBRARY OF CONGRESS CATALOG CARD NUMBER 72-97478

ISBN 0-8247-6050-6

PRINTED IN THE UNITED STATES OF AMERICA

CONTENTS

PREFACE

Within the past two decades, the availability, diversity, and utilization of structural composite materials have increased at a remarkable rate. The field has undergone enormous advances, both in the design and in the fabrication of high-performance composites. Much of this progress is due to the relatively recent development of high-modulus fibers and to the urgent need for light-weight high-strength structures by aerospace and related industries. The availability of these materials has in turn encouraged the development of theoretical models capable of describing composite material behavior under static and dynamic loads.

The purpose of this book is to present in a systematic way an account of various analysis techniques. It was, of course, impossible to treat all (or even most) of the areas in which research is currently active. Therefore, the selection of topics reflects, to a certain degree, the prejudices and personal interests of the authors.

Considerable space has been devoted to statistical theories of fibrous composite tensile strength, and to continuum theories for wave propagation in particle- and fiber-reinforced composites, and in laminated materials. The elastic properties of these media are also characterized. Attention has also been given to nonstatistical theories for strength of fiber-reinforced composites and laminated media. On the other hand, fabrication techniques, fatigue, and wave propagation in media with random inhomogeneities are not discussed.

The book may be divided into three parts. The first (Chapters 1-5) deals with the elastic properties of fiber- and particle-reinforced composites, and of laminated media. Both statistical and nonstatistical theories of strength for fibrous composites and laminated materials are discussed in Chapters 6-11. For ready reference, an appendix containing the pertinent statistical and probabilistic concepts is also included. Finally, Chapters 12-15 treat elastic and shock wave propagation in fiber- and particle-reinforced composites, and in laminated plates.

Although the book is primarily intended for the analyst rather than the designer, numerous comparisons of theoretical models with experimental data are provided. This information may be utilized to generate design data. It is hoped that the book will prove useful to both students and research workers in applied mechanics, aerospace, and civil engineering.

For their kindness and help in various matters, we should like to thank Miss Andreé Coste, Miss Corazon I. de Juan, Dr. R. A. Kruger, Miss Lynne S. Lance, Mr. J. W. Pritchett, Dr. T. D. Riney, Dr. J. M. Walsh, and Mrs. Adrienne E. Wheat. In addition, Vytas Svalbonas wishes to dedicate this book to his wife Zigrida for her continued encouragement and help.

Analysis of Structural Composite Materials

CHAPTER 1

WHY COMPOSITES?

I. INTRODUCTION

Composite materials have been in existence for centuries. In fact, there are extant fragments of laminated wood dating about 2780 B.C., the time of the third Egyptian dynasty. In the broadest sense, well-known materials such as glass, plywood, concrete, and dispersion-hardened metal alloys are composites. These have been used in a wide variety of man-made structures: from houses and bridges to furniture, from laminated tennis racquets and skis to Damascus swords, and from optical lenses to arch dams. However, the idea of designing a composite for a specific purpose is comparatively recent.

In the past, the engineer was forced to choose among available materials to design a particular structure. Soon, however, he may be in a position to prescribe material properties in the expectation that an appropriate composite will be fabricated. For example, consider the case of a proposed laminate composite for future building applications. It has been suggested [1] that such a composite might have an outside layer consisting of a structural and insulating sandwich covered with an attractive weather-resistant film. In addition, it could have an inside skin consisting of a metallic sheet to provide either radiant heat or cooling (depending upon the direction of electric current) along with a fluorescent material for internal lighting. The task of fabricating such a composite is formidable. As another example, it has been proposed that costs may be reduced by replacing solid titanium with titanium clad steel for applications requiring strength and corrosion resistance. Bonding of these materials has to date been difficult. Thus, while one may readily visualize the advantages of replacing high-strength castings or forgings with light-weight fiber-reinforced composites or honeycomb sandwich materials, fabrication techniques for complex structures are, at present, not well developed. Evidently, there is a hiatus between proposing an ideal composite and the ability to produce it and put it to use.

Lest the impression be given that composites are completely theoretical materials, let us examine where they may be and have been put to use. The main structural problems which these materials have overcome are those involving strength, weight, and stiffness. Thus they have been used in the construction of ships, submarines, aircraft radomes, wing and empennage structures, and helicopter blades. The use of composites increases the capacity of a commercial plane for passengers or cargo; additional fuel, armament, or sophisticated electronic equipment may be carried by military aircraft. Composite helicopter blades provide lower inertia loads while maintaining stiffness and providing good damping and fatigue properties.

Materials which can provide strength and are at the same time corrosion-resistant, rot-proof, thermally insulating, dielectric, and nonmagnetic are obviously desirable in ship construction. This accounts for the widespread use of fiberglass in the fabrication of small boats. In spacecraft, every pound of structural weight saved by the use of composites provides an additional pound of payload. Composites are also utilized in heat shield construction for reentry vehicles, where the material's thermal behavior properties are of paramount importance. As a further example of space applications, the wall construction of a proposed lunar shelter (shown in Fig. 1) may be contrasted with the hypothetical general construction laminate discussed above. This construction composite combines fabrics, plastics, and metal fibers to perform its required functions [2]. Elastic recovery of the flexible foam, compressed in a container for storage, erects and maintains the cylindrical shape of the shelter without internal pressurization. The foam also provides primary meteoroid protection. Since

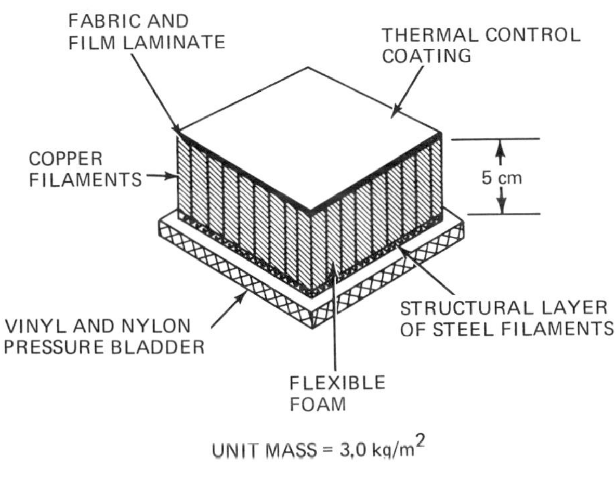

FIG. 1. Wall construction of lunar shelter. (Reproduced from Ref. [2].)

the foam is an excellent insulator, copper filaments* are required for proper thermal balance between the inner and outer surfaces. Steel wires provide the load-carrying strength for the internal pressure bladder, and the fabric outer surface, which provides a meteoroid bumper and container, has a suitable thermal control coating. The filamentary or matrix materials used in the shelter are all within the current state of the art.

Filamentary composites have not been thus far utilized to any great extent outside the aerospace and allied fields, due to the relative newness of the technology, limited design methods, occasional low reliability, high cost, and low producibility. All these problems are characteristic of an immature technology; future research and development will presumably overcome the present obstacles. Indeed, some of these problems are already being eliminated; Figure 2 illustrates the spectacular decline of the costs of certain composite materials. Present limited design methods and lack of reliability and confidence are also due to the paucity of test data on composite materials. This, too, is changing; even standard building materials are feeling the impact of new technology, as, for example, in the creation and study of fiberglass-reinforced concrete [4-7] or concrete-polymer composites [8-10]. The aforementioned hypothetical construction laminate is somewhat more sophisticated than materials presently in common use. Nonetheless, even though the wall unit shown in Fig. 3 is not as elaborate as the earlier proposal, it is here now and in current use.

* In this book, the words "filament" and "fiber" will be employed interchangeably.

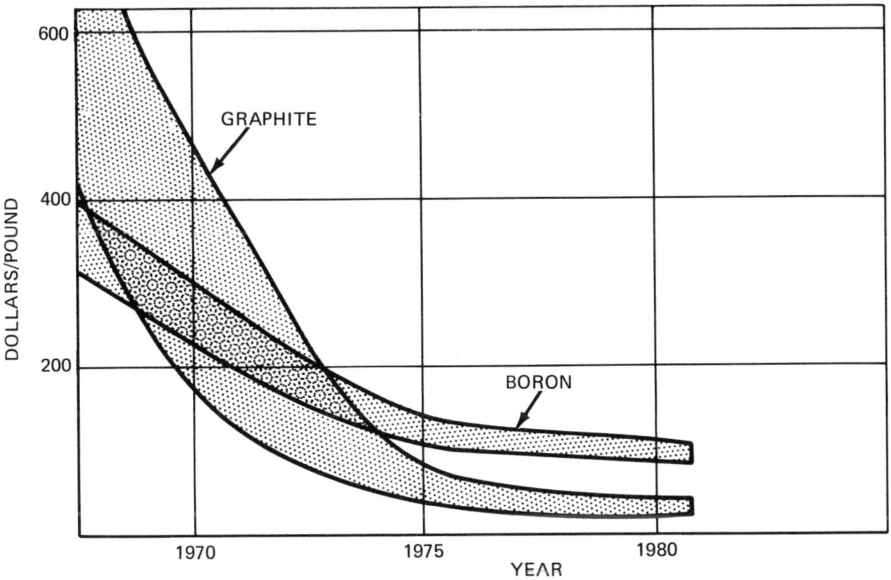

FIG. 2. Projected cost of boron and graphite filaments in the 1970s. (Reproduced from Ref. [3].)

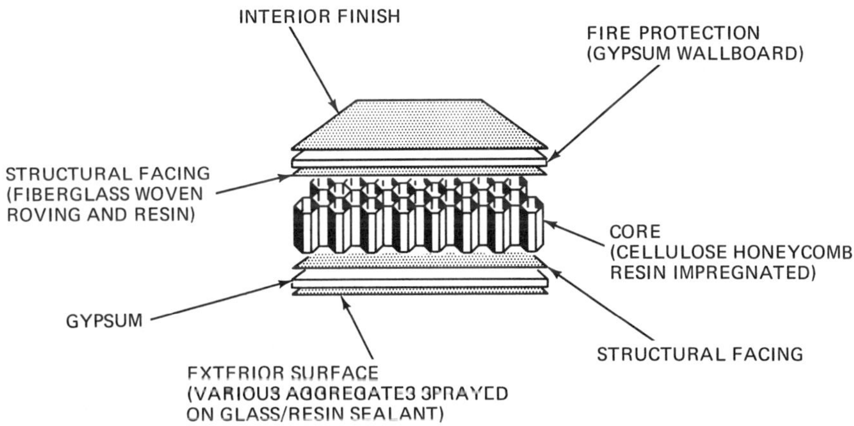

FIG. 3. Fiber-shell is a composite material which is the basis for TRW's building system. Company offers the materials in three ways: precut panels, unfinished shells, and completely finished modules. (Reproduced from Ref. [11].)

II. SCOPE OF THE BOOK

The above remarks are intended to show that composites are not theoretical materials. The principal obstacles to their widespread use are fabrication problems of various kinds. These problems, however, are deemed to be outside the scope of the present text, wherein we will concentrate upon the mathematical prediction and analysis of the behavior of a given composite. The various approaches to an understanding of composite behavior fall into two categories: those that are empirical and those that are analytical.

In the empirical approach, the composite is viewed as if it were a single material. If the composite is approximately isotropic, then its average properties may be substituted into the usual design formulas. A more sophisticated approach is required for anisotropic materials. Empirical descriptions are necessary whenever there is insufficient data to sustain an analytical model. However, since one knows nothing regarding the internal behavior of the composite, this viewpoint has major limitations. Specifically, for inhomogeneous materials, test data cannot be safely applied to novel situations; large safety factors are necessary to compensate for ignorance of the real material behavior. The empirical approach gives no basis upon which to recommend, quantitatively, any change in design which might result in a more efficient utilization of the composite [12].

The analytical approach, on the other hand, begins with a study of the individual constituent materials. The aim here is to find how the behavior of the composite depends on the composition, size, volume fraction, distribution, and orientation of the constituents. General relationships of this sort will then provide predictions of the behavior of a composite resulting from an arbitrary combination of these constituents. In common with the empirical approach, the analytical approach aims at predicting "effective" material properties that may be utilized in conventional design formulas. However, in this case, a clear distinction is made between the behavior of the "real" and "effective" materials. In practice, of course, the two approaches complement each other. Much effort is being expended today in developing models for materials that heretofore could only be treated empirically. But even with better analytical models, empirical verification of composite behavior will be required at least in the foreseeable future.

The present book will be primarily concerned with the structural theory of composite materials. In order to further define the scope, it is necessary to introduce some basic composite vocabulary. "Phase" denotes one of the materials comprising the composite. Thus, a two-phase composite is a composite made up from two materials. The dominant or containing phase is called the "matrix", while the minor or contained phase is the "reinforcement." Thus, one can have particle, flake, or fiber reinforcement of a composite. The laminates, or filled composites (e.g., filled honeycomb) cannot be truly characterized in this manner. The discussion will be limited to particle- and fiber-reinforced composites, and laminates.

Important examples of particle-reinforced composites are dispersion-hardened alloys, cermets, and concrete. The distribution of the particles is usually random, and the particulate composites are macroscopically isotropic. The size and nature of the particles are determined by the intended application.

On the other hand, fiber-reinforced composites are usually anisotropic. It is a well-known fact that materials are stronger in fiber form than in bulk form. This property cannot be usefully exploited unless the fibers are held together by a matrix material. In a fiber-reinforced composite, the fibers carry the bulk of the applied load, and the matrix serves to bind the fibers together, to space them, to distribute the load to individual fibers, and to protect them from mechanical or chemical damage. In selecting a matrix material for a fiber-reinforced composite, it is necessary to ensure that the fibers are properly bonded and are in total contact with the matrix. Furthermore, the matrix must not react chemically with the fibers and should have adequate shear strength and ductility. Great care must also be taken in handling the fibers to avoid surface damage, since this may result in significant weakening. The major advantage of fiber-reinforced composites lies in their anisotropy. They offer the greatest improvements in uniaxial strength and stiffness-to-weight ratios. In Fig. 4, the typical specific strengths achievable for fibrous composites are shown as a function of specific modulus.

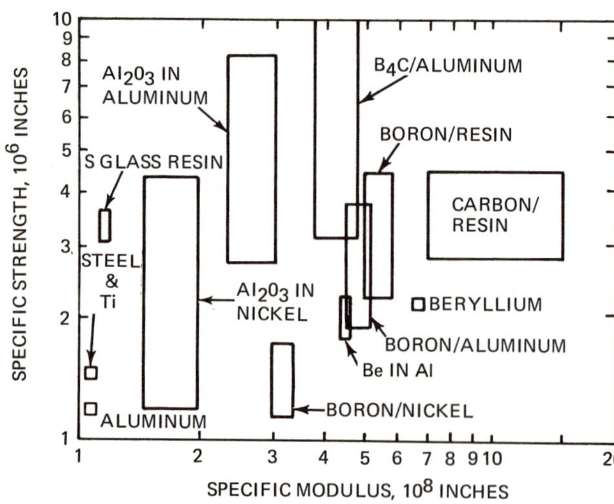

FIG. 4. Potential room temperature strength properties of conventional alloys and advanced fiber reinforced composites. (Reproduced from Ref. [13].)

This book will deal almost exclusively with two-phase composites. In the first part (Chapters 2-5), describing macroscopic material properties, both the reinforcement and the matrix phases will be treated in a rather general way. Various analytical models will be discussed along with relevant experimental data. A major portion of the second part (Chapters 6-11), dealing with fracture mechanics of unidirectionally fiber-reinforced and laminated composites, is devoted to tensile strength analysis. The reinforcing filaments of interest, e.g., glass and graphite, are usually brittle, with a wide scatter in their tensile strength values. Thus, statistical representations become useful to describe these materials. For ready reference, an appendix containing the pertinent statistical concepts is provided. Finally, in the third and last part (Chapters 12-15), the dynamic response of composites is considered.

REFERENCES

1. W. Lubars, "Laminar Composites," Materials in Design Eng., 58, 109-118, September 1963.

2. R. Heldenfels, "Applications of Composite Materials in Space Vehicle Structures," Proc. Fifth Symp. Naval Structural Mechanics, Pergamon, New York, 1967, pp. 157-174.

3. A. M. Lovelace and S. W. Tsai, "Composites Enter the Mainstream of Aerospace Vehicle Design," Astronautics and Aeronautics, 8 (7), 56-61 (1970).

4. E. Nawy, G. Neuwerth, and C. Phillips, "Behavior of Fiber-Glass Reinforced Concrete Beams," ASCE J. Structural Div., 97 (8), 2203-2215 (1971).

5. I. A. Rubinsky and A. Rubinsky, "A Preliminary Investigation of the Use of Fiber Glass for Prestressed Concrete," Magazine of Concrete Research, 6 (17), 71-78, September 1954.

6. J. Wines and G. Hoff, "Laboratory Investigation of Plastic-Glass Fiber Reinforcement for Reinforced and Prestressed Concrete, Report #1," U. S. Army Waterways Experiment Station, February 1966.

7. J. Wines, R. Dietz, and J. Hawley, "Laboratory Investigation of Plastic-Glass Fiber Reinforcement for Reinforced and Prestressed Concrete, Report #2," U. S. Army Waterways Experiment Station, February 1966.

8. L. Kukacka, P. Colombo, M. Steinberg, and B. Manowitz, "Concrete-Polymer Composites," ASCE J. Structural Div., 97 (9), 2217-2227 (1971).

9. M. Steinberg, L. Kukacka, P. Colombo, J. J. Kelsch, and B. Manowitz, "Concrete-Polymer Materials, First Topical Report," Brookhaven National Lab., BNL-50134 (T-509) and USBR General Report 41, December 1968.

10. M. Steinberg, L. Kukacka, P. Colombo, A. Auskern, and B. Manowitz, "Concrete-Polymer Materials, Second Topical Report," Brookhaven National Lab., BNL-50218 (T-560) and REC OCE 70-1, December 1969.

11. R. B. Aronson,"Industrialized Housing, a New Kind of Cave," Machine Design, 43 (26), 20-22 and 24-26, October 28, 1971.

12. J. Campbell, "Design of Composites," Materials in Design Eng., 58, 118-126, September 1963.

13. W. Sutton and H. Rauch, Sr., "Review of Current Developments in New Refractory Fibers and Their Utilization as High-Temperature Reinforcements," General Electric Space Sciences Lab. Report, R66 SD 46, August 1966.

CHAPTER 2

SIMPLE ANALYTIC MODELS

I. INTRODUCTION

In order to analyze the mechanical behavior (deformation, failure strength, etc.) of composite materials, a mathematical model of the material must first be formulated. In this and several of the following chapters, we shall present the various models that can be utilized to analyze the material properties of the composite, and thus predict its stress-strain behavior. These models shall be introduced in order of increasing complexity and accuracy of prediction of results.

II. NETTING-TYPE ANALYSES

The simplest analytic model is obtained by only considering the fibers and ignoring any effect of the binder phase. It is perhaps obvious that such a model has certain drawbacks. For example, if the composite is a sheet with all the fibers aligned in one direction, the netting analysis predicts zero strength in the direction at right angles to the fibers. This model was first developed by Cox [1] and applied to fibrous felts and paper. The mathematical analysis for this model is based upon the following assumptions: (a) long, straight, thin fibers (diameter neglected), (b) load applied only at the fiber ends (no interfacial forces), and (c) no bending stiffness for the fibers.

Consider a planar mat of fibers subjected to tensile strains ε_{11} and ε_{22} in two directions at right angles to each other (Fig. 1) and to a shear strain γ_{12} between these directions. The strain of a fiber inclined at an arbitrary angle θ to the direction x_1 is [2]

$$\varepsilon_{11} \cos^2\theta + \varepsilon_{22} \sin^2\theta + \gamma_{12} \sin\theta \cos\theta$$

The stress in the fiber is assumed to be proportional to this strain. If the load in the fiber is L, then the contribution of the fiber to the loads in the directions x_1 and x_2 will be $L \cos\theta$ and $L \sin\theta$, respectively.

Let $f(\theta)$ be the distribution function, i.e., the fraction of fibers inclined at angle θ to the direction

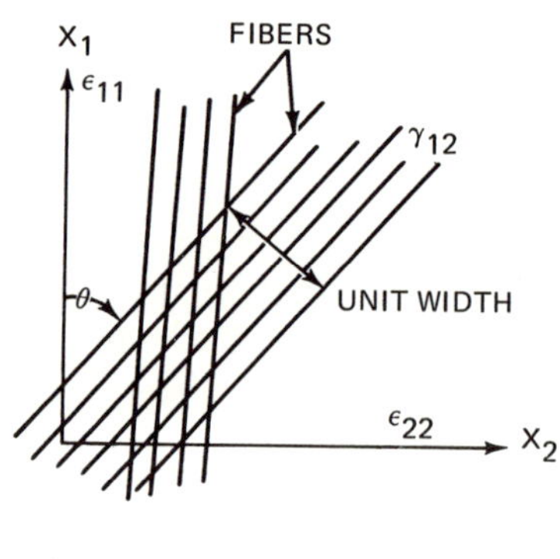

FIG. 1. Planar mat of fibers.

x_1 in the unit width transverse to their direction such that

$$\int_0^{\pi} f(\theta)\, d\theta = 1 \tag{1}$$

The fractions of fibers intersecting lines of unit width perpendicular to the directions x_1 and x_2 are then $f(\theta) \cos\theta$ and $f(\theta) \sin\theta$, respectively. Therefore, the loads per unit width of edge, i.e., σ_{11} (along x_1, on the edge perpendicular to x_1), σ_{22} (along x_2, on the edge perpendicular to x_2), and τ_{12} (along x_1, on the edge perpendicular to x_2, or vice versa) are

$$\begin{aligned}
\sigma_{11} &= E_f V_f \int_0^{\pi} \Big(\varepsilon_{11} \cos^2\theta + \varepsilon_{22} \sin^2\theta \\
&\qquad + \gamma_{12} \cos\theta \sin\theta\Big) \cos^2\theta\, f(\theta)\, d\theta \\
\sigma_{22} &= E_f V_f \int_0^{\pi} \Big(\varepsilon_{11} \cos^2\theta + \varepsilon_{22} \sin^2\theta \\
&\qquad + \gamma_{12} \cos\theta \sin\theta\Big) \sin^2\theta\, f(\theta)\, d\theta \\
\tau_{12} &= E_f V_f \int_0^{\pi} \Big(\varepsilon_{11} \cos^2\theta + \varepsilon_{22} \sin^2\theta \\
&\qquad + \gamma_{12} \cos\theta \sin\theta\Big) \sin\theta \cos\theta\, f(\theta)\, d\theta
\end{aligned} \tag{2}$$

where E_f is the fiber modulus, and V_f is the fiber packing density, i.e., the ratio of fiber volume to the volume of the composite. Alternately, Eqs. (2) can be rewritten as

$$\begin{aligned}\sigma_{11} &= C_{11}\varepsilon_{11} + C_{12}\varepsilon_{22} + C_{16}\gamma_{12} \\ \sigma_{22} &= C_{12}\varepsilon_{11} + C_{22}\varepsilon_{22} + C_{26}\gamma_{12} \\ \tau_{12} &= C_{16}\varepsilon_{11} + C_{26}\varepsilon_{22} + C_{66}\gamma_{12}\end{aligned} \tag{3}$$

where

$$\begin{aligned}C_{11} &= E_f V_f \int_0^{\pi} \cos^4\theta\ f(\theta)\ d\theta \\ C_{16} &= E_f V_f \int_0^{\pi} \cos^3\theta\ \sin\theta\ f(\theta)\ d\theta \\ C_{22} &= E_f V_f \int_0^{\pi} \sin^4\theta\ f(\theta)\ d\theta \\ C_{26} &= E_f V_f \int_0^{\pi} \sin^3\theta\ \cos\theta\ f(\theta)\ d\theta \\ C_{12} = C_{66} &= E_f V_f \int_0^{\pi} \cos^2\theta\ \sin^2\theta\ f(\theta)\ d\theta\end{aligned} \tag{4}$$

The analysis for three-dimensional structures is carried out in a similar manner. The strain ε on a single fiber at polar angles θ, ϕ (Fig. 2) is

$$\varepsilon = \varepsilon_{11} \cos^2\theta + \varepsilon_{22} \cos^2\phi \sin^2\theta + \varepsilon_{33} \sin^2\phi \sin^2\theta$$

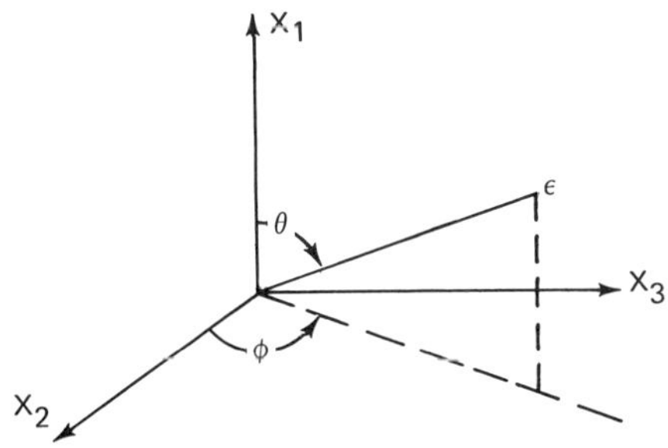

FIG. 2. Three-dimensional coordinate system.

$$+ \gamma_{23} \sin\phi \cos\phi \sin^2\theta + \gamma_{31} \sin\phi \sin\theta \cos\theta$$

$$+ \gamma_{12} \cos\phi \sin\theta \cos\theta \qquad (5)$$

Assuming $f(\theta, \phi)$ as the distribution function, the expressions for the stresses σ_{ij} can be written in a fashion similar to that for the two-dimensional case [3]:

$$\begin{Bmatrix} \sigma_{11} \\ \sigma_{22} \\ \sigma_{33} \\ \tau_{23} \\ \tau_{31} \\ \tau_{12} \end{Bmatrix} = \begin{bmatrix} C_{11} & C_{12} & C_{13} & C_{14} & C_{15} & C_{16} \\ & C_{22} & C_{23} & C_{24} & C_{25} & C_{26} \\ & & C_{33} & C_{34} & C_{35} & C_{36} \\ & & & C_{44} & C_{45} & C_{46} \\ \text{SYM.} & & & & C_{55} & C_{56} \\ & & & & & C_{66} \end{bmatrix} \begin{Bmatrix} \varepsilon_{11} \\ \varepsilon_{22} \\ \varepsilon_{33} \\ \gamma_{23} \\ \gamma_{31} \\ \gamma_{12} \end{Bmatrix} \qquad (6)$$

where the elastic constants C_{ij} are similarly given by

$$C_{11} = E_f V_f \int_0^{2\pi} \int_0^{\pi/2} \cos^4\theta \, f(\theta, \phi) \, d\theta \, d\phi$$

etc.

Thus, from a knowledge of the distribution functions $f(\theta)$ or $f(\theta, \phi)$, the elastic constants C_{ij} for the composite may be calculated in a simple manner. For the isotropic, two-dimensional case with a random distribution of fibers, the distribution function $f(\theta)$ used is

$$f(\theta) = 1/\pi, \quad 0 \leq \theta \leq \pi \tag{7}$$

Equation (7) represents a rectangular probability density function, and implies that the fibers are uniformly distributed in the interval $0 \leq \theta \leq \pi$. Then, from Eq. (4), the elastic constants are found to be

$$C_{11} = C_{22} = (3/8)E_f V_f, \quad C_{12} = (1/8)E_f V_f$$

$$C_{16} = C_{26} = 0$$

This yields for the composite

$$\begin{aligned} &\text{Young's modulus} \quad E_c = C_{11} - \left(\frac{C_{12}^2}{C_{22}}\right) = E_f(V_f/3) \\ &\text{Shear modulus} \quad G_c = C_{12} = E_f(V_f/8) \\ &\text{Poisson's ratio} \quad \nu_c = \left(\frac{E_c}{2G_c}\right) - 1 = 1/3 \end{aligned} \tag{8}$$

For the three-dimensional random array, the distribution function $f(\theta, \phi)$ used is

$$f(\theta, \phi) = (\sin\theta)/2\pi, \quad 0 \leq \theta \leq \pi/2, \quad 0 \leq \phi \leq 2\pi \tag{9}$$

This gives the following values for the elastic constants of the composite:

$$\begin{aligned} &\text{Young's modulus} \quad E_c = E_f(V_f/6) \\ &\text{Shear modulus} \quad G_c = E_f(V_f/15) \\ &\text{Poisson's ratio} \quad \nu_c = 1/4 \end{aligned} \tag{10}$$

Although the netting-type analyses do not take into account the effects of the matrix, they are useful for the analysis and design of a large class of spun or filament-wound pressure vessels [4]. By the very nature of their geometry, these filament-wound structures eliminate major drawbacks of the simple netting analysis model, and are often designed by its use. The major advantage of the filament-winding technique of construction is that a high structural efficiency and strength optimization can be achieved. This is made possible by orienting the fibers in the direction of principal loads and thereby utilizing most of the potential strength of the material. As an example, consider a cylindrical pressure vessel. The ratio of the hoop stresses to the longitudinal stresses is 2:1. When fabricating a cylinder of an isotropic material, the cylinder is naturally overdesigned for the longitudinal loads. The over-design condition can be corrected in filament-wound cylinders by placement of the strands in such a manner that the load-carrying potential of the strands is the same as the load requirements. This balanced strength can be achieved by winding parallel to the axis of the vessel until the required longitudinal strength is obtained, and then by winding perpendicular to the vessel axis for hoop strength. This, however, is not a practical approach for cylinders with integral domes at each end, since openings in the ends must be provided for fittings and mandrel shafts, etc. Therefore, the longitudinal windings are offset at some helix angle θ to the vessel axis, depending upon the vessel and dome geometry. A helical path is generated on the surface of the closed-ended cylinder. This generation of one complete cycle around the vessel is called a loop. Depending upon the helix angle and the path taken across the domes, the completion of one loop may or may not end at the starting point. It is evident, however, that for most specific paths, some Nth loop will terminate at the starting point. At this point, the winding becomes repetitive, and the full generation of the N loops is called a "pattern." A four-loop pattern is shown in Fig. 3. For an efficient structure, it is desirable to minimize "crossover" points, since these points are sources of stress concentrations. Hence, a low number of loops is always sought in a winding pattern.

For the cylinder, the total load L carried by the reinforcement longitudinal windings at angle θ is resolved into a longitudinal load $L \cos \theta$ and a hoop load $L \sin \theta$. This decrease in the longitudinal load carried by the reinforcement, from L to $L \cos \theta$, requires additional longitudinal windings. On the other hand, the superimposed hoop windings can be reduced, since a portion of the hoop load ($L \sin \theta$) is carried by the longitudinal windings. By increasing the helix angle θ to 54.7 deg, the addition of hoop windings becomes unnecessary. It can be shown, however, that no weight advantage results from all-helical winding or a combination wrap [4]. Also, the total thickness is independent of helix angle. A more detailed mathematical description of filament-wound pressure vessels is delayed until Chapter 5, where it will be considered along with other laminated structures.

III. LAW OF MIXTURES

Although the netting analysis is adequate for the design of a certain class of structures, it leads to erroneous results in many cases. As an example, it predicts zero strength transverse to the direction of the reinforcing fibers. In the netting analysis, the interfacial stresses between the fiber and the matrix are assumed to be nonexistent. In actuality, interfacial stresses enable the matrix to transfer and distribute

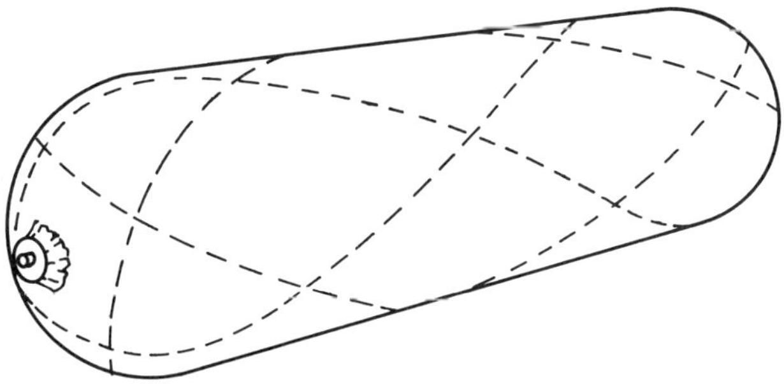

FIG. 3. A 4-loop winding pattern.

loads between the fibers. These limitations of the netting analysis were recognized by Gordon [5] and Arridge [3], who proposed several modifications to account for the fiber-matrix interaction.

Consider a composite made up of a cylindrical fiber with Young's modulus E_f surrounded by a tubular matrix with Young's modulus E_m (Fig. 4). Assuming equal strain in the components under a uniform axial load, one has

$$\varepsilon_f = \varepsilon_m = \varepsilon_c, \quad \sigma_f V_f + \sigma_m(1 - V_f) = \sigma_c \tag{11}$$

where the subscripts f and m refer to the fiber and the matrix, respectively, and V_f denotes the fraction of

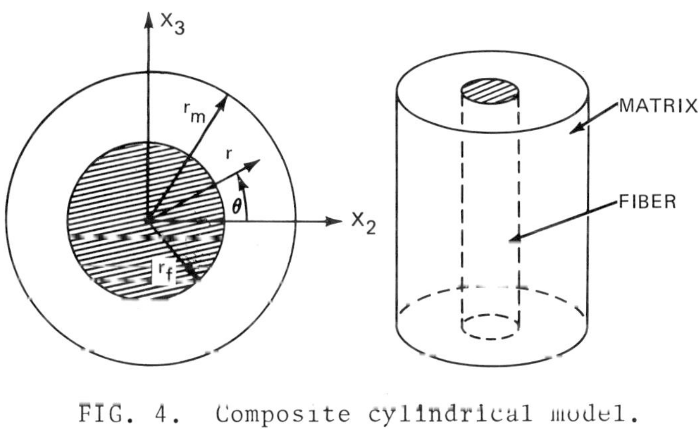

FIG. 4. Composite cylindrical model.

the total composite cross-sectional area occupied by the fiber. Equations (11) may be combined with Hooke's law to yield the following expression for the elastic modulus E_c of the composite:

$$E_c = V_f E_f + (1 - V_f)E_m \tag{12}$$

This is the well-known law of mixtures formula. It will be shown in Chapters 3 and 4 that it serves as a good approximation to the Young's modulus for a unidirectional composite in the direction of the fibers. In the derivation of Eq. (12), the radial forces generated due to Poisson's ratio differences in the two materials were neglected. Equation (12) may be rewritten in the form

$$E_c = V_f[(k - 1)/k]E_f + (E_f/k) \tag{13}$$

where k is the ratio of the fiber modulus to the matrix modulus. Equation (13) can be interpreted to imply that the fiber modulus is reduced in the ratio $[(k - 1)/k]$ by the presence of the matrix. It is now assumed that this reduction in fiber modulus applies to the three-dimensional isotropic problem also. Thus, the stresses σ_{ij} on a unit cube consist of two independent components, i.e., $[(k - 1)/k]C_{ijk\ell}\,\varepsilon_{k\ell}$ from the fibers and $(2G_m\varepsilon_{ij} + \lambda_m e\,\delta_{ij})$ from the matrix, where G_m and λ_m are the Lamé constants for the matrix, δ_{ij} is the Kronecker delta

$= 1$ if $i = j$

$= 0$ if $i \neq j$

e is the dilatation $(\varepsilon_{11} + \varepsilon_{22} + \varepsilon_{33})$, and the $C_{ijk\ell}$ are the elastic constants derived by Cox [1]. Therefore,

$$\sigma_{ij} = \frac{k - 1}{k} C_{ijk\ell}\varepsilon_{k\ell} + \frac{E_f}{k(1 + \nu_m)}\varepsilon_{ij} + \frac{\nu_m E_f e\,\delta_{ij}}{k(1 + \nu_m)(1 - 2\nu_m)} \tag{14}$$

where ν_m denotes the Poisson's ratio for the matrix. Equation (14) enables the calculation of the total stress in a fiber-reinforced material. Thus, for a three-dimensional random network, we obtain, using Eq. (6)

$$\sigma_{11} = \frac{k - 1}{k} \frac{E_f V_f}{5} \left(\varepsilon_{11} + \frac{1}{3} \varepsilon_{22} + \frac{1}{3} \varepsilon_{33}\right) + \frac{E_f}{k(1 + \nu_m)} \varepsilon_{11} + \frac{\nu_m E_f (\varepsilon_{11} + \varepsilon_{22} + \varepsilon_{33})}{k(1 + \nu_m)(1 - 2\nu_m)} \tag{15}$$

with similar expressions for σ_{22} and σ_{33}. To find the modulus in a single direction, we must describe a uniaxial tension problem. Assuming $\sigma_{22} = \sigma_{33} = 0$ and solving for ε_{11}, the following expression for Young's modulus in the x_1-direction is obtained:

$$E_{c1} = a - \left[\frac{2b^2}{(a + b)}\right]$$

where

$$a = \frac{k - 1}{5k} E_f V_f + \frac{E_f}{k(1 + \nu_m)} + \frac{\nu_m E_f}{k(1 + \nu_m)(1 - 2\nu_m)}$$

$$b = \frac{k - 1}{15k} E_f V_f + \frac{\nu_m E_f}{k(1 + \nu_m)(1 - 2\nu_m)} \tag{16}$$

For $\nu_m = 1/4$, the modulus becomes

$$E_{c1} = \frac{E_f V_f}{6} + \frac{E_f (6 - V_f)}{6k} \tag{17}$$

which, in the limit as $k \to \infty$, reduces to the result of Cox [Eq. (10)].

Arridge [3] also introduces a second type of approximation. The stress on a unit cube can also be written as

$$\sigma_{ij} = V_f \sigma'_{ij} + (1 - V_f)\sigma''_{ij} \tag{18}$$

where σ'_{ij} is the stress due to the fibers and σ''_{ij} is the stress due to the matrix. Utilizing Hooke's law, we obtain

$$\sigma_{ij} = V_f C_{ijk\ell}\varepsilon_{k\ell} + (1 - V_f)\left[\frac{E_f \varepsilon_{ij}}{k(1 + \nu_m)} + \frac{\nu_m E_f e\ \delta_{ij}}{k(1 + \nu_m)(1 - 2\nu_m)}\right] \tag{19}$$

Proceeding as before, this leads to the following expression for Young's modulus in the x_1-direction (with $\nu_m = 1/4$):

$$E_{c1} = \frac{E_f V_f}{6} + \frac{E_f(1 - V_f)}{k} \tag{20}$$

which again reduces to Cox's results for a random felt, in the limit as $k \to \infty$.

Although the two equations (17) and (20) both reduce to Cox's results in the limit at $k \to \infty$ and both give approximately the same results when V_f is small, for a larger concentration of fibers, however, the answers will vary significantly. This is a consequence of the simplifying assumptions introduced in both approximations. Arridge proposes the following expression for Young's modulus:

$$E_{c1} = \frac{E_f V_f}{6} + \frac{E_f(6 - BV_f)}{6k} \tag{21}$$

where the constant B would be found for each combination of materials by suitable experimentation.

IV. EFFECT OF MATRIX YIELDING

Few authors, after Cox and Arridge, considered composites with a random array of fibers, and most of the later work has been concerned with the analysis of unidirectional fiber-reinforced composites. Shaffer [6,7] derived the elastic constants for a unidirectional composite based upon the law of mixtures, but also including the effects of matrix yielding. In order to obtain the elastic modulus in the direction of the reinforcement (direction x_1), the composite is represented by two bars (Fig. 5). A little reflection reveals that this model is basically the same as that shown in Fig. 4, and is based upon the same assumptions:

(a) Stress is linearly proportional to strain, in both phases, up to yield. Here, however, the matrix is assumed to be ideally elastic-plastic.

(b) Both phases are securely bonded together and are strained an equal amount under load.

(c) The fibers are straight, continuous, and aligned with the axis of the applied load.

(d) Both phases are homogeneous and isotropic.

On the basis of this model, we previously derived an expression for the elastic modulus:

$$E_{c1} = V_f E_f + (1 - V_f) E_m \qquad (22)$$

where E_{c1} is the elastic modulus of the composite parallel to the direction of the fibers (direction x_1), E_f is the Young's modulus of the fibers, E_m is the Young's modulus of the matrix, $V_f = A_f$ is the fiber packing density, and A_f and A_m are the fiber and the matrix cross-sectional area fractions, respectively.

The force L applied at the end plates (Fig. 5) is

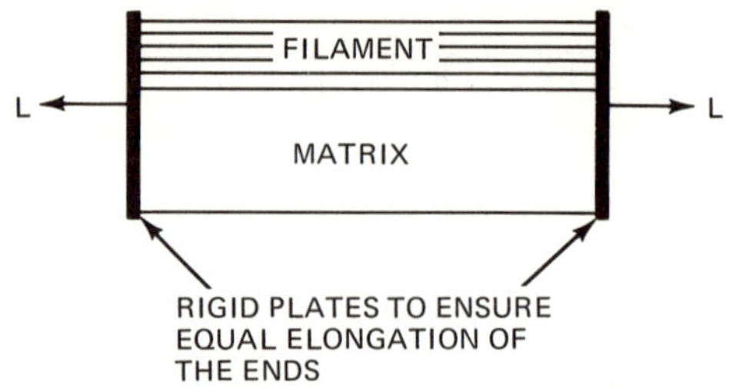

FIG. 5. Model of composite with fibers parallel to the applied force.

decomposed into L_f, the force in the fibers, and L_m, the force in the matrix:

$$L = L_f + L_m \tag{23}$$

Thus, for a unit composite area, we obtain from Eqs. (11), (22), and (23)

$$\frac{L_f}{L_f + L_m} = \frac{1}{1 + (E_m/E_f)(A_m/A_f)}$$

and

$$\frac{\sigma_f}{\sigma_{c1}} = \frac{1}{A_f + (E_m/E_f)(1 - A_f)} = \frac{E_f\sigma_m}{E_m\sigma_{c1}} \tag{24}$$

Since the modulus of elasticity of the matrix is less than that of the fibers, the stress in the filament bar is greater than that in the matrix bar. To continue into the area of matrix yielding, it is necessary to introduce an additional assumption, i.e.,

$$Y_f/Y_m > E_f/E_m \tag{25}$$

where Y_f and Y_m denote the yield stresses of the fibers and the matrix, respectively. This assumption merely states that the matrix yields before the fibers. At the instant of matrix yield,

$$\sigma_{c1} = A_f\sigma_f + A_m\sigma_m, \quad \sigma_m = Y_m, \quad \sigma_f = (E_f/E_m)\sigma_m \tag{26}$$

Elimination of σ_f and σ_m from Eqs. (26) yields

$$\sigma_{c_1} = Y_m\{1 + A_f[(E_f/E_m) - 1]\}$$

Also,

$$\varepsilon_m = \varepsilon_{c1} = \varepsilon_f = Y_m/E_m$$

Combining the above two equations, we have

$$E_{c_1} = \sigma_{c_1}/\varepsilon_{c_1} = E_m[A_f(E_f/E_m) + (1 - A_f)] \tag{27}$$

Equation (27) is identical with Eq. (22). It describes the point of incipient matrix yield. Once the matrix yields, the stress in the matrix remains at a constant value Y_m while the fiber stress continues to increase. Therefore, from Eq. (26)

$$\sigma_f = (L/A_f) - Y_m(A_m/A_f) \tag{28}$$

Finally, at incipient fiber yield ($\sigma_f = Y_f$), the composite stress is

$$\sigma_{c_1Y} = Y_mA_m + Y_fA_f = Y_m + A_f(Y_f - Y_m) \tag{29}$$

Therefore, the second slope in the composite stress-strain diagram (Fig. 6) is

$$E'_{c_1} = \frac{\sigma_{c_1Y} - \sigma_{c_1}}{\varepsilon_{c_1Y} - \varepsilon_{c_1}} = \frac{A_fY_f - A_f(E_f/E_m)Y_m}{(Y_f/E_f) - (Y_m/E_m)} = A_fE_f \tag{30}$$

From Eqs. (22) and (30), it is evident that the first slope

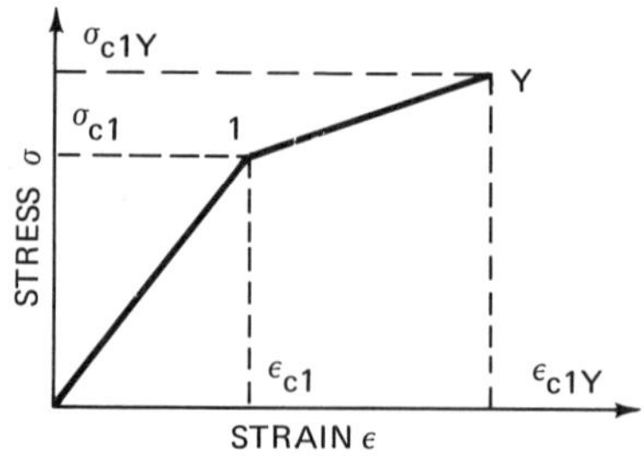

FIG. 6. Stress-strain diagram for a composite material.

(E_{c1}) is greater than the second slope (E'_{c1}) in the stress-strain diagram for the composite.

Now let us consider the elastic properties of the composite material in the x_2-direction, i.e., transverse to the reinforcing fibers. A unit thickness of the composite in this direction is shown in Fig. 7.

The cross section contains a uniform distribution of fibers of diameter d_f spaced a distance h apart. The triangular region is repeated throughout the composite. From the geometry of the region, we obtain

$$A_f = (\pi/2\sqrt{3})\ d_f^2/h^2 \tag{31}$$

and

$$s = (\sqrt{3}/2)h - d_f \tag{32}$$

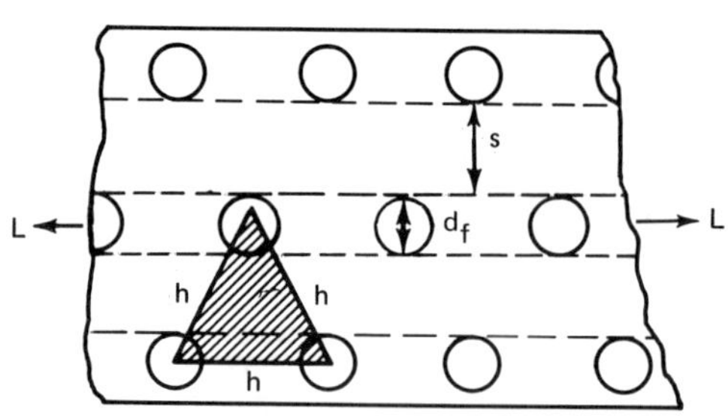

FIG. 7. Transverse cross-section of unidirectional fiber composite.

The distance s between adjacent layers of fibers is positive as long as A_f is less than 0.68. The analysis will first be given for the case when $s > 0$ and then modified to describe the situation when $s < 0$.

When $s > 0$, the stress caused by an external load L follows two distinct paths through the composite, one strictly through the matrix bars of width s, and the other through bars of width d_f, containing both fibers and matrix. This situation can be represented by the two-bar model shown in Fig. 8. In the combined fiber-matrix bar, the length of the fiber phase ℓ_f is given by

$$\ell_f/h = \left(\pi d_f^2/4\right)/hd_f = \pi d_f/4h \tag{33}$$

The modulus E of the fiber-matrix bar, made up in fact, of two bars in series, is obtained by considering the stress distribution in the equivalent bar:

$$\sigma_f = \sigma_m = \sigma, \quad u_f + u_m = u \tag{34}$$

where u is the displacement of the bar. Therefore

$$(\sigma/E_f)\ell_f + (\sigma/E_m)(h - \ell_f) = (\sigma/E)h$$

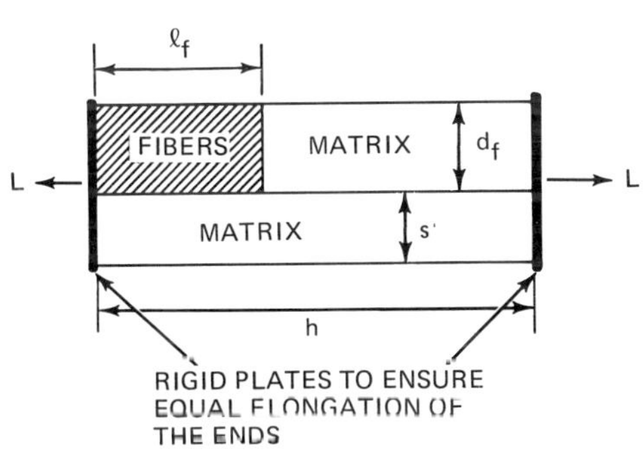

FIG. 8. Composite model with fibers transverse to the applied load.

or

$$E_m/E = 1 - (\ell_f/h)[1 - (E_m/E_f)] \tag{35}$$

Combining Eqs. (33) and (35), we obtain

$$E_m/E = 1 - (\pi d_f/4h)[1 - (E_m/E_f)] \tag{36}$$

The modulus of the equivalent bar (E) is greater than that of the resin bar (E_m) since $E_f > E_m$. Therefore, the stress in the equivalent bar will be greater than that in the matrix bar, and the matrix portion of the equivalent bar will be the first to yield. At the moment of incipient yield, we have

$$\sigma_{c2}(d_f + s) = \sigma d_f + \sigma_m s, \quad \sigma = Y_m, \quad \sigma_m = (E_m/E)\sigma \tag{37}$$

where σ_{c2} and σ denote the stresses in the composite and the equivalent bars, respectively. Substituting for E from Eq. (36) in Eq. (37), we obtain

$$\sigma_{c2} = \frac{Y_m}{d_f + s}\left\{d_f + s\left[1 - \frac{\pi d_f}{4h}\left(1 - \frac{E_m}{E_f}\right)\right]\right\} \tag{38}$$

The strain ε_{c2} in the composite is given by

$$\varepsilon_{c2} = \frac{Y_m}{E} = \frac{Y_m}{E_m}\left[1 - \frac{\pi d_f}{4h}\left(1 - \frac{E_m}{E_f}\right)\right] \tag{39}$$

Therefore, the modulus of elasticity for the composite in the x_2-direction is

$$E_{c2} = \frac{\sigma_{c2}}{\varepsilon_{c2}} = \frac{E_m}{d_f + s}\,\frac{d_f + s\left\{1 - (\pi d_f/4h)\left[1 - (E_m/E_f)\right]\right\}}{1 - (\pi d_f/4h)\left[1 - (E_m/E_f)\right]} \tag{40}$$

If $s = 0$, then only the equivalent bar exists, and

$$E_{c2} = E$$

If we continue to load the composite, the stress in the equivalent bar remains Y_m, and the stress in the matrix bar increases until it also reaches Y_m. The composite stress and strain are then

$$\sigma_{c2Y} = Y_m, \quad \varepsilon_{c2Y} = Y_m/E_m$$

The composite stress-strain curve can again be represented by a two-slope curve (Fig. 6), with the second slope given by

$$E'_{c2} = \frac{\sigma_{c2Y} - \sigma_{c2}}{\varepsilon_{c2Y} - \varepsilon_{c2}} = \frac{E_m s}{d_f + s} \tag{41}$$

We mentioned previously that if $s = 0$, then $E_{c2} = E$. It should be noted that this is only a dividing line between $s > 0$ and $s < 0$. For $s < 0$, the model consists of only the equivalent bar. In this case, E_{c2} and E prior to matrix yield are given by

$$E = E_{c2} = \frac{E_m}{1 - A_f\left[1 - (E_m/E_f)\right]} \tag{42}$$

where A_f is the fiber cross-sectional area per unit composite area [not the expression of Eq. (31)].

In the above analysis, the matrix was assumed to be ideally elastic-plastic. Realistically once the matrix yields, it may fracture. Therefore, the above results are strictly applicable only for strains of the order of the elastic strains. Also, it must be emphasized that this analysis is inappropriate for brittle matrices.

In order to completely characterize a thin, unidirectionally fiber-reinforced composite plate, we require besides the elastic moduli determined above, the shear modulus G_{12}, and the Poisson's ratio ν_{12}. More accurate models for the determination of E_1, E_2, G_{12} and ν_{12}, as well as the larger question regarding the number of material constants required to completely characterize the elastic behavior of a fiber-reinforced composite (not necessarily unidirectional), form the subject matter of Chapter 3.

REFERENCES

1. H. L. Cox, "The Elasticity and Strength of Paper and other Fibrous Materials," British J. Appl. Phys., 3, 72-79 (1952).

2. A.E.H. Love, A Treatise on the Mathematical Theory of Elasticity, 4th ed., Dover Publications, New York, 1927.

3. R.G.C. Arridge, "Orientation Effects in Fiber Reinforced Composites Where the Modulus of the Fibers is No More Than an Order of Magnitude Greater Than That of the Matrix," Presented at the 18th Annual Meeting of the Reinforced Plastics Division of the Society of the Plastics Industry, 1963.

4. R. C. Stedfeld and C. T. Hoover, "The Filament-Wound Pressure Vessel," Allison Research and Engineering Co. Internal Report, TR-41, 1962.

5. J. E. Gordon, "On the Present and Potential Efficiency of Structural Plastics," J. Royal Aero. Soc., 56, 704-728 (1952)

6. B. W. Shaffer, "Stress-Strain Relations of Reinforced Plastics Parallel and Normal to Their Internal Filaments," AIAA J., 2, 348-352 (1964).

7. B. W. Shaffer, "Elastic-Plastic Stress Distribution within Reinforced Plastics Loaded Normal to Its Filaments," AIAA J., 6, 2316-2324 (1968).

CHAPTER 3

ELASTICITY ANALYSES

I. INTRODUCTION

In Chapter 2, we considered several simple analytic models for analyzing composite material properties. Only two moduli were found for several models, and no consideration was given as to whether these were sufficient to describe the composite behavior. In order to more thoroughly investigate the number of required material constants for the elastic description of composites, we start here with the three-dimensional stress-strain relationships for an orthotropic material (three planes of elastic symmetry):

$$\begin{aligned}
\varepsilon_{11} &= (1/E_1)\left[\sigma_{11} - (\nu_{12}\sigma_{22} + \nu_{13}\sigma_{33})\right] \\
\varepsilon_{22} &= (1/E_2)\left[\sigma_{22} - (\nu_{21}\sigma_{11} + \nu_{23}\sigma_{33})\right] \\
\varepsilon_{33} &= (1/E_3)\left[\sigma_{33} - (\nu_{31}\sigma_{11} + \nu_{32}\sigma_{22})\right] \\
\gamma_{23} &= \tau_{23}/G_{23}, \quad \gamma_{12} = \tau_{12}/G_{12}, \quad \gamma_{31} = \tau_{31}/G_{31}
\end{aligned} \tag{1}$$

where

$$\nu_{13}/E_1 = \nu_{31}/E_3, \quad \nu_{12}/E_1 = \nu_{21}/E_2, \quad \nu_{23}/E_2 = \nu_{32}/E_3$$

Let us now consider a unidirectionally fiber-reinforced composite (Fig. 1). In this case, the elastic properties in the x_2-direction will be the same as in the x_3-direction, and the above equations reduce to the following set containing six independent elastic constants (E_1, E_2, ν_{12}, G_{12}, ν_{23}, G_{23}) [1]:

$$\begin{aligned}
\varepsilon_{11} &= (1/E_1)\left[\sigma_{11} - \nu_{12}(\sigma_{22} + \sigma_{33})\right] \\
\varepsilon_{22} &= (1/E_2)\left[\sigma_{22} - (\nu_{21}\sigma_{11} + \nu_{23}\sigma_{33})\right] \\
\varepsilon_{33} &= (1/E_2)\left[\sigma_{33} - (\nu_{21}\sigma_{11} + \nu_{23}\sigma_{22})\right] \\
\gamma_{12} &= \tau_{12}/G_{12}, \quad \gamma_{23} = \tau_{23}/G_{23}, \quad \gamma_{31} = \tau_{31}/G_{12}
\end{aligned} \tag{2}$$

where

$$E_2 = E_3, \quad \nu_{21} = \nu_{31}, \quad \nu_{12} = \nu_{13}, \quad \nu_{23} = \nu_{32}$$

$$G_{12} = G_{31}$$

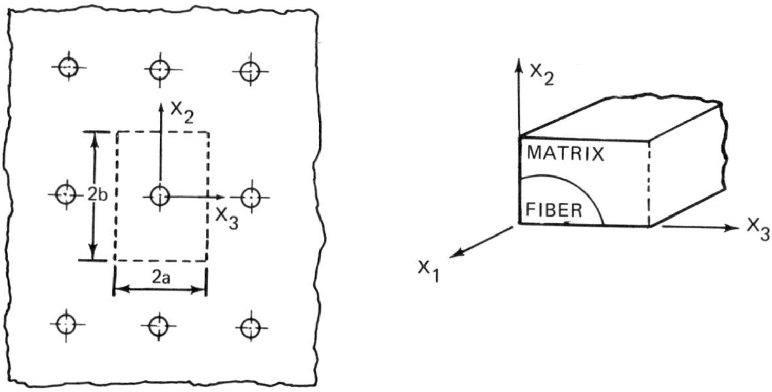

FIG. 1. Unidirectionally fiber-reinforced composite.

If the fibers in the Fig. 1 were arranged in such a fashion that the x_2-x_3 plane could be considered isotropic, an additional equation would become available [1]:

$$G_{23} = E_2/2(1 + \nu_{23}) \tag{3}$$

reducing the number of independent elastic constants in Eq. (2) to five. If the unidirectional composite were such that it was very thin in the x_3 direction (plane stress), two-dimensional orthotropic equations could be employed in place of Eqs. (1)

$$\varepsilon_{11} = (1/E_1)(\sigma_{11} - \nu_{12}\sigma_{22}), \quad \varepsilon_{22} = (1/E_2)(\sigma_{22} - \nu_{21}\sigma_{11})$$

$$\gamma_{12} = \tau_{12}/G_{12}, \qquad E_2\nu_{12} = E_1\nu_{21} \tag{4}$$

Equations (4) utilize only four independent material constants (E_1, E_2, ν_{12}, G_{12}).

The number of constants could even be further reduced for specific lay-ups in these composite sheets (see Fig. 2).

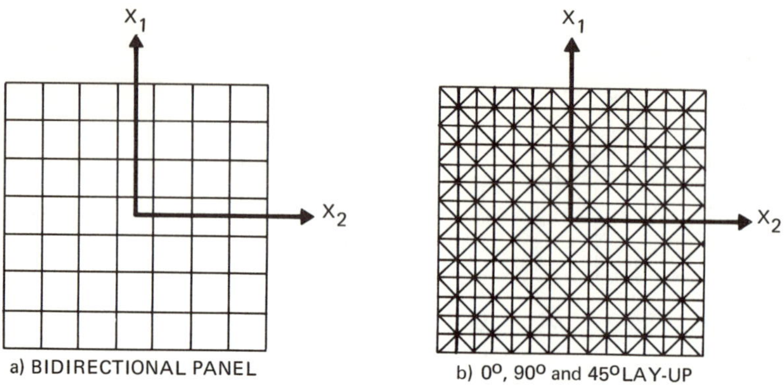

FIG. 2. Various sheet composite lay-ups.

If we consider the bidirectional panel of Fig. 2a, Eqs. (4) become

$$\varepsilon_{11} = (1/E_1)(\sigma_{11} - \nu_{12}\sigma_{22}), \quad \varepsilon_{22} = (1/E_2)(\sigma_{22} - \nu_{12}\sigma_{11})$$

$$\gamma_{12} = \tau_{12}/G_{12}, \quad E_1 = E_2, \quad \nu_{21} = \nu_{12} \tag{5}$$

involving only three independent elastic constants (E_1, ν_{12}, G_{12}). The pseudoisotropic lay-up pictured in Fig. 2b can be described by only two independent elastic constants [1], since the following identities are valid in Eqs. (4):

$$E_2 = E_1, \quad \nu_{21} = \nu_{12}, \quad G_{12} = E_1/2(1 + \nu_{12}) \tag{6}$$

Thus it is seen that in general the unidirectionally reinforced composite can have four independent elastic constants (two-dimensional orthotropy), five independent elastic constants (transverse isotropy), or six independent elastic constants (tetragonal symmetry).

In this chapter, we shall consider models and analyses which are consistent in the number of independent material constants. We shall consider these in order of complexity and completeness of analysis, from models still basically in the "mechanics of materials" category to those requiring the use of the theory of elasticity or numerical analysis methods.

II. MECHANICS-OF-MATERIALS MODELS

A typical model in this category is that due to Ekvall [2,3]. Ekvall's analysis, being more sophisticated than the models discussed in Chapter 2, attempts to take into account the actual cross-section shape of the reinforcing fibers. Specifically considered are square or rectangular fibers and round fibers. This group of models attempts to find the material properties necessary for an elastic analysis of a two-dimensionally orthotropic composite structure (four independent elastic constants). The models of Chapter 2 obtained only E_1 and E_2, the Young's moduli parallel and normal to the reinforcing fibers. The present analysis will derive values for the four independent material constants, E_1, E_2, ν_{12}, and G_{12}, and obtain the fifth by using the orthotropic identity

$$\nu_{12}E_2 = \nu_{21}E_1 \tag{7}$$

The assumptions necessary for this model are as follows:

(a) The composite is composed of unidirectional layers one fiber diameter (d_f) thick.

(b) The fibers in the layer are evenly spaced.

(c) The fiber and matrix materials are securely bonded together and act as a composite.

(d) Both phases obey Hooke's law.

(e) There is no interaction (chemical or otherwise) between the two phases that would change their individual material properties.

If we consider the properties of the composite in the longitudinal (parallel to the fibers) direction, we have already derived the classic law of mixtures formula:

$$E_1 = V_f E_f + V_m E_m \tag{8}$$

where V_f and V_m are the fiber and matrix volume ratios respectively, and $V_f + V_m = 1$. Reasoning similar to this leads to the following equation for Poisson's ratio for the composite:

$$\nu_{12} = V_f \nu_f + V_m \nu_m \tag{9}$$

This follows from the earlier statement that longitudinal strains in the composite and each of the phases are equal. Equation (9) thus states that each phase contributes to the composite Poisson's ratio in proportion to its volume. We note here that Eqs. (8) and (9) are true regardless of the reinforcing fiber shape, since the longitudinal strain in the fibers or matrix is independent of this parameter as long as the distributions of the phases are continuous throughout the composite cross sections. Considering now the properties in the transverse (normal to the fibers) direction, we will first derive the necessary equations for square or rectangular fibers. For these fibers, the Ekvall model in the transverse direction is shown in Fig. 3. Since the fibers are square or rectangular, the changes in stiffness in the x_3-direction are neglected. The strains in the x_2-direction are

$$\varepsilon_{m2} = (1/E_m)(\sigma_{m2} - \nu_m \sigma_{m1}), \quad \varepsilon_{f2} = (1/E_f)(\sigma_{f2} - \nu_f \sigma_{f1})$$

$$\varepsilon_{c2} = V_f \varepsilon_{f2} + V_m \varepsilon_{m2} \tag{10a}$$

The stresses are

$$\sigma_{c2} = \sigma_{m2} = \sigma_{f2} \tag{10b}$$

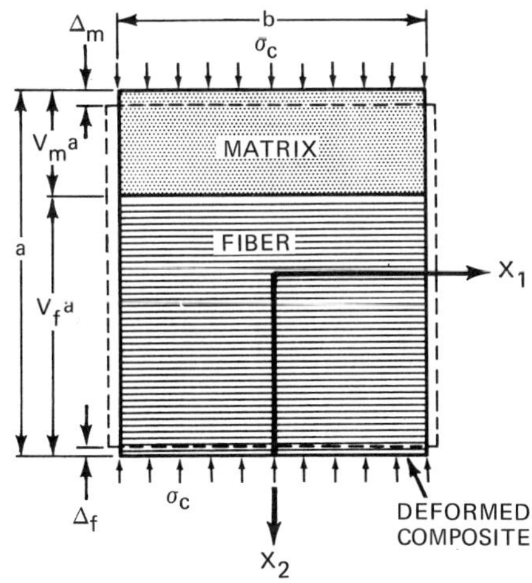

FIG. 3. Equivalent composite for analyzing the transverse stiffness (from Ref. [2]).

In the x_1-direction, we have

$$\varepsilon_{m_1} = (1/E_m)(\sigma_{m_1} - \nu_m \sigma_{m_2}), \quad \varepsilon_{f_1} = (1/E_f)(\sigma_{f_1} - \nu_f \sigma_{f_2})$$

$$\varepsilon_{m_1} = \varepsilon_{f_1} = \varepsilon_{c_1}, \quad \sigma_{c_1} = \sigma_{m_1} V_m + \sigma_{f_1} V_f = 0 \qquad (11)$$

Combining Eqs. (11), there follows

$$(1/E_m)(\sigma_{m_1} - \nu_m \sigma_{m_2}) = (1/E_f)(\sigma_{f_1} - \nu_f \sigma_{f_2})$$

Young's modulus in the x_2-direction can then be found by manipulating Eqs. (10) to form

$$\varepsilon_{c_2} = \frac{\sigma_{c_2}}{E_2} = \frac{V_f}{E_f}(\sigma_{f_2} - \nu_f \sigma_{f_1}) + \frac{V_m}{E_m}(\sigma_{m_2} - \nu_m \sigma_{m_1})$$

Using Eqs. (11) to eliminate stresses in the above equation, we obtain

$$\frac{1}{E_2} = \frac{V_m}{E_m} + \frac{V_f}{E_f} - \frac{V_f}{E_f} \frac{\left[(E_f/E_m)\nu_m - \nu_f\right]^2}{(V_f/V_m)(E_f/E_m) + 1} \qquad (12a)$$

Occasionally, some physical approximations are possible with Eq. (12a) for some combinations of materials. Thus if ν_f is small compared to $\nu_m E_f/E_m$, and 1 is small compared to $V_f E_f / V_m E_m$, then

$$E_2 = E_f E_m / \left[V_f E_m + V_m E_f \left(1 - \nu_m^2\right)\right] \tag{12b}$$

In addition, if $1 - \nu_m^2 \approx 1$, then

$$E_2 = E_f E_m / (V_f E_m + V_m E_f) \tag{12c}$$

A chart showing the relative magnitudes of the analytical predictions of elastic moduli for steel-epoxy composites as given by Eqs. (8) and (12b) is presented in Fig. 4. It can be seen that $E_2 << E_1$ for almost all fiber volume ratios. By using Eqs. (10) and (11), Poisson's ratio ν_{21} can be obtained from the definition

$$\nu_{21} = -\varepsilon_{c_1}/\varepsilon_{c_2}$$

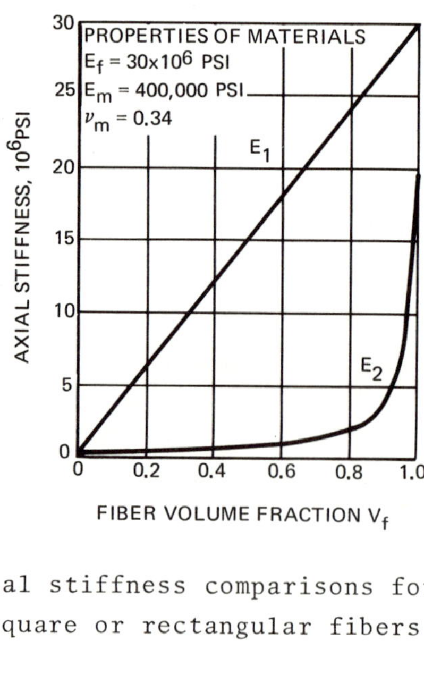

FIG. 4. Axial stiffness comparisons for steel-epoxy composites with square or rectangular fibers (from Ref. [2]).

Substitution from Eqs. (10) and (11) and appropriate simplification of the resulting equation yields the orthotropic identity. This serves as a check upon the consistency of the assumptions and derivations used in obtaining the three material constants thus far.

The remaining constant to be determined for the square or rectangular fibers is the shear stiffness. The equivalent model for this problem is presented in Fig. 5. Since the shear load is uniform around the edge, it is immaterial in which direction the fibers are oriented. The shear stiffness of the composite is

$$G_{12} = \tau_{12}/(\Delta_f + \Delta_m)$$

Rewriting this equation in terms of the component properties gives

$$G_{12} = \frac{\tau_{12}}{(V_f\tau_f/G_f) + (V_m\tau_m/G_m)}$$

In order for interface equilibrium to exist, the shear stresses must be equal in the fibers, matrix, and composite.

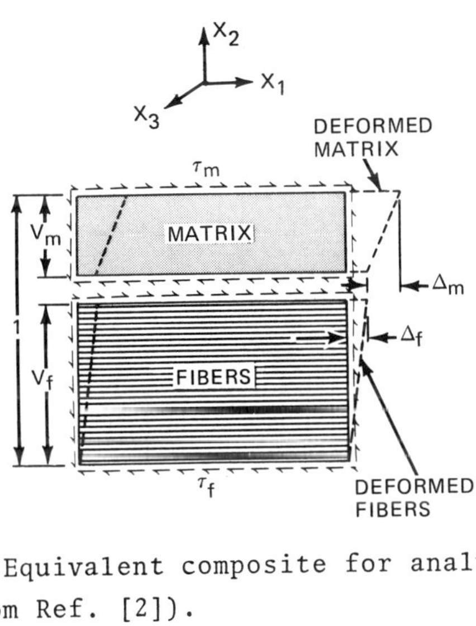

FIG. 5. Equivalent composite for analyzing the shear stiffness (from Ref. [2]).

Therefore we have

$$G_{12} = G_f G_m / (V_m G_f + V_f G_m) \tag{13}$$

Thus we have found the relationships for the four independent material constants necessary for the elastic analysis of the square or rectangular fiber model, Eqs. (8), (9), (12), and (13). The analysis for the round fiber model is more difficult, since variation of fiber thickness with distance must be considered. However, Eqs. (8) and (9) for E_1 and ν_{12} still hold. In the transverse direction, the model for finding the transverse Young's modulus is presented in Fig. 6. Since, due to the shape of the fiber, the composite stiffness varies across the model width, we must apply a nonuniform stress distribution to obtain equal deflections at every point. Assuming that Eq. (12b) applies to the differential element $dx_3 \times 1 \times 1$, the stiffness of this element is

$$E_2 = E_f E_m / \left[V_{m2} E_f \left(1 - \nu_m^{\,2}\right) + V_{f2} E_m\right]$$

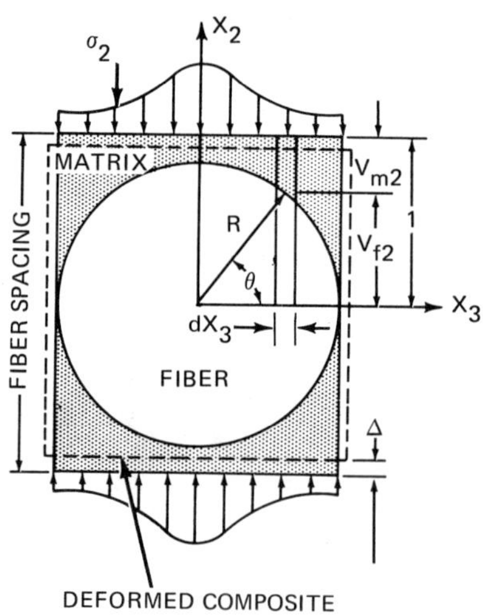

FIG. 6. Typical element for analyzing the transverse stiffness (from Ref. [2]).

To get the average stiffness, we must integrate this equation from zero to R and divide by R, where R is not the actual physical fiber radius but is the ratio of fiber radius to fiber spacing in x_2 direction. Therefore

$$E_2 = \frac{1}{R}\int_0^R \frac{E_m E_f \, dx_3}{V_{m2} E_f \left(1 - \nu_m^2\right) + V_{f2} E_m}$$

Transforming variables

$$V_{f2} = R \sin\theta, \quad V_{m2} = (1 - R \sin\theta), \quad dx_3 = R \sin\theta \, d\theta$$

we obtain

$$E_2 = E_f \int_0^{\pi/2} \frac{E_m \sin\theta \, d\theta}{E_f\left(1 - \nu_m^2\right) + R(\sin\theta)\left[E_m - E_f\left(1 - \nu_m^2\right)\right]} \tag{14}$$

If the effect of ν_m^2 is negligible, we have

$$E_2 = E_f \int_0^{\pi/2} \frac{\sin\theta \, d\theta}{(E_f/E_m) + R(\sin\theta)[1 - (E_f/E_m)]}$$

A plot of Eq. (14) for several fiber spacings is presented in Fig. 7 for steel-epoxy composites. The variation of the stiffness along the width is dramatically demonstrated in the figure.

The calculation of the shear stiffness for round fiber composites is made by use of the models in Figs. 8 and 9. It is assumed that there is no warping in the x_2-x_3 plane, and thus the shear distribution at any distance x_2 is as shown in Fig. 8. Equilibrium gives

$$2R\tau_{12} = \tau_{f2}(2R \sin\theta) + \tau_{m2}(2R - 2R \sin\theta)$$

or

$$\tau_{12} = \tau_{f2} \sin\theta + \tau_{m2}(1 - \sin\theta)$$

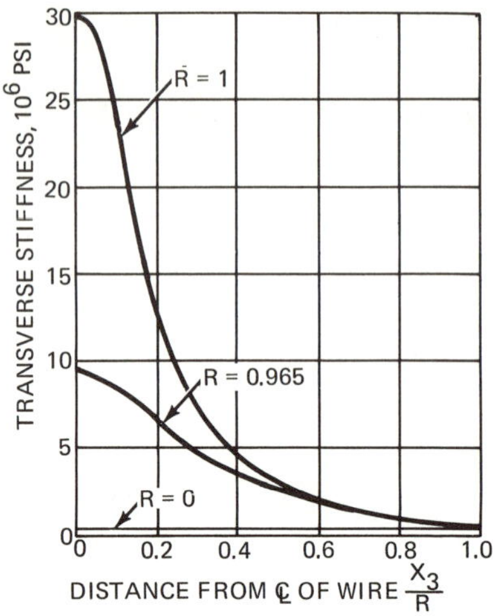

FIG. 7. Transverse stiffness of steel epoxy composites for several fiber spacings (from Ref. [2]).

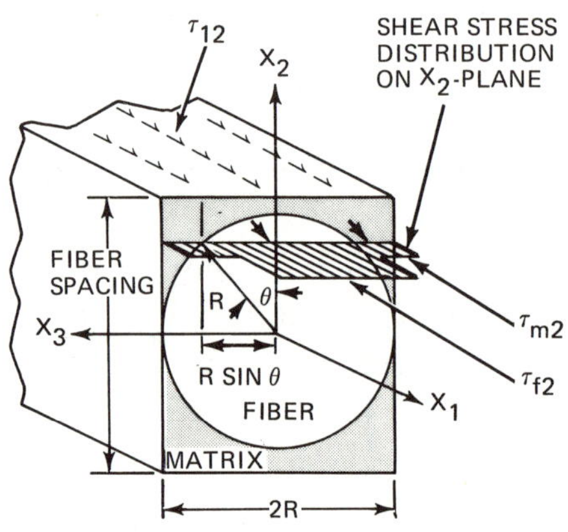

FIG. 8. Typical element for analyzing shear stiffness (from Ref. [2]).

The equality of matrix and fiber displacements at their interface (any plane of x_2-constant; see Fig. 8) results in

$$\tau_{m_2} = (G_m/G_f)\tau_{f_2}$$

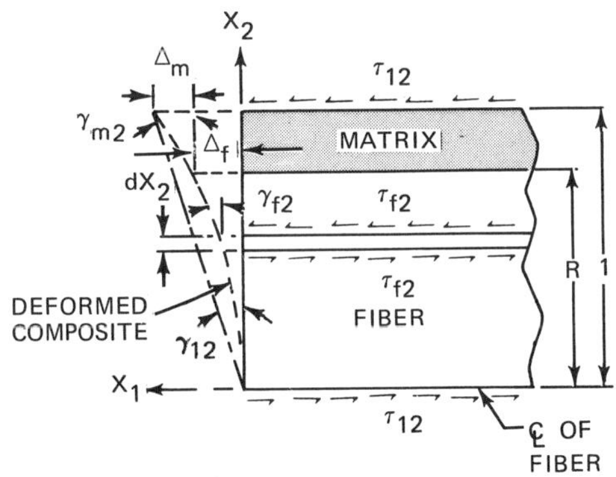

FIG. 9. Shear deformation of typical element shown in Fig. 8 (from Ref. [2]).

Combining the last two equations to eliminate τ_{m_2} yields

$$\tau_{f_2} = \frac{\tau_{12}}{(\sin\theta) + (G_m/G_f)(1 - \sin\theta)} \tag{15}$$

From Fig. 9, which shows the deformation picture at a cross section through the center of the fiber, the fiber deformation can be written as

$$\Delta_f = \int_0^R \gamma_{f_2}\, dx_2 = \int_0^R (\tau_{f_2}/G_f)\, dx_2$$

Utilizing Eq. (15) and transforming variables gives

$$\Delta_f = \frac{\tau_{12} R}{G_f} \int_0^{\pi/2} \frac{\sin\theta\, d\theta}{(G_m/G_f) + (\sin\theta)[1 - (G_m/G_f)]}$$

$$= \frac{\tau_{12} R}{G_f}\, \Phi \tag{16}$$

where Φ has been set equal to the integral. Since for this cross section, the shear stress in the matrix is constant from $x_2 = R$ to $x_2 = 1$ and equal to τ_{12}, the shear deformation in the matrix is

$$\Delta_m = (\tau_{12}/G_m)(1 - R)$$

Finally, we have

$$\gamma_{12} = \frac{\Delta_f + \Delta_m}{1} = \tau_{12}\left(\frac{R\Phi}{G_f} + \frac{1 - R}{G_m}\right)$$

The shear stiffness of the composite is therefore

$$G_{12} = \frac{G_m G_f}{R\Phi G_m + (1 - R)G_f} \tag{17}$$

This is in a form similar to the value for square or rectangular fibers [Eq. (13)] with $V_m = 1 - R$ and $V_f = R\Phi$. Equations (13) and (17) are presented graphically for steel-epoxy composites in Fig. 10.

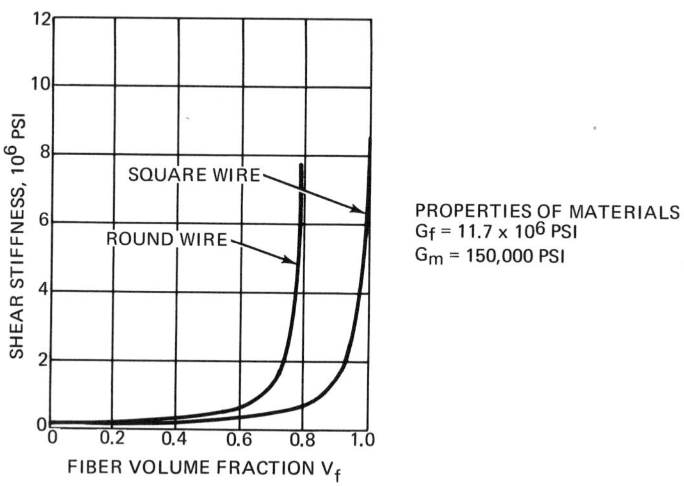

FIG. 10. Shear fiber stiffness comparisons for steel-epoxy composites (from Ref. [2]).

One drawback of these models is the assumption of regularity in geometry. In working with a representative volume of the composite, it is assumed that:

(a) All fibers have the same diameter.

(b) All fibers are equally spaced.

(c) All fibers are fully aligned and straight.

(d) The matrix has no voids.

(e) The composite geometry is perfect and regular.

In general a fiber composite does not satisfy these strict requirements; the fiber spacing, for instance, is random, and the matrix almost always contains some voids. Ekvall [2,3] attempted to accomodate his formulas to account for some of these discrepancies. The Ekvall models previously discussed assumed the model layers to be equivalent to one fiber diameter. If this is not the case and the actual layers are thicker, or if they overlap, the formulas need to be modified. In the longitudinal direction, there is no change. In the transverse direction, if the layers are thicker, there are actually two layers, one of matrix material of thickness $(s - d_f)$ and the other of thickness d_f for which the modulus E_2 was calculated by the previous formulas. Applying then the law of mixtures to the "composite" layer, we have

$$E_2' = (d_f/s)E_2 + [(s - d_f)/s]E_m \tag{18}$$

where s denotes the fiber spacing in the x_3-direction. Similarly,

$$G_{12}' = (d_f/s)G_{12} + [(s - d_f)/s]\, G_m \tag{19}$$

For round fibers, especially with random spacing, it is possible for the layers to overlap, that is, for there to be more than a single fiber in a layer of thickness d_f. For the calculation of the transverse Young's modulus, E_2, this can be taken into account by modifying R. Equating the fiber volume ratio for uniform spacing to the fiber volume ratio of the actual composite, we find

$$V_f = \frac{\pi d_f^2}{4sd_f} = \frac{\pi d_f}{4s} = \frac{\pi R}{4}$$

Therefore

$$R = (4/\pi)V_f \tag{20}$$

Here, s denotes the average spacing of the fibers in the x_2-x_3 plane. Using Eq. (20) in Eq. (14) will provide the necessary modification. The adjusted shear stiffness may be obtained by considering that the overlap area in the layers was considered twice in the analysis. Therefore, the adjusted shear stiffness would be

$$G'_{12} = G_{12} - [(s - d_f)/s]G_{12} = (d_f/s)G_{12} \tag{21}$$

A comparison between Ekvall's analytic results and experimental data is presented in Fig. 11. In this figure,

$$E_{12} = E_2/\nu_{21} = E_1/\nu_{12} \tag{22}$$

and the effect of voids in the matrix is taken into account by reducing V_m so that

$$V_f + V_m < 1 \tag{23}$$

the difference being due to the void volume. This method of accounting for voids cannot be used for G_{12}, since this value is independent of volume ratios for the Ekvall model.

Although these models are based on strength-of-materials theory their results are relatively accurate as shown in Fig. 11. It will be seen in later sections that the simple law of mixtures formula for E_1 is rarely improved upon even by the use of the more elaborate theories and models. Of course, it should be noted that measurements of matrix voids and the layer overlaps in a multilayer composite are rarely accurate, or even practical. Thus the use of such

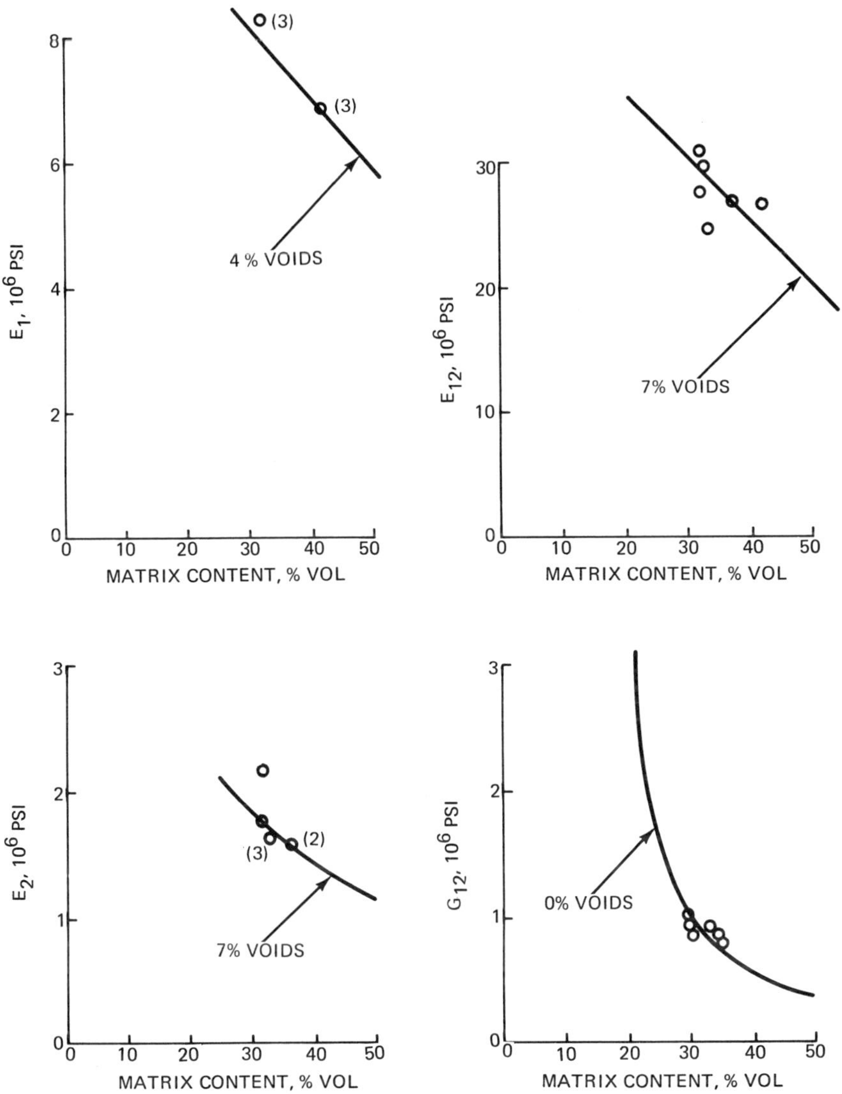

FIG. 11. Comparison of theoretical and experimental principal elastic properties (Ekvall Model). One test per data point unless indicated by ( ) (from Ref. [3]).

factors to adjust the basic formulas by more than small amounts must be viewed as being of questionable validity.

III. ELASTICITY THEORY SOLUTIONS--FIBER-REINFORCED COMPOSITES

Many composite material property derivations utilize the theory of elasticity as the basis of their solutions [1,4-9]. Some combine variational principles to find bounds for the elastic constants. These will be separately discussed in Chapter 4. The following sections will be concerned with elasticity-based analyses (both closed-form and numerical solutions) which do not use minimum-energy theorems to obtain property bounds. Most of these analyses are similar in theoretical bases and differ mainly in application to fiber or particulate composites or in the composite packing geometry considered. Thus, rather than considering all of these analyses in detail, we will select a representative one for detailed discussions.

The model chosen for a deeper examination is due to Whitney and Riley [1]. The basic assumptions of this unidirectionally fiber-reinforced composite model are:

(a) Fibers are circular in cross section and continuous through the composite.

(b) The fiber packing makes the plane of the composite cross section isotropic (transversely isotropic construction).

(c) Both fiber and matrix are isotropic.

(d) The composite material is macroscopically homogeneous and obeys Hooke's law.

The "repeating element" chosen in this analysis is an encased cylinder (Fig. 4 of Chapter 2). The term "repeating element" is used to mean that the entire composite is made up of these elements, thus making the composite elastic behavior equal to that of a single repeating element. The geometrical choice of a repeating element is relatively simple, depending upon the amount of symmetry and regularity assumed or existing in a composite. If there is little regularity, the most convenient element may be the encased cylinder to be discussed here. If greater regularity exists, hexagonal and rectangular patterns may be identified. The inherent weakness in the use of a repeating element lies in the assumptions required to describe the boundary stress state. Due to interaction among the repeating elements, the boundary stress state may be quite complex, whereas most analyses assume the state to be uniform to obtain closed-form solutions for the repeating element. These assumptions will be further discussed when we consider other models.

The stress-strain relations for this transversely isotropic composite are given by Eqs. (2) and (3). We note that there are five independent constants to be found: the longitudinal modulus E_1, the Poisson's ratios ν_{12} and ν_{23}, the transverse modulus E_2, and the shear modulus G_{12}. These constants can be found by defining and solving several elasticity problems for the repeating element.

A. Longitudinal Modulus [1]

An axial force is applied to the repeating element (Fig. 4, Chapter 2) such that the axial strain of the composite ε_1 is a constant. A "modified plane strain" problem is then solved, wherein it is required that the radial and tangential displacements be functions of r and θ only, while the axial displacement is a linear function of x_1. The particular loading problem considered here is axisymmetric (not a function of θ), and thus with the use of an Airy stress function, the compatibility equation may be written as

$$\nabla^4\phi = \frac{d^4\phi}{dr^4} + \frac{2}{r}\frac{d^3\phi}{dr^3} - \frac{1}{r^2}\frac{d^2\phi}{dr^2} + \frac{1}{r^3}\frac{d\phi}{dr} = 0 \qquad (24)$$

The solution of this equation is given by

$$\phi = C_1 \log r + C_2 r^2 \log r + C_3 r^2 + D \tag{25}$$

where C_1, C_2, C_3, and D are constants. The fiber and matrix stresses are

$$\begin{aligned}
\sigma_{fr} &= (1/r)\, d\phi_f/dr = (A/r^2) + B(1 + 2 \log r) + 2C \\
\sigma_{f\theta} &= d^2\phi_f/dr^2 = -(A/r^2) + B(3 + 2 \log r) + 2C \\
\sigma_{mr} &= (1/r)\, d\phi_m/dr = (F/r^2) + J(1 + 2 \log r) + 2H \\
\sigma_{m\theta} &= d^2\phi_m/dr^2 = -(F/r^2) + J(3 + 2 \log r) + 2H
\end{aligned} \tag{26}$$

Since the fiber stress must be finite at $r = 0$, it follows that $A = B = 0$. By utilizing the stress Beltrami-Michell compatibility equations, without resorting to the Airy stress function, it is found that $J = 0$. Thus it can be shown [4] that

$$\begin{aligned}
&\sigma_{fr} = \sigma_{f\theta} = 2C, \quad \sigma_{mr} = (F/r^2) + 2H \\
&\sigma_{m\theta} = -(F/r^2) + 2H
\end{aligned} \tag{27}$$

The axial stresses σ_{f1} and σ_{m1} can be found from stress-strain relationships:

$$\begin{aligned}
\varepsilon_{f1} &= \varepsilon_1 = (1/E_f)[\sigma_{f1} - \nu_f(\sigma_{fr} + \sigma_{f\theta})] \\
\varepsilon_{m1} &= \varepsilon_1 = (1/E_m)[\sigma_{m1} - \nu_m(\sigma_{mr} + \sigma_{m\theta})]
\end{aligned} \tag{28}$$

Combining Eqs. (27) and (28) and solving for axial stresses, one obtains

$$\sigma_{f1} = E_f\varepsilon_1 + 4\nu_f C, \quad \sigma_{m1} = E_m\varepsilon_1 + 4\nu_m H \tag{29}$$

The radial displacements are given by

$$u_{fr} = r\varepsilon_{f\theta} = (r/E_f)[2C - \nu_f(2C + 4\nu_f C + E_f\varepsilon_1)]$$
$$u_{mr} = r\varepsilon_{m\theta} = (r/E_m)\left\{2H - (F/r^2) - \nu_m[2H + 4\nu_m H + E_m\varepsilon_1 + (F/r^2)]\right\} \quad (30)$$

The boundary conditions to be satisfied for this problem are as follows:

At $r = r_f$,

$$\sigma_{fr} = \sigma_{mr} \quad \text{and} \quad u_{fr} = u_{mr}$$

i.e.,

$$2C = (F/r_f^2) + 2H$$
$$E_m[2C(1 - \nu_f - 2\nu_f^2) - \nu_f E_f \varepsilon_1] = E_f[2H(1 - \nu_m - 2\nu_m^2) - (F/r_f^2)(1 + \nu_m) - \nu_m E_m \varepsilon_1] \quad (31)$$

At $r = r_m$,

$$\sigma_{mr} = 0$$

i.e.,

$$2H = -F/2r_m^2 \quad (32)$$

The elastic modulus E_1 may now be obtained by solving the above equations for the fiber and matrix stresses and then equating the strain energy, applied to the composite cylinder, to the sum of the strain energies of the fiber and matrix:

$$\frac{1}{2}\int_V \sigma_1\varepsilon_1 \, dV = \frac{1}{2}\int_{V_f} (\sigma_{f1}\varepsilon_1 + \sigma_{f\theta}\varepsilon_{f\theta} + \sigma_{fr}\varepsilon_{fr}) \, dV_f + \frac{1}{2}\int_{V_m} (\sigma_{m1}\varepsilon_1 + \sigma_{m\theta}\varepsilon_{m\theta} + \sigma_{mr}\varepsilon_{mr}) \, dV_m \quad (33)$$

Thus, using the stress-strain relationships, the left-hand side of Eq. (33) can be expressed in terms of E_1 and ε_1, while the right-hand side is expressed in terms of the component stresses. These substitutions finally yield

$$E_1 = \left\{2(\nu_f - \nu_m)^2 E_f E_m (1 - V_f) V_f\right\} / \left\{E_m(1 - V_f)\left(1 - \nu_f - 2\nu_f^2\right) + \left[V_f\left(1 - \nu_m - 2\nu_m^2\right) + (1 + \nu_m)\right] E_f\right\} + E_m + (E_f - E_m)V_f \tag{34}$$

where V_f denotes the fiber volume fraction. It is to be noted that the first term in Eq. (34) is usually insignificant for most composites, and thus the equation reduces to the law of mixtures:

$$E_1 = E_m + (E_f - E_m)V_f \tag{35}$$

B. Poisson's Ratios [1]

The major Poisson's ratio ν_{12} for the composite can be found from the results of Section III.A. The radial displacement for the composite can be written as

$$\Delta_r = r_m \nu_{12} \varepsilon_1 \tag{36}$$

From the previous solution, it is also equal to

$$\Delta_r = u_{mr}\Big|_{r=r_m} = (r_m/E_m)(\sigma_{m\theta} - \nu_m \sigma_{m1}) \tag{37}$$

Utilizing the matrix stress solutions and combining Eqs. (36) and (37), it is found that

$$\nu_{12} = \nu_m - \left\{2(\nu_m - \nu_f)(1 - \nu_m^2)E_f V_f\right\} / \left\{E_m(1 - V_f)\right.$$

$$(1 - \nu_f - 2\nu_f^2) + E_f\Big[V_f(1 - \nu_m - 2\nu_m^2)$$

$$+ (1 + \nu_m)\Big]\Big\} \qquad (38)$$

The Poisson's ratio ν_{23} is not solved in Ref. [1], but is approximated by the law of mixtures relationship.

$$\nu_{23} = \nu_f V_f + \nu_m V_m \qquad (39)$$

C. Transverse Modulus [1]

A radial pressure and an axial stress are applied to the repeating element (Fig. 4, Chapter 2). The radial pressure is

$$p = (\sigma_2 + \sigma_3)/2 \qquad (40)$$

while the axial stress is such that the axial strain is kept at zero under the combined loading. The change in volume of one cubic unit of the cylinder is then

$$1 + \Delta V = (1 + \varepsilon_2)(1 + \varepsilon_3) \qquad (41a)$$

or, neglecting second-order terms,

$$\Delta V = \varepsilon_2 + \varepsilon_3 \qquad (41b)$$

Using Eqs. (2), we have

$$\Delta V = \Big[(1 - \nu_{23})/E_2\Big](\sigma_2 + \sigma_3) - (2\nu_{12}/E_1)\sigma_1 \qquad (42)$$

Utilizing Eq. (40) and the fact that $\varepsilon_1 = 0$, there follows from Eqs. (2) and (42)

$$\Delta V = 2p\left(\frac{1 - \nu_{23}}{E_2} - \frac{2\nu_{12}^2}{E_1}\right) \qquad (43)$$

The "plane strain" bulk modulus K_{23} will be defined as

$$K_{23} = \frac{p}{\Delta V} = \frac{E_2}{2\left[1 - \nu_{23} - \left(2\nu_{12}^2 E_2/E_1\right)\right]} \tag{44}$$

It should be noted that the stress equations (27) from the previous solution still hold for the current problem. Since $\varepsilon_{f1} = \varepsilon_{m1} = 0$, we obtain from Hooke's law

$$\sigma_{f1} = \nu_f(\sigma_{fr} + \sigma_{f\theta}), \quad \sigma_{m1} = \nu_m(\sigma_{mr} + \sigma_{m\theta}) \tag{45}$$

The boundary conditions for this problem are as follows:

At $r = r_f$,

$$\sigma_{fr} = \sigma_{mr} \quad \text{and} \quad u_{fr} = u_{mr}$$

i.e.,

$$2C = \left(F/r_f^2\right) + 2H$$

$$(2C/E_f)\left(1 - \nu_f - 2\nu_f^2\right) = (2H/E_m)\left(1 - \nu_m - 2\nu_m^2\right) - \left[F/\left(r_f^2 E_m\right)\right](1 + \nu_m) \tag{46a}$$

At $r = r_m$,

$$\sigma_{rm} = -p$$

i.e.,

$$-p = \left(F/r_m^2\right) + 2H \tag{46b}$$

At this stage, the solution steps are identical to those in Section III.A. The stresses are solved from Eqs. (27) and (46) and the energy of the composite is equated to the sum of the two component energies. Thus we can obtain for the transverse modulus

$$E_2 = \left[2K_{23}E_1(1 - \nu_{23})\right]/\left(E_1 + 4\nu_{12}^2 K_{23}\right) \tag{47}$$

where

$$K_{23} = \left\{\left[\frac{E_f}{2\left(1 - \nu_f - 2\nu_f^2\right)} + G_m\right]\frac{E_m}{2\left(1 - \nu_m - 2\nu_m^2\right)}\right.$$

$$\left. + \left[\frac{E_f}{2\left(1 - \nu_f - 2\nu_f^2\right)} - \frac{E_m}{2\left(1 - \nu_m - 2\nu_m^2\right)}\right] G_m V_f\right\}$$

$$\div \left\{\frac{E_f}{2\left(1 - \nu_f - 2\nu_f^2\right)} + G_m - \left[\frac{E_f}{2\left(1 - \nu_f - 2\nu_f^2\right)}\right.\right.$$

$$\left.\left. - \frac{E_m}{2\left(1 - \nu_m - 2\nu_m^2\right)}\right] V_f\right\} \tag{48}$$

D. Shear Modulus [10]

A pure shear, $\tau_{12} = \tau$, is applied to the surface of the composite cylinder repeating element. The equations of equilibrium in displacement form with no body forces or dilatation reduce to (in Cartesian form)

$$\nabla^2 u_1 = 0, \quad \nabla^2 u_2 = 0, \quad \nabla^2 u_3 = 0 \tag{49}$$

where ∇^2 is the three-dimensional Laplacian and (u_1, u_2, u_3) denote the displacements in the (x_1, x_2, x_3) directions. A displacement solution can be assumed to be in the form (in cylindrical coordinates)

$$u_r = C_1 x_1 \cos\theta, \quad u_\theta = -C_1 x_1 \sin\theta$$

$$u_1 = f(r)\cos\theta \tag{50}$$

where x_1 denotes distance along the x_1-axis. In order to satisfy the first of Eqs. (49), we have

$$\nabla^2 u_1 = \left(\frac{d^2 f}{dr^2} + \frac{1}{r}\frac{df}{dr} - \frac{f}{r^2}\right)\cos\theta = 0 \tag{51}$$

Solution of Eq. (51) yields

$$f(r) = \left[C_2 r + (C_3/r)\right]$$

Therefore

$$u_1 = \left[C_2 r + (C_3/r)\right]\cos\theta \tag{52}$$

Using the strain displacement and the isotropic Hooke's laws, we find

$$\varepsilon_1 = \varepsilon_r = \varepsilon_\theta = \gamma_{r\theta} = 0$$

$$\gamma_{\theta 1} = -\left[C_2 + (C_3/r^2) + C_1\right]\sin\theta$$

$$\gamma_{r1} = \left[C_2 - (C_3/r^2) + C_1\right]\cos\theta \tag{53}$$

and

$$\sigma_1 = \sigma_r = \sigma_\theta = \tau_{r\theta} = 0$$

$$\tau_{\theta 1} = -G\left[C_2 + (C_3/r^2) + C_1\right]\sin\theta$$

$$\tau_{r1} = G\left[C_2 - (C_3/r^2) + C_1\right]\cos\theta \tag{54}$$

Therefore in the repeating cylindrical element, the fiber and matrix stresses can be written as

$$\tau_{f\theta 1} = -G_f\left[A + (B/r^2) + C\right]\sin\theta$$

$$\tau_{fr1} = G_f\left[A - (B/r^2) + C\right]\cos\theta$$

$$\tau_{m\theta 1} = -G_m\left[D + (F/r^2) + H\right]\sin\theta$$

$$\tau_{mr1} = G_m\left[D - (F/r^2) + H\right]\cos\theta \tag{55}$$

The boundary conditions to be applied for the problem are the regularity of stresses at $r = 0$ $(B \equiv 0)$ and the following:

At $r = r_f$,

(i) $u_{fr} = u_{mr}$, $u_{f\theta} = u_{m\theta}$, and $u_{f1} = u_{m1}$,

i.e.,

$$C = H \quad \text{and} \quad A = D + \left(F/r_f^2\right) \tag{56}$$

(ii) $\tau_{fr1} = \tau_{mr1}$

i.e.,

$$G_f(A + C) = G_m\left[D - \left(F/r_f^2\right) + H\right] \tag{57}$$

At $r = r_m$,

$$\tau_{mr1} = \tau\cos\theta \quad \text{and} \quad u_1 = 0$$

i.e.,

$$D - F/r_m^2 + H = \tau/G_m \quad \text{and} \quad Dr_m^2 + F = 0 \tag{58}$$

Again the energy equivalence after solving for the unknown constants is identical to that in Section III.A. Thus the shear modulus for the composite can be obtained as

$$G_{12} = G_m\frac{(G_f + G_m) + (G_f - G_m)V_f}{(G_f + G_m) - (G_f - G_m)V_f} \tag{59}$$

E. Experimental Comparison

An experimental comparison with the Whitney-Riley model for boron-epoxy composites is presented in Table 1. As can be seen, the theoretical predictions are good both for the transverse and longitudinal moduli. The predictions for the shear modulus, however, are significantly different from experiment. This may be due, at least partly, to the fact that there is no consideration of fiber or repeating element interaction in the analysis. It can be speculated that this lack may be especially felt in the prediction of the shear modulus. The nonuniformity of the applied stress state resulting from a more rigorous analysis would require a computer solution. Such an analysis will be discussed in Section III.F.

It should be noted that there are also Poisson's ratio comparisons available in Ref. [1]. However, there seems some doubt as to the validity of the experimental values. Further comparisons of this model with other models will be presented in Chapter 4.

F. Other Fiber-Reinforced Composite Models

One of the basic advantages of the Whitney-Riley model is that closed-form solutions resulting in simple formulas can be obtained. This very advantage, important to a designer, necessitated some simplifying assumptions which may have also caused errors in the predictions. The next model [5] we shall discuss, while being on the same theoretical basis as the Whitney-Riley model, utilizes a finite-difference solution. Thus certain assumptions can be relaxed, and results improved, at the cost of losing simple designer-oriented formulas.

The model of Ref. [5] allows the longitudinal modulus E_1 and the major Poisson ratio ν_{12} to be predicted from

TABLE 1

Experiment vs Whitney-Riley Theoretical Model for Boron-Epoxy Composite

($E_f = 60 \times 10^6$ psi, $\nu_f = 0.2$, $E_m = 0.6 \times 10^6$ psi, $\nu_m = 0.35$)

	E_1 (psi)			E_2 (psi)			G_{12} (psi)		
Vol % fiber	Experiment	Theoretical	% Dif.	Experiment	Theoretical	% Dif.	Experiment	Theoretical	% Dif.
20	11.7×10^6	12.5×10^6	6.84	--	--	--	--	--	--
55	30.1×10^6	33.3×10^6	7.80	--	--	--	--	--	--
60	35.7×10^6	36.2×10^6	1.40	3.10×10^6	3.08×10^6	0.65	--	--	--
65	35.5×10^6	39.2×10^6	10.41	3.40×10^6	3.55×10^6	4.41	--	--	--
70	34.5×10^6	42.2×10^6	23.30	3.88×10^6	4.20×10^6	8.25	1.77×10^6	1.09×10^6	38.4
75	--	--	--	4.90×10^6	5.00×10^6	2.04	2.43×10^6	1.25×10^6	48.5

the law of mixtures discussed previously, and concentrates on better predictions for the transverse modulus E_2 and the shear modulus G_{12}. The important assumptions of this model are:

(a) No variation of stress in the direction of the fiber is allowed.

(b) A rectangular fiber packing array is assumed as shown in Fig. 1; the repeating element is shown in dashed lines.

(c) The individual fibers have cross sections symmetric about each of the coordinate axes, x_2 and x_3. Other than this requirement, the fiber shape is arbitrary (e.g., circle, diamond, ellipse, sphere).

(d) Perfect fiber-matrix bonding is assumed.

The moduli G_{12} and E_2 are predicted from the solutions of elasticity problems similar to those discussed earlier in connection with the Whitney-Riley model. However, since the solution technique is numerical relaxation using finite differences, the problem formulation can be more rigorous. To obtain E_2, the composite is subjected to transverse normal loads applied at a distance from the repeating element being analyzed. This induces a complex state of stress within the composite and the stress distribution along the sides of the repeating element will not be uniform, although the average stresses must be equal to the average applied stresses. A similar problem with applied shear loads is posed in order to obtain the shear modulus G_{12}. Due to the greater flexibility of the numerical solution scheme as opposed to closed-form solutions, both the above problems can be formulated and solved in displacement form, and interactions between adjacent fibers which induce the non-uniform stresses at the boundaries of the repeating element can be correctly taken into account.

The results of such solutions, using the computer

programs of Ref. [5], showing the effect of various parameters on the transverse modulus E_2 are presented in Figs. 12 and 13. The base E_f for Fig. 13 is 60×10^6 psi, corresponding to typical boron fibers. Various current composite combinations are noted. Similar information, as well as the effects of various fiber shapes, for the shear modulus G_{12} are presented in Figs. 14-16. In Fig. 16, the relevant component properties for glass-epoxy are $G_f = 4.0 \times 10^6$ psi and $G_m = 0.2 \times 10^6$ psi, while the base G_f for Fig. 15 is 24×10^6 psi. The improvement of these results over the Whitney-Riley model can be seen from Table 2, where the epoxy shear modulus is taken as 0.2×10^6 psi. The difference between theory and experiment is much smaller than in the previous model.

Another model requiring a computer solution is presented

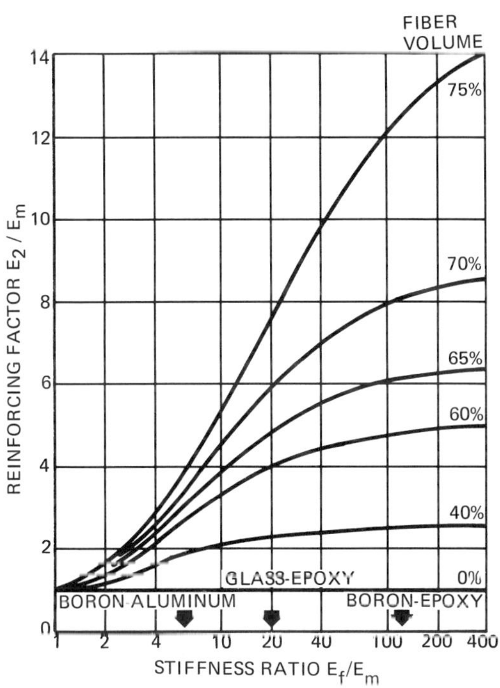

FIG. 12. Composite transverse stiffness for circular fibers in a square array (from Ref. [5]).

TABLE 2

Composite Shear Modulus G_{12} (psi)

Percent fiber volume	Glass-epoxy			Boron-epoxy		
	Experiment	Theoretical	% Dif.	Experiment	Theoretical	% Dif.
70	1.2×10^6	1.1×10^6	9.1	1.5×10^6	1.4×10^6	7.1

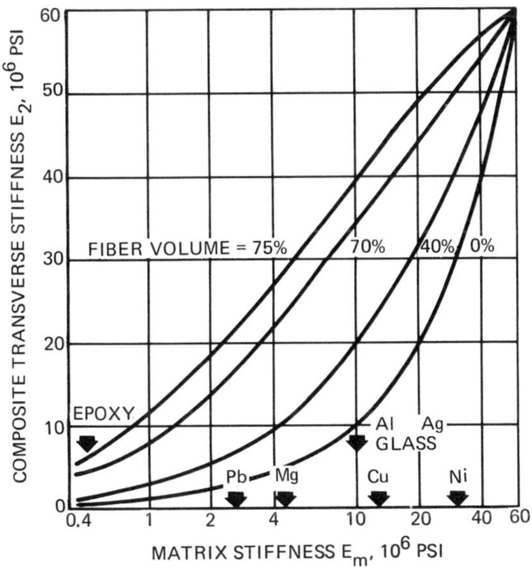

FIG. 13. Composite transverse stiffness for boron fibers as a function of matrix Young's modulus and fiber volume (from Ref. [5]).

in Refs. [6, 7]. With this model, both the rectangular and the hexagonal packing arrays (see Fig. 17) are considered. The material constants that are sought are those for an orthotropic material, Eqs. (1). Equations (1) contain nine independent moduli. This number of independent moduli could be reduced depending upon fiber geometry and packing arrays.

The relevant assumptions made in the model are:

(a) The components are isotropic.

(b) The matrix-fiber bond is perfect.

(c) There are no voids or boundary layer effects at the fiber surface.

(d) Each component is homogeneous and elastic.

(e) Fibers are of uniform size and arranged in

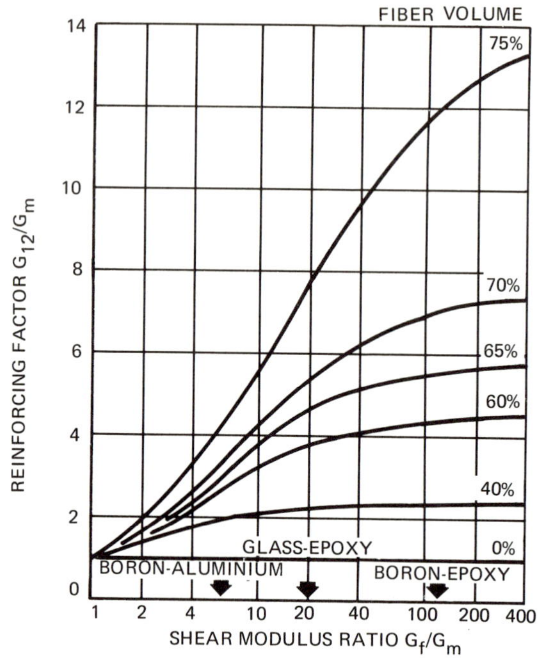

FIG. 14. Composite shear modulus for circular fibers in a square packing array (from Ref. [5]).

a uniform array. The cases analyzed are circular fibers in rectangular or hexagonal arrays.

(f) The composite stress tensor has the same magnitude and orientation throughout every cross section, i.e., the tensor is homogeneous.

The elasticity problems posed to find the necessary moduli are similar to those previously discussed. The method of solution chosen is expansion of the variables (Airy stress function and displacements) in infinite Fourier series, with the boundary conditions being satisfied in the least-square sense. Three least-square type methods were actually used. After the elasticity equations were satisfied by use of the Fourier series expansions in

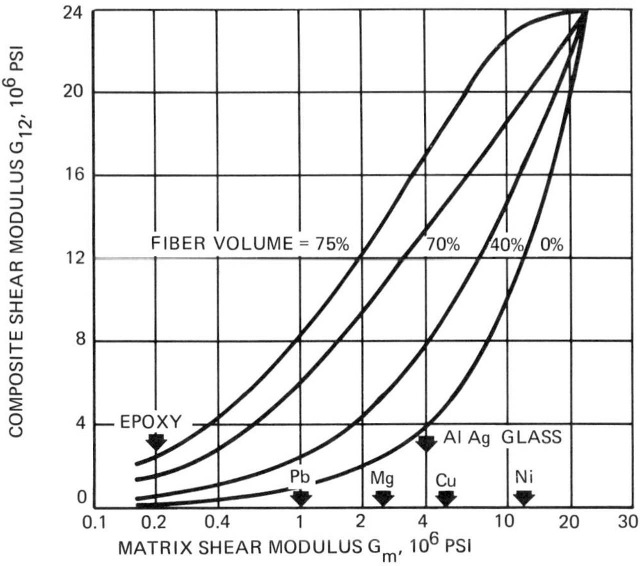

FIG. 15. Composite shear modulus for boron fibers as a function of matrix shear modulus and fiber volume (from Ref. [5]).

the interior, the boundary conditions were satisfied by any of the following methods:

(a) The square of an error over a particular portion of the boundary is minimized with respect to a Fourier series parameter.

(b) The square of the errors integrated along the boundary is minimized by variation of Fourier series parameters.

(c) Combination of the above two methods.

Independent computations were made for almost all of the nine moduli, a cross-check against expected symmetry being used to decide upon the truncation point which was satisfactory in considering the Fourier series expansions.

Unfortunately, no comparison of this model with experi-

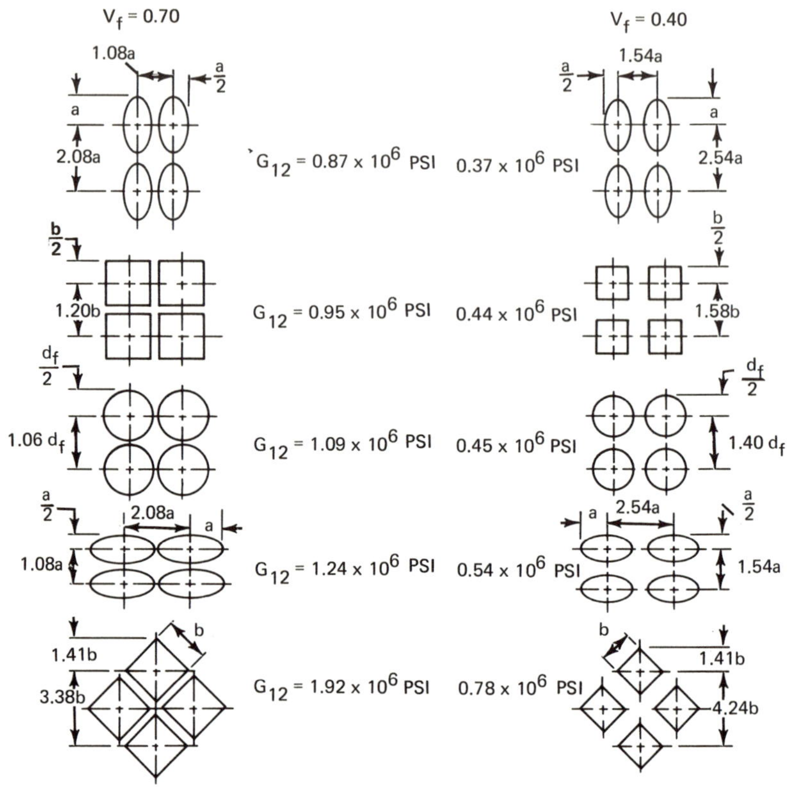

FIG. 16. Shear modulus G_{12} for glass-epoxy composites subjected to an applied shear stress (reproduced from Ref. [5]).

mental data is available in Ref. [6] or [7]. A comparison with another model [10,11] based on the variational techniques to be discussed in Chapter 4 is, however, made and is provided in Fig. 18. It can be seen that the current solution falls within the bounds of Refs. [10,11]. It was also found that the longitudinal stiffness and the major Poisson's ratio were linear functions of the fiber volume content, thus verifying the law of mixtures for these moduli.

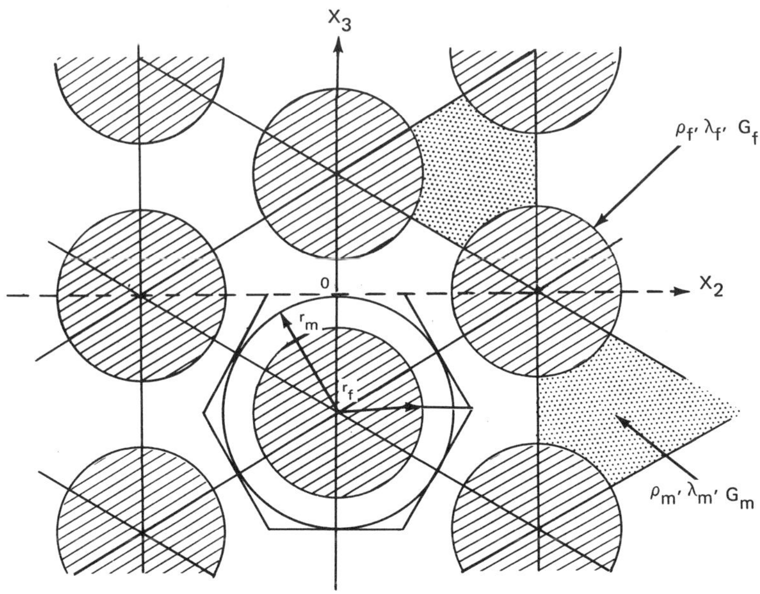

FIG. 17. Hexagonal array.

A model similar to that discussed above is available in Ref. [8]. The main difference between the two formulations is that the latter work utilizes point matching of the boundary conditions at specific points in order to determine the Fourier series coefficients. Again the elasticity equations in the interior can be satisfied by choosing appropriate Fourier series representations of the Airy stress function in polar coordinates. The boundary conditions are then satisfied at discrete points, using enough points to determine all the unknown Fourier series coefficients of the truncated series. This method is perhaps more sensitive to the choice of point locations than the least-square method of Ref. [7]. Unfortunately, no experimental verification is provided for this analysis.

Recently, finite-element techniques have been gaining

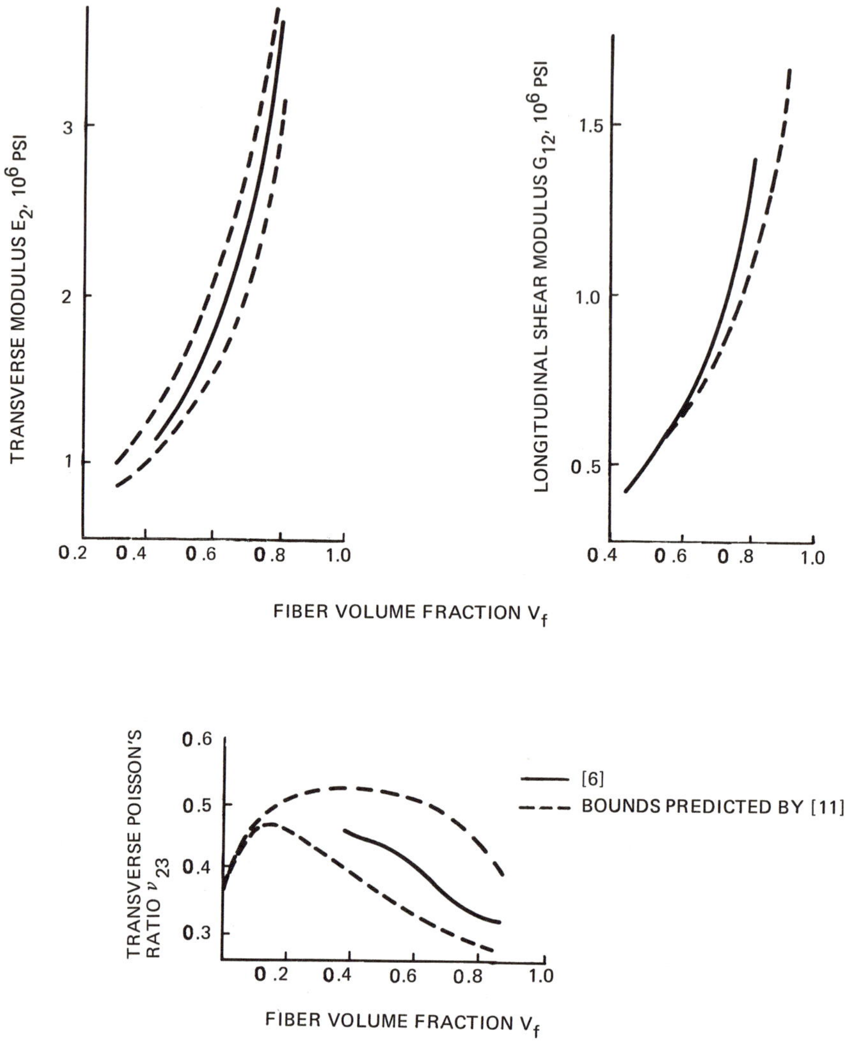

FIG. 18. Predictions for E_2, G_{12}, ν_{23} (from Ref. [6]).

popularity for the determination of the elastic moduli. Since finite elements (like the finite differences) are a numerical tool, and treat the composite problem as any other problem in mechanics, we will not go into finite-element theory in this text. The reader is referred to the books by Zienkiewicz and Cheung [12] and Przemieniecki [13]. The first application of this technique to composite problems was apparently made by Foye [14] to find the elastic moduli of unidirectional composites. The method has since seen increasing application [15-19]. It is difficult to currently evaluate these efforts, partly due to the fact that only a few moduli have been found with this technique. Some of these results will be compared with the values of moduli obtained from other methods in Chapter 4.

G. Anisotropic Reinforcing Filaments

In the previous analyses, the fibers were assumed to be isotropic. With the advent and increased usage of graphite fibers, however, this assumption becomes somewhat questionable. Graphite fibers are highly anisotropic, with the longitudinal stiffness being an order of magnitude greater than the transverse stiffness. Thus additional or corrected analyses are necessary [20,21] to take the anisotropic effects into account.

The simplest correction was suggested by Whitney [20]. He proposed to utilize the appropriate properties of the fiber when calculating the associated directional properties of the composite. If the fiber is transversely isotropic, the independent elastic moduli for the fiber are E_{f_1}, E_{f_2}, $\nu_{f_{12}}$, $G_{f_{12}}$ and $G_{f_{22}}$, where 1 and 2 refer to the longitudinal and transverse directions, respectively. For an isotropic fiber, we have

$$E_{f_1} = E_{f_2}, \quad G_{f_{12}} = G_{f_{22}}, \quad G_{f_{12}} = \frac{E_{f_1}}{2(1 + \nu_{f_{12}})} \qquad (60)$$

Utilizing the "Whitney correction method," the longitudinal stiffness E_1 and major Poisson's ratio ν_{12} of the com posite can still be obtained by the rule of mixtures as

$$E_1 = V_f E_{f1} + V_m E_m \quad \text{and} \quad \nu_{12} = V_f \nu_{f12} + V_m \nu_m \tag{61}$$

The derivation of the transverse composite stiffness E_2 in Section III.C was based on a plane-strain problem. Therefore, only the fiber properties E_{f2} and G_{f22} need be used. Similarly, in the derivation of the composite shear modulus G_{12} direct substitution of G_{f12} can be made.

The above simple correction is utilized by Whitney [20] in the automated solution of Ref. [5] for the prediction of transverse Young's and shear moduli, and the results are compared with experimental data in Figs. 19 and 20. The experimental component properties are

$$E_{f1} = 24 \times 10^6 \text{ psi}, \quad E_m = 0.6 \times 10^6 \text{ psi}$$

$$E_{f2} = 2 \times 10^6 \text{ psi}, \quad \nu_m = 0.3$$

$$\nu_{f12} = 0.3, \quad G_m = 0.23 \times 10^6 \text{ psi}$$

$$G_{f12} = 4.0 \times 10^6 \text{ psi}, \quad \nu_{f22} = 0.15$$

Note that E_{f2} and G_{f12} are only approximate due to experimental difficulty. It is clear from Figs. 19 and 20 that fiber anisotropy characteristics have significant effect on the composite elastic moduli.

The above correction was quite approximate in that it assumed only transverse component properties would affect the transverse properties of the overall composite. This assumption was eliminated by Chen and Cheng [21]. They utilized a procedure similar to the one outlined earlier for isotropic fibers [6], for the analysis of anisotropic fibers. The basic results found are as follows:

(a) The composite shear stiffness G_{12} can be

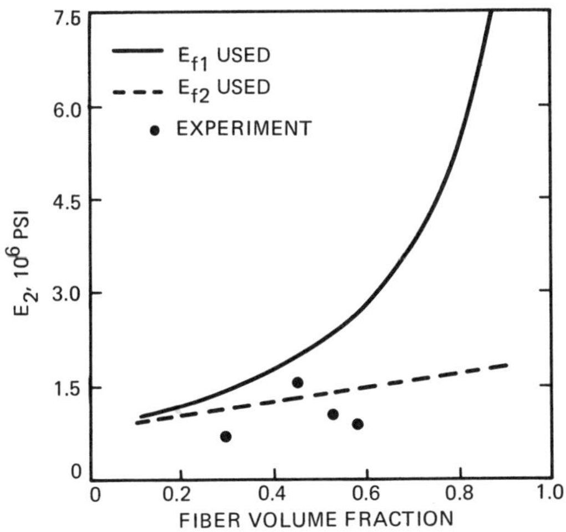

FIG. 19. Transverse modulus as a function of fiber volume for anisotropic and isotropic fiber composites (from Ref. [20]).

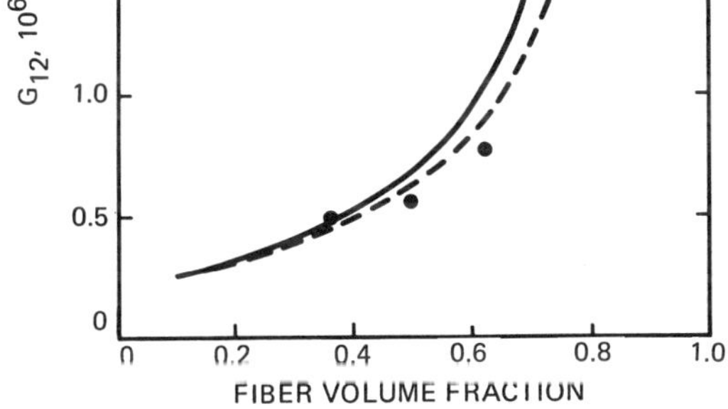

FIG. 20. Shear modulus as a function of fiber volume for anisotropic and isotropic fiber composites (from Ref. [20]).

calculated by using existing isotropic solutions and utilizing $G_f = G_{f12}$.

(b) The rule of mixtures is satisfactory for the calculation of the longitudinal modulus E_1 and the major Poisson's ratio ν_{12} [see Eq. (60)].

(c) Changes in the values of ν_{f22} and ν_{f12} have little effect on the composite Poisson's ratios or the transverse modulus E_2.

(d) Changes in E_{f2} have appreciable effects on the composite transverse modulus, slight effect on the composite Poisson's ratios, and practically no effect on the composite longitudinal modulus.

Lest the above conclusions be used to completely justify the "Whitney correction" technique, Fig. 21 presents the predictions of the above two methods for a hypothetical composite with the following component properties:

$$E_{f1} = 50 \times 10^6 \text{ psi}, \quad E_m = 6 \times 10^5 \text{ psi}$$

$$E_{f2} = 2 \times 10^6 \text{ psi}, \quad G_m = 2.22 \times 10^5 \text{ psi}$$

$$G_{f12} = 10 \times 10^6 \text{ psi}, \quad \nu_m = 0.35$$

$$\nu_{f12} = 0.20, \quad \nu_{f22} = 0.15$$

As can be seen, the predictions for G_{12} are essentially the same, but those for E_2 are not in agreement. Comparisons of the current theory with the experimental data of Ref. [20] are also shown in Figs. 22 and 23. It is seen that the theory and experiment are in good agreement for both G_{12} and E_2.

H. Fiber Composite Thermal Properties

Until this point, our primary concern has been with

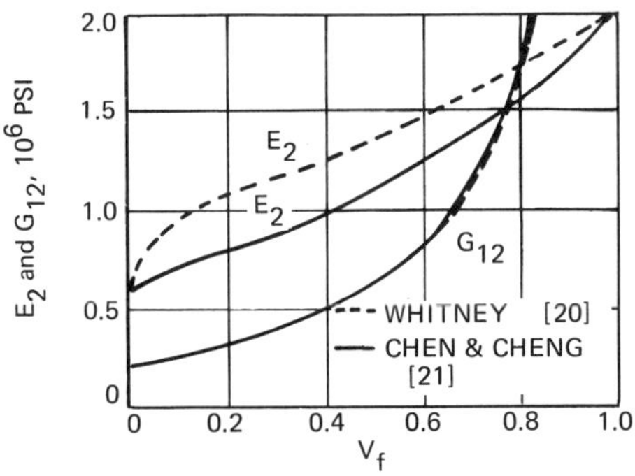

FIG. 21. Numerical results for composites with anisotropic fibers (from Ref. [21]).

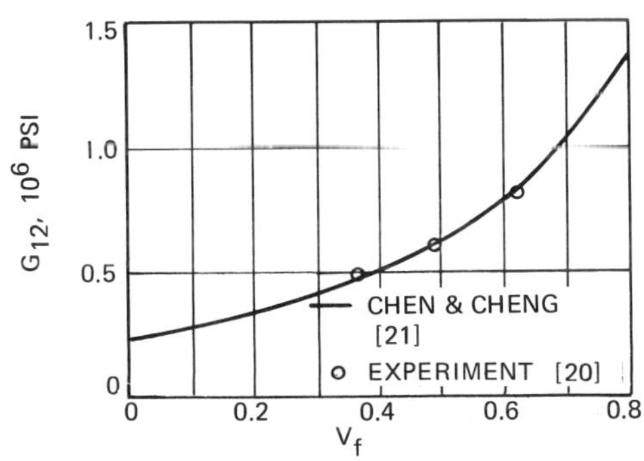

FIG. 22. Experimental comparison, G_{12} (from Ref. [21]).

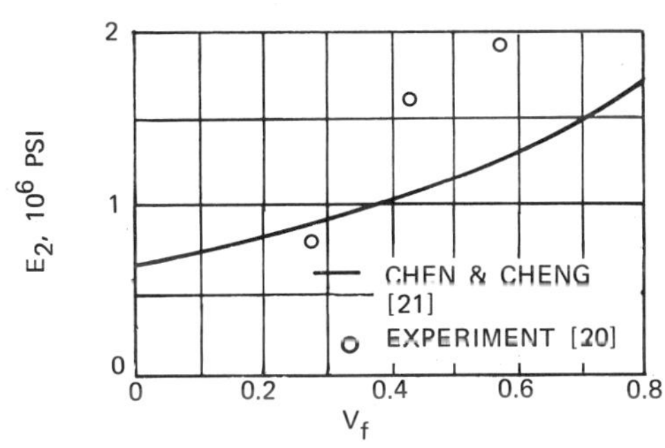

FIG. 23. Experimental comparison, E_2 (from Ref. [21]).

the elastic material properties of a composite. Since thermal problems are often encountered in service life, prediction of thermal constants is also necessary. One of the earliest thermal models is due to Greszczuk [22]. This model is essentially a mechanics-of-materials type, very similar to the round-filament Ekvall [2] model (Fig. 24) and with the same basic assumptions.

Applying a temperature T to the Greszczuk round-filament model and considering the longitudinal (x_1) direction, we can write from equilibrium

$$\sigma_{m1} V_m + \sigma_{f1} V_f = 0, \quad \sigma_{m2} = \sigma_{f2} = 0 \tag{62a}$$

Strain compatibility requires

$$\varepsilon_{m1} = \varepsilon_{f1} = \varepsilon_{c1} \tag{62b}$$

Application of Hooke's laws yields

$$\varepsilon_{f1} = (\sigma_{f1}/E_f) + \alpha_f T, \quad \varepsilon_{m1} = (\sigma_{m1}/E_m) + \alpha_m T$$

$$\varepsilon_{c1} = \alpha_1 T \tag{63}$$

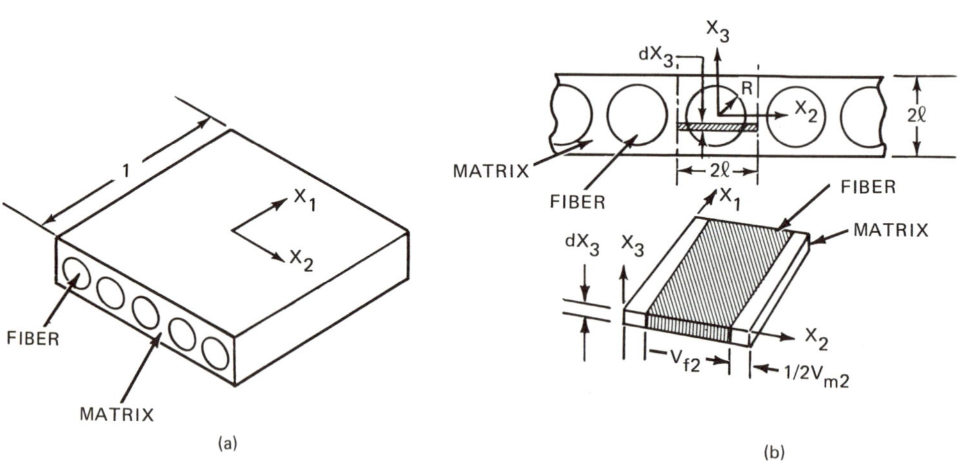

FIG. 24. Model for determining α_1 and α_2.

where α_f, α_m, and α_1 denote the coefficients of linear expansion for the fiber, the matrix, and the composite (x_1-direction) respectively. Combining Eqs. (62) and (63) we obtain

$$\alpha_1 = (\alpha_m V_m E_m + \alpha_f V_f E_f)/(E_m V_m + E_f V_f) \tag{64}$$

In the transverse direction, due to the round shape of the fiber, as mentioned earlier, the composite stiffness E_2 varies across the model width. Thus at any point x_3 (in the x_3-direction), for an element dx_3 to be in equilibrium, and for compatibility to be satisfied, we have

$$\sigma_{m1} V_{m2} + \sigma_{f1} V_{f2} = 0 \tag{65a}$$

$$\sigma_{m2} = \sigma_{f2} = \sigma_2 = 0 \tag{65b}$$

$$\varepsilon_{m2} V_{m2} + \varepsilon_{f2} V_{f2} = \varepsilon_2(x_3) \tag{65c}$$

$$\varepsilon_{m1} = \varepsilon_{f1} = \varepsilon_{c1} \tag{65d}$$

where V_{f2} and V_{m2} denote the fiber and the matrix volume fractions at height x_3 respectively. Hooke's law for this element is given by

$$\varepsilon_{f2} = (\sigma_{f2}/E_f) - \nu_f(\sigma_{f1}/E_f) + \alpha_f T \tag{66a}$$

$$\varepsilon_{m2} = (\sigma_{m2}/E_m) - \nu_m(\sigma_{m1}/E_m) + \alpha_m T \tag{66b}$$

$$\varepsilon_{f1} = (\sigma_{f1}/E_f) - \nu_f(\sigma_{f2}/E_f) + \alpha_f T \tag{66c}$$

$$\varepsilon_{m1} = (\sigma_{m1}/E_m) - \nu_f(\sigma_{m2}/E_m) + \alpha_m T \tag{66d}$$

$$\varepsilon_2(x_3) = \alpha_2(x_3) T \tag{66e}$$

Combining Eqs. (65a), (65b), (65d), (66c) and (66d) gives

$$\sigma_{m1} = (\alpha_f - \alpha_m)E_m E_f V_{f2} T/(E_m V_{m2} + E_f V_{f2})$$
$$\sigma_{f1} = -(\alpha_f - \alpha_m)E_m E_f V_{m2} T/(E_m V_{m2} + E_f V_{f2}) \tag{67}$$

Combining Eqs. (65b), (65c), (66a), (66e), and (67), we can obtain $\alpha_2(x_3)$, the thermal expansion of the element dx_3 in the x_2-direction, as

$$\alpha_2(x_3) = \alpha_m V_{m2} + \alpha_f V_{f2} - \left(\frac{\nu_m}{E_m} - \frac{\nu_f}{E_f}\right) \cdot \frac{(\alpha_f - \alpha_m)E_f E_m V_{m2} V_{f2}}{E_m V_{m2} + E_f V_{f2}} \tag{68}$$

To obtain the average α_2, it is convenient to define the average strain ε_2 as follows:

$$\varepsilon_2 = (1/2\ell)\int_{-\ell}^{\ell} \varepsilon_2(x_3)\, dx_3 = \alpha_2 T \tag{69}$$

where α_2 in this equation refers to the average value. Thus $\varepsilon_2(x_3)$ at any x_3 may be thought to consist of two parts, (1) the average thermal strain ε_2, and (2) $\varepsilon_2'(x_3)$ ∿ the local variation from ε_2:

$$\varepsilon_2(x_3) = \alpha_2(x_3)T = \varepsilon_2 + \varepsilon_2'(x_3) \tag{70}$$

The local strain $\varepsilon_2'(x_3)$ causes a local stress $\sigma_2(x_3)$ in the dx_3 element. Neglecting terms involving Poisson's ratios, $\sigma_2(x_3)$ may be approximately written as

$$\sigma_2(x_3) \approx E_2(x_3)\varepsilon_2'(x_3) \tag{71}$$

where $E_2(x_3)$ is given by Eq. (12a) for $0 \leq x_3 < R$, and $E_2(x_3) = E_m$ for $R \leq x_3 \leq \ell$. Now in order to satisfy Eq. (65b), we must require that

$$\int_{-\ell}^{\ell} E_2(x_3)\varepsilon_2'(x_3)dx_3 \simeq \int_{-\ell}^{\ell} \sigma_2(x_3)dx_3 = 0 \tag{72}$$

Therefore we obtain on multiplying Eq. (70) with $E_2(x_3)$ and integrating from $-\ell$ to ℓ

$$T \int_{-\ell}^{\ell} \alpha_2(x_3)E_2(x_3)\, dx_3 = \int_{-\ell}^{\ell} \varepsilon_2 E_2(x_3)\, dx_3$$

$$= \alpha_2 T \int_{-\ell}^{\ell} E_2(x_3)\, dx_3$$

Hence

$$\alpha_2 = \frac{\int_{-\ell}^{\ell} \alpha_2(x_3)E_2(x_3)\, dx_3}{\int_{-\ell}^{\ell} E_2(x_3)\, dx_3} \tag{73}$$

Due to the complexity of expressions for $\alpha_2(x_3)$ and $E_2(x_3)$, Eq. (73) cannot be evaluated analytically, and it is necessary to resort to numerical integration. Unfortunately, no comparisons between theory and experiment are available for α_1 and α_2 as given by Eqs. (64) and (73).

Another important composite property which must be predicted is the thermal conductivity. A numerical method for predicting the transverse thermal conductivity is discussed in Ref. [23]. Since the equations for the thermal composite problem reduce to those for the prediction of the shear modulus [5], the computer program utilized in Ref. [5] is directly applicable. Only substitution of the component thermal data need be made in place of the elastic modulus input data used previously. A comparison of the predictions of this program for κ_2 with the experimental data of Ref. [24] is shown in Fig. 25. As can be seen, correlation of results is good. The longitudinal thermal conductivity, as with most longitudinal properties for unidirectionally reinforced composites, can be adequately predicted by the law of mixtures:

$$\kappa_1 = V_f\kappa_f + (1 - V_f)\kappa_m \tag{74}$$

IV. PARTICLE COMPOSITES

The composites heretofore considered have all consisted of unidirectional fibers reinforcing a given matrix material. While this may be the dominant type of composite in terms of new usage, it is only a minor family of the types of possible composites. One need only to look at concrete to find an example of the much larger and diverse particle composite family. In terms of analysis, the two families are substantially different. The unidirectional fiber composite is well defined and the main geometric variables are fiber packing and cross-section shape. For a particle composite, one needs to consider the grain shape and the distribution. A particle composite may consist of N - 1 phases embedded in the Nth (matrix) phase. Such a composite, regardless of local irregularities, can be considered to be homogeneous and isotropic in the large.

A majority of the efforts in finding the effective properties of particle composites consist in stating universal bounds by the use of variational techniques. As

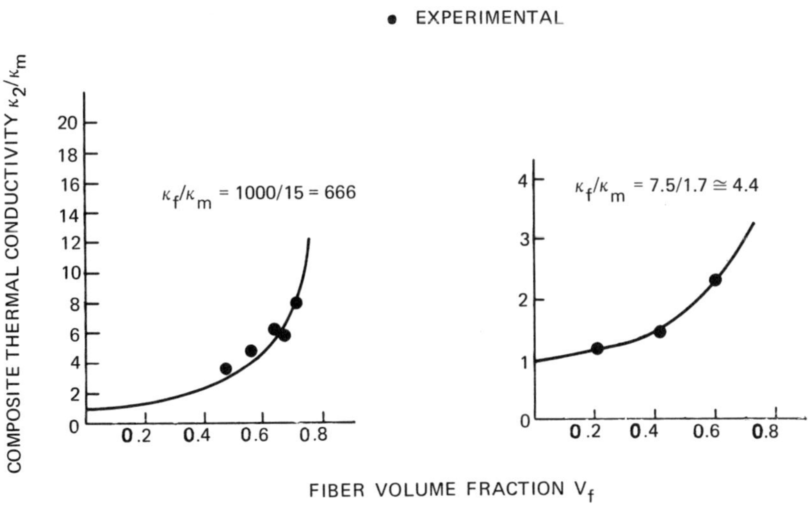

FIG. 25. Comparison between theory and experimental data. Experimental data from Thornburg and Pears [24] for glass-plastic (κ_f/κ_m = 4.4) and graphite fabric-plastic (κ_f/κ_m = 666) composites are shown (from Ref. [23]).

mentioned previously, these efforts will be discussed in Chapter 4. Presently, we will discuss an elasticity model which finds exact formulas for predicting overall moduli. The model to be considered is the "self-consistent model" (as defined below) adopted for composites by Hill [25-27]. This model was originally proposed by Hershey [28] for polycrystalline aggregates, and the argument is basically the following. The environment of each particle in a composite involves the matrix and many other particles. The interaction of a local particle with its environment creates local deformations. The local strain (for elastic deformations) can be obtained from the local deformation by use of linear differential equations, and the local stress at equilibrium also involves linear differential equations. Thus, if all the particles of the same phase were superimposed and their strains added together, the conglomerate strain and stress would still be obtained from the conglomerate displacement by use of the same differential equations as the local strain and stress. The average strain pattern in the neighborhood of a particle is therefore the same as the strain pattern for an overall particle immersed in a continuous matrix with the average properties of the composite. Under any overall loading uniform at infinity, the inclusion strain is then readily calculable [25-27]. The basic idea of the self-consistent model (and actually its definition) is the adoption of this strain as the average over all the inclusions of a particular phase, thus allowing the composite moduli to be determined uniquely.

As an example, we adopt the one presented by Budiansky [29,30]. Consider the problem of a single spherical inclusion in an infinite isotropic solid of bulk modulus K_c and Poisson's ratio ν_c. If at infinity and the cavity boundary we apply normal traction rates $\dot{\sigma}$ and $\dot{\sigma}_i$, respectively, letting $\dot{e}$ be the resulting dilatation rate at infinity, it can be shown [29,30] that the cavity displacements are consistent with a uniform dilatation rate $\dot{e}_i$ at the cavity. Furthermore, we have

$$\dot{\sigma}_i - \dot{\sigma} = K_c[(2 - 4\nu_c)/(1 + \nu_c)](\dot{e} - \dot{e}_i) \qquad (75)$$

The key approximation now made in the self-consistent model is to assume that the above equation holds for the relation between the average hydrostatic stress rate $\dot{\sigma}_i$ and average dilatation rate $\dot{e}_i$ in the ith component $(i = 1, 2, \ldots, N - 1)$, with K_c and ν_c being the unknown macroscopic properties of the composite. Thus $\dot{\sigma}$ and $\dot{e}$ become the mean hydrostatic stress and dilatation rates, respectively. Since

$$\dot{\sigma} = \sum_{i=1}^{N} V_i \dot{\sigma}_i, \quad \dot{e} = \sum_{i=1}^{N} V_i \dot{e}_i \qquad (76)$$

where V_i is the ith-component volume fraction, it is readily seen that Eq. (75) holds for the matrix (Nth component) as well. Substituting

$$\dot{\sigma}_i = K_i\dot{e}_i, \quad \dot{\sigma} = K_c\dot{e} \tag{77}$$

and Eq. (76) into Eq. (75), we can obtain the equation for the composite bulk modulus as

$$\sum_{i=1}^{N} V_i\left[1 - \frac{1 + \nu_c}{3 - 3\nu_c} + \frac{K_i}{K_c}\left(\frac{1 + \nu_c}{3 - 3\nu_c}\right)\right]^{-1} = 1 \tag{78}$$

In an entirely analogous fashion, by using pure shear states, we may obtain the composite shear modulus G_c,

$$\sum_{i=1}^{N} V_i\left[1 - \frac{8 - 10\nu_c}{15 - 15\nu_c} + \frac{G_i}{G_c}\left(\frac{8 - 10\nu_c}{15 - 15\nu_c}\right)\right]^{-1} = 1 \tag{79}$$

Equations (78) and (79) are actually simultaneous equations with the common factor ν_c being obtained from the isotropic identity

$$\nu_c = (3K_c - 2G_c)/(6K_c + 2G_c)$$

With $N = 2$, Eqs. (78) and (79) are the same as those obtained by Hill [27] for spherical particles reinforcing a matrix material.

The self-consistent particulate composite model may also be utilized in obtaining composite thermal properties [30]. If, for instance, $\dot{\sigma} = 0$ and

$$\dot{e}_i = (\dot{\sigma}_i/K_i) + \beta_i\dot{T} \tag{80}$$

is substituted into Eq. (75), we obtain

$$\dot{e}_i = \frac{[(1 + \nu_c)/(3 - 3\nu_c)]K_i\beta_i}{[(1 + \nu_c)/(3 - 3\nu_c)]K_i + K_c[(2 - 4\nu_c)/(3 - 3\nu_c)]}\dot{T}$$

$$+ \frac{K_c[(2 - 4\nu_c)/(3 - 3\nu_c)]}{[(1 + \nu_c)/(3 - 3\nu_c)]K_i + K_c[(2 - 4\nu_c)/(3 - 3\nu_c)]} \dot{e} \tag{81}$$

Utilizing Eq. (75), the second of Eqs. (76), and $\dot{e} = \beta_c \dot{T}$ in the above equation finally gives the volumetric thermal expansion coefficient β_c of the composite:

$$\beta_c = \sum_{i=1}^{N} V_i\left(\frac{K_i}{K_c}\right)\beta_i \left[1 - \frac{1 + \nu_c}{3 - 3\nu_c} + \frac{K_i}{K_c}\left(\frac{1 + \nu_c}{3 - 3\nu_c}\right)\right]^{-1} \tag{82}$$

Utilizing $N = 2$, the above equation can be reduced to Levin's [31] result:

$$(\beta_c - \beta_1)/(\beta_2 - \beta_1) = \left(K_c^{-1} - K_1^{-1}\right)/\left(K_2^{-1} - K_1^{-1}\right) \tag{83}$$

Other thermal composite constants similarly found by Budiansky [30] are as follows:

Heat capacity per unit volume, at constant volume, C_v:

$$C_v = \sum_{i=1}^{N} V_i C_{vi} + \left(\frac{1 + \nu_c}{3 - 3\nu_c}\right) K_c \beta_c^2 T \left\{\sum_{i=1}^{N} V_i \left(\frac{K_i}{K_c}\right)^2 \left(\frac{\beta_i}{\beta_c}\right)^2 \left[1 - \left(\frac{1 + \nu_c}{3 - 3\nu_c}\right) + \frac{K_i}{K_c}\left(\frac{1 + \nu_c}{3 - 3\nu_c}\right)\right]^{-1} - 1\right\} \tag{84}$$

Heat capacity per unit volume, at constant pressure, C_p:

$$C_p = C_v + K_c \beta_c^2 T \tag{85}$$

Thermal conductivity, κ_c:

$$\sum_{i=1}^{N} V_i \left[\frac{2}{3} + \frac{1}{3}\left(\frac{\kappa_i}{\kappa_c}\right)\right]^{-1} = 1 \tag{86}$$

While it has been noted that the self-consistent model gives unreliable results at extreme ratios of the stiffness of the phases, no comparison with experiment is available in Refs. [27, 29, 30, 31]. The model predictions are found to be within the bounds obtained by analyses such as those of Ref. [10]. The ideas of the self-consistent model are not restricted to any particular composite reinforcement geometry. They apply equally well to particles, fibers, and plate like reinforcements. While we chose to introduce the model in the discussion of particle-reinforced composites, it should be pointed out that the same ideas have also been applied to composites reinforced by aligned fibers [26].

As mentioned previously, all the analyses in this chapter were aimed at providing a single formula or numerical results for a given elastic composite property. The variational approaches to be discussed in the next chapter aim to eliminate uncertainty in the analytical models of the composite by obtaining bounds on the actual composite properties. The usefulness of these bounds depends, of course, upon their relative closeness. The basic advantage of the bounding approach is that less specific geometric models need be used, and thus certain geometric idealizations and assumptions of the models presented in this chapter can be avoided. The drawback, of course, is the difficulty in finding bounds close enough to be useful without falling back again on geometric model idealizations. These bounding analyses will form the basis of discussion for Chapter 4.

REFERENCES

1. J. M. Whitney and M. B. Riley, "Elastic Stress-Strain Properties of Fiber-Reinforced Composite Materials," Air Force Materials Laboratory Report AFML-TR-65-238, December 1965.

2. J. C. Ekvall, "Elastic Properties of Orthotropic Monofilament Laminates," ASME Paper No. 61-AV-56, Presented at the ASME Aviation Conference, March 1961.

3. J. C. Ekvall, "Structural Behavior of Monofilament Composites," Presented at the AIAA 6th Structures and Materials Conference, April 1965.

4. J. M. Whitney, "Elastic Moduli of Composite Materials Reinforced with Orthotropic Filaments," Air Force Materials Laboratory Report AFML-TR-65-411, January 1966.

5. S. W. Tsai, D. F. Adams, and D. R. Doner, "Effect of Constituent Material Properties on the Strength of Fiber-Reinforced Composite Materials," Air Force Materials Laboratory Report AFML-TR-66-190, August 1966.

6. C. H. Chen and S. Cheng, "Mechanical Properties of Fiber-Reinforced Composites," J. Composite Materials, 1, 30-41 (1967).

7. G. Pickett, "Elastic Moduli of Fiber-Reinforced Plastic Composites," in Fundamental Aspects of Fiber Reinforced Plastic Composites, Interscience, New York, 1968, pp. 13-27.

8. A. W. Leissa and W. E. Clausen, "Application of Point Matching to Problems in Micromechanics," in Fundamental Aspects of Fiber-Reinforced Plastic Composites, Interscience, New York, 1968, pp. 29-44.

9. T. S. Chow and J. J. Hermans, "The Elastic Constants of Fiber-Reinforced Materials," J. Composite Materials, 3, 382-396 (1969).

10. Z. Hashin and B. W. Rosen, "The Elastic Moduli of Fiber-Reinforced Materials," J. Appl. Mechanics, 31, 223-232 (1964).

11. N. F. Dow and B. W. Rosen, "Evaluations of Filament-Reinforced Composites for Aerospace Structural Applications," NASA Report CR-207, April 1965.

12. O. Zienkiewicz and Y. Cheung, The Finite-Element Method in Structural and Continuum Mechanics, McGraw-Hill, London, 1967.

13. J. Przemieniecki, Theory of Matrix Structural Analysis, McGraw-Hill, New York, 1968.

14. R. L. Foye, "An Evaluation of Various Engineering Estimates of the Transverse Properties of Unidirectional Composites," Science of Advanced Materials and Process Engineering Series, Vol. 10, Western Periodicals Co., pp. G-31 to G-42, 1966.

15. R. R. Rizzo, "More on the Influence of End Constraints on Off-Axis Tensile Tests," J. Composite Materials, 3, 202-219 (1969).

16. R. B. Enie and R. R. Rizzo, "Three-Dimensional Laminate Moduli," J. Composite Materials, 4, 150-154 (1970).

17. D. F. Adams, "Inelastic Analysis of a Unidirectional Composite Subjected to Transverse Normal Loading," RAND Corp. Report RM-6245-PR, May 1970.

18. R. R. Rizzo and A. A. Vicario, "A Finite-Element Analysis of Laminated Anisotropic Tubes: Part 1--A Characterization of the Off-Axis Tensile Specimen," J. Composite Materials, 4, 344-359 (1970).

19. G. Isakson and A. Levy, "Finite-Element Analysis of Interlaminar Shear in Fibrous Composites," J. Composite Materials, 5, 273-276 (1971).

20. J. M. Whitney, "Elastic Moduli of Unidirectional Composites with Anisotropic Filaments," J. Composite Materials, 1, 188-193 (1967).

21. C. H. Chen and S. Cheng, "Mechanical Properties of Anisotropic Fiber-Reinforced Composites," J. Appl. Mechanics, 37, 186-189 (1970).

22. L. B. Greszczuk, "Thermoelastic Properties of Filamentary Composites," presented at the AIAA 6th Structures and Materials Conference, April 1965.

23. S. G. Springer and S. W. Tsai, "Thermal Conductivities of Unidirectional Materials," J. Composite Materials, 1, 166-173 (1967).

24. J. D. Thornburgh and C. D. Pears, "Prediction of the Thermal Conductivity of Filled and Reinforced Plastics," paper no. 65/WA/HT-4, ASME Winter Annual Meeting, 1965.

25. R. Hill, "Continuum Micro-Mechanics of Elastoplastic Polycrystals," J. Mech. Phys. Solids, 13, 89-101 (1965).

26. R. Hill, "Theory of Mechanical Properties of Fiber-Strengthened Materials III: Self-Consistent Model," J. Mech. Phys. Solids, 13, 189-198 (1965).

27. R. Hill, "A Self-Consistent Mechanics of Composite Materials," J. Mech. Phys. Solids, 13, 213-222 (1965).

28. A. V. Hershey, "The Elasticity of an Isotropic Aggregate of Anisotropic Cubic Crystals," J. Appl. Mechanics, 21, 236-240 (1954).

29. B. Budiansky, "On the Elastic Moduli of Some Heterogeneous Materials," J. Mech. Phys. Solids, 13, 223-227 (1965).

30. B. Budiansky, "Thermal and Thermoelastic Properties of Isotropic Composites," J. Composite Materials, 4, 286-295 (1970).

31. V. M. Levin, "Thermal Expansion Coefficients of Heterogeneous Materials," Mekhanika Tverdovo Tela, 2(1), 88-94 (1967) [Eng. trans.: Mechanics of Solids 2(1), 58-61 (1967)].

CHAPTER 4

BOUNDS ON ELASTIC PROPERTIES BY ENERGY METHODS

I. INTRODUCTION

In the preceding chapter, several mathematical models were introduced in order to obtain estimates of the composite material properties. All of the models had one common feature, i.e., a specific idealized geometry and packing arrangement had to be assumed. In many composite materials, this regularity in geometry does not exist, and the material cannot be identified as having a hexagonal, or a rhombic, or a square packing arrangement. Thus the models of Chapter 3, in the strictest sense, do not truly represent a real composite. One way to avoid this idealization problem is through the use of the variational techniques to be presented in this chapter, which may but do not specifically need to use a precise geometric model.

II. ELEMENTARY BOUNDS

The earliest bounds on the properties of a multiphase material were actually first derived for polycrystals. Voigt [1] assumed that under a uniform state of strain on the boundary, each constituent would also be in the same state of uniform strain. Using Hooke's law, this leads to the following estimates for composite bulk and shear moduli:

$$K_V = \sum_{i=1}^{n} V_i K_i \tag{1a}$$

$$G_V = \sum_{i=1}^{n} V_i G_i \tag{1b}$$

where V_i are the phase volume fractions, and n denotes the number of phases. The corresponding assumption on stress states is due to Reuss [2]. Under a uniform stress state at the composite boundary, the various constituents are assumed to be in a state of uniform stress. This yields the following values for composite bulk and shear moduli:

$$1/K_R = \sum_{i=1}^{n} (V_i/K_i) \tag{2a}$$

$$1/G_R = \sum_{i=1}^{n} (V_i/G_i) \tag{2b}$$

It should be noted that neither of the above assumptions is correct. The implied stresses in the Voigt assumption are such that the tractions at the constituent interfaces would not be in equilibrium, while the implied Reuss strains are such that displacement continuity at constituent interfaces is violated. It can be shown that the Voigt values always exceed the Reuss ones. Thus Voigt and Reuss estimates serve as upper and lower bounds, respectively, on the composite elastic properties. The difference between the two values is small for relatively small differences between the phases but becomes rather large when one of the n phases is relatively rigid or weak.

Bounds similar to Eqs. (1) and (2) were derived by Paul [3] for multiphase materials by use of the minimum energy theorems. The only assumptions necessary were isotropy and homogeneity of the phases giving rise to a macroscopically quasiisotropic, quasihomogeneous composite. We shall discuss this derivation in some detail as an example of the use of the energy theorems in obtaining bounds on elastic properties.

The internal energy of a body under deformation may be written as

$$U = \frac{1}{2}\int_V \left(\sigma_1\varepsilon_1 + \sigma_2\varepsilon_2 + \sigma_3\varepsilon_3 + \tau_{12}\gamma_{12} + \tau_{23}\gamma_{23} + \tau_{31}\gamma_{31}\right) dV \qquad (3)$$

If, through the use of appropriate Hooke's laws, the above equation is written containing stress components only, we may define the result as the complementary energy U^σ, and if only strain components are used, the result may be called the strain energy U^ε. For a linear Hookean material, of course, the two energies are equal:

$$U^\sigma = U^\varepsilon \qquad (4)$$

The necessary energy theorems can now be stated. The

theorem of minimum complementary energy can be phrased as follows: Let a stress state be specified within the body which satisfies the traction boundary conditions and the equations of equilibrium throughout the body. The complementary energy obtained from this state, U_0^σ, will then always be greater than or equal to the actual complementary energy U^σ in the body:

$$U^\sigma \leq U_0^\sigma \tag{5a}$$

It should be noted that the strains from the assumed stress state will not, in general, satisfy compatibility requirements. The second required theorem, the theorem of minimum potential (or strain) energy, may be stated as follows: Let a strain state be specified which satisfies the displacement boundary conditions and strain compatibility throughout the body. The strain energy obtained from this state U_0^ε will then always be greater than or equal to the actual strain energy U^ε in the body:

$$U^\varepsilon \leq U_0^\varepsilon \tag{5b}$$

In this case, the stresses calculated from the assumed strain state will not, in general, satisfy equilibrium.

By defining appropriate stress and strain states, reflecting the degree of complexity desired and the amount of specific geometric information one wishes to employ, and utilizing the above theorems, various bounds on the composite properties may be found. To find the composite Young's modulus E_c, for example, Paul [3] assumes that the composite is loaded in uniform tension, σ, and chooses the internal stress state as

$$\sigma_1^0 = \sigma, \quad \sigma_2^0 = \sigma_3^0 = \tau_{12}^0 = \tau_{32}^0 = \tau_{13}^0 = 0 \tag{6}$$

The complementary energy is therefore (for a two-phase material)

$$U_0^\sigma = \frac{\sigma^2}{2} \int_V \frac{dV}{E} = \frac{\sigma^2}{2} \left(\frac{V_1}{E_1} + \frac{V_2}{E_2} \right) V \tag{7a}$$

$$U^\sigma = \frac{\sigma^2}{2} \frac{V}{E_c} \tag{7b}$$

Upon using Eq. (7a) as an upper bound to the actual complementary energy, Eq. (7b), we obtain the following lower bound upon Young's modulus:

$$1/E_c \leq (V_1/E_1) + (V_2/E_2) \tag{8}$$

To obtain an upper bound, Paul [3] assumes the elongation of the composite specimen of length ℓ to be $\varepsilon\ell$ in the x_1-direction, and chooses the internal strain state to be

$$\varepsilon_1^0 = \varepsilon, \quad \varepsilon_2^0 = \varepsilon_3^0 = -m\varepsilon \tag{9a}$$

$$\gamma_{12}^0 = \gamma_{32}^0 = \gamma_{13}^0 = 0 \tag{9b}$$

where m is an unspecified constant. The strain energy for this assumed state can be written as [3]

$$U_0^\varepsilon = \frac{\varepsilon^2}{2} \int_V \left(\frac{1 - \nu - 4\nu m + 2m^2}{1 - \nu - 2\nu^2} \right) E dV \tag{10a}$$

$$U^\varepsilon = \frac{1}{2} \varepsilon^2 E_c V \tag{10b}$$

where ν is the Poisson's ratio. If the Poisson's ratios of the two phases are equal, then $\nu_1 = \nu_2 = \nu = m$, and substituting Eqs. (10a) and (10b) into Eq. (5b), we obtain an upper bound for the composite Young's modulus:

$$E_c \leq V_1E_1 + V_2E_2 \tag{11}$$

If the phase Poisson's ratios are unequal, the unknown constant m can be found by minimization of the strain energy equation (10a)

$$\partial U_0^{\varepsilon}/\partial m = 0 \tag{12a}$$

Thus

$$m = \frac{\nu_1(1 + \nu_2)(1 - 2\nu_2)V_1E_1 + \nu_2(1 + \nu_1)(1 - 2\nu_1)V_2E_2}{(1 + \nu_2)(1 - 2\nu_2)V_1E_1 + (1 + \nu_1)(1 - 2\nu_1)V_2E_2} \tag{12b}$$

The upper bound on Young's modulus is then given by

$$E_c \leq \frac{1 - \nu_1 + 2m(m - 2\nu_1)}{1 - \nu_1 - 2\nu_1^2} E_1V_1 + \frac{1 - \nu_2 + 2m(m - 2\nu_2)}{1 - \nu_2 - 2\nu_2^2} V_2E_2 \tag{13}$$

Note that Eq. (12b) reduces to $m = \nu$ and Eq. (13) reduces to Eq. (11) for $\nu_1 = \nu_2 = \nu$. It should be noted that in finding these bounds on the composite Young's modulus, no specific geometric information was utilized as to any packing arrangement, etc.

The analogous bounds for the shear modulus G_c can be found if the composite is considered as undergoing uniform shear strain under a simple uniform shear stress. Thus

$$\frac{1}{(V_1/G_1) + (V_2/G_2)} \leq G_c \leq V_1G_1 + V_2G_2 \tag{14}$$

The predicted bounds are plotted against experimental results in Fig. 1. The material considered is an alloy system of tungsten carbide and cobalt, with the constituent properties being

$$E_1 = 30 \times 10^6 \text{ psi}, \; E_2 = 102 \times 10^6 \text{ psi}$$

$$\nu_1 = 0.3, \; \nu_2 = 0.22$$

The experimental results are taken from Refs. [4] and [5]. As can be seen from the figure, the bounds predicted for this alloy are substantially far apart. As the difference between the two phases increases, the bounds move even farther apart. If one phase is very rigid, or weak, the upper or lower bounds could approach infinity or zero in the limit, respectively. Thus these bounds are really useful only in the small range where the phase moduli differ by a factor of two or less.

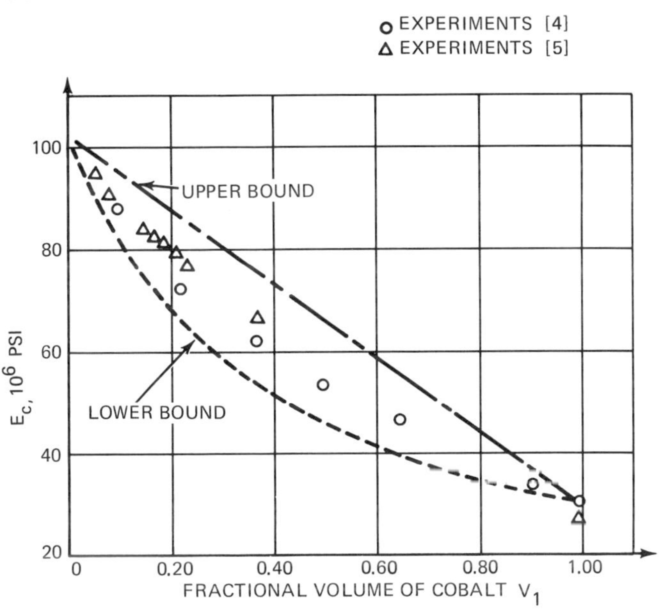

FIG. 1. Comparison of theory with experiment (from Ref. [3]).

III. IMPROVED BOUNDS FOR COMPOSITE PROPERTIES--ARBITRARY PHASE GEOMETRY

It was seen in the preceding section that the use of complementary energy and strain energy theorems leads to bounds which are often too far apart to be useful. To improve these bounds, Hashin and Shtrikman [6-8] and Hashin [9] formulated new variational theorems on the basis of polarization tensors. The polarization tensors were introduced into the theory of elasticity by Eshelby [10], and the name was coined by Kröner [11] in analogy to the polarization vector in electrostatics. These variational formulations are briefly summarized below [6,8] using indicial notation.*

Let σ^0_{ij} and ε^0_{ij} be known stress and strain tensors in a deformed elastic body of volume V and surface S. The stresses and strains are related by Hooke's law:

$$\sigma^0_{ij} = C^0_{ijk\ell}\varepsilon^0_{k\ell}, \quad i,j,k,\ell = 1,2,3$$

For the present homogeneous and isotropic case, this equation reduces to

$$\sigma^0_{ij} = \lambda_0 \varepsilon^0_{kk}\, \delta_{ij} + 2G_0 \varepsilon^0_{ij} \tag{15a}$$

or in operator form

$$\sigma^0_{ij} = L_0(\varepsilon^0_{ij}) \tag{15b}$$

where λ_0 and G_0 are the Lamé constants and δ_{ij} is the Krönecker delta. The strains are given by

*In this chapter, we will sometimes find it convenient to employ the tensor definition of shear strain. The tensor shear strain will be denoted by ε_{ij} $(i \neq j)$. Note that $\varepsilon_{ij} = \frac{1}{2}\gamma_{ij}$, where γ_{ij} denotes the engineering shear strain.

$$\varepsilon^0_{ij} = \frac{1}{2}(u^0_{i,j} + u^0_{j,i}) \tag{15c}$$

Here, u^0_i denotes the displacement in the ith direction. Now, a part or the whole of the body is changed to a material of different moduli, λ and G, which may vary in space, and the surface displacements $u^0_i(S)$ are held fixed. The resulting unknown stresses and strains in the new body are denoted by σ_{ij} and ε_{ij}. The following definitions are now introduced:

$$\sigma_{ij} = L_0(\varepsilon_{ij}) + p_{ij} \tag{16a}$$

$$u'_i = u_i - u^0_i \tag{16b}$$

$$\varepsilon'_{ij} = \varepsilon_{ij} - \varepsilon^0_{ij} \tag{16c}$$

Here, p_{ij} denotes the stress polarization tensor, and it and ε'_{ij} can be treated as unknowns in Eqs. (15) and (16).

The variational principles involving ε'_{ij} and p_{ij} are now formulated. The integral

$$U_p = U_0 - \frac{1}{2}\int_V \left[p_{ij}H(p_{ij}) - p_{ij}\varepsilon'_{ij} - 2p_{ij}\varepsilon^0_{ij}\right] dV \tag{17}$$

where

$$U_0 = \frac{1}{2}\int_V \sigma^0_{ij}\varepsilon^0_{ij}\, dV, \quad H = (L - L_0)^{-1}$$

$$L(\varepsilon_{ij}) = \lambda\varepsilon_{kk}\delta_{ij} + 2G\varepsilon_{ij}$$

subject to the subsidiary condition (in the absence of body forces)

$$L_0(\varepsilon'_{ij})_{,j} + p_{ij,j} = 0 \tag{18}$$

and the boundary condition

$$u'_i(S) = 0 \tag{19}$$

is stationary for

$$p_{ij} = L(\varepsilon_{ij}) - L_0(\varepsilon_{ij}) \tag{20}$$

It can further be shown [6,8] that the stationary value of U_p is an absolute maximum when

$$\lambda > \lambda_0, \quad G > G_0 \tag{21a}$$

and an absolute minimum when

$$\lambda < \lambda_0, \quad G < G_0 \tag{21b}$$

Furthermore, the stationary value of U_p is equal to the strain energy U stored in the changed body. When K_0 and G_0 are very small compared to K and G, the above variational principle reduces in the limit to the principle of minimum complementary energy. When K_0 and G_0 are very large in comparison to K and G, the principle of minimum potential energy is obtained [8]. For prescribed surface tractions, the strain polarization tensor can be used to obtain other variational principles [6].

Hashin and Shtrikman have applied these variational principles to both polycrystals [7] and multiphase materials [8]. The application is similar to that discussed in Section II, except for substantially more complicated mathematics, and we will not delve into the details here. The resulting bounds, however, deserve investigation.

For multiphase materials, the Hashin-Shtrikman bounds on the bulk modulus K_c are

$$K_1^* = K_1 + \frac{A_1}{1 + B_1 A_1} < K_c < K_n + \frac{A_n}{1 + B_n A_n} = K_2^* \qquad (22)$$

where

$$B_1 = -3/(3K_1 + 4G_1), \quad B_n = -3/(3K_n + 4G_n)$$

$$A_1 = \sum_{i=2}^{n} V_i / \left\{ [1/(K_i - K_1)] - B_1 \right\}$$

$$A_n = \sum_{i=1}^{n-1} V_i / \left\{ [1/(K_i - K_n)] - B_n \right\}$$

Subscripts 1 and n denote the smallest and the largest phase moduli, respectively, and V_i is the volume fraction of the ith phase. The bounds on the shear modulus G_c are

$$G_1^* = G_1 + \frac{1}{2} \frac{D_1}{1 + F_1 D_1} < G_c < G_n + \frac{1}{2} \frac{D_n}{1 + F_n D_n} = G_2^* \qquad (23)$$

where

$$F_1 = -\frac{3(K_1 + 2G_1)}{5G_1(3K_1 + 4G_1)}, \quad F_n = -\frac{3(K_n + 2G_n)}{5G_n(3K_n + 4G_n)}$$

$$D_1 = \sum_{i=2}^{n} \frac{V_i}{\left\{1/[2(G_i - G_1)]\right\} - F_1}$$

$$D_n = \sum_{i=1}^{n-1} \frac{V_i}{\left\{1/[2(G_i - G_n)]\right\} - F_n}$$

Here K_1^* and G_1^* are the lower bounds and K_n^* and G_n^* are the upper bounds. The Voigt and Reuss bounds, or those found by Paul [3] for two-phase materials, were given by

Eqs. (1a), (1b), (2a), (2b), (8), (11), and (14), respectively. The comparison of these sets of bounds (for $n = 2$) with the experimental results of Ref. [4] is presented in Fig. 2. It can be seen that the Hashin-Shtrikman bounds are much closer than the previous predictions. For Fig. 2, Young's modulus was calculated by

$$E_c = 9K_cG_c/(3K_c + G_c) \tag{24}$$

It should be noted that for the amount of information utilized here, that is, the phase moduli and the volume fractions only, the bounds of Eq. (22) on the bulk modulus (for $n = 2$) cannot be improved. Thus for further improvement, additional information as to phase distribution would be required. This statement will be shown to be true in the following section where the bounds on bulk modulus for a matrix with spherical inclusions will be shown to duplicate the current results. This conclusion is not, however, true

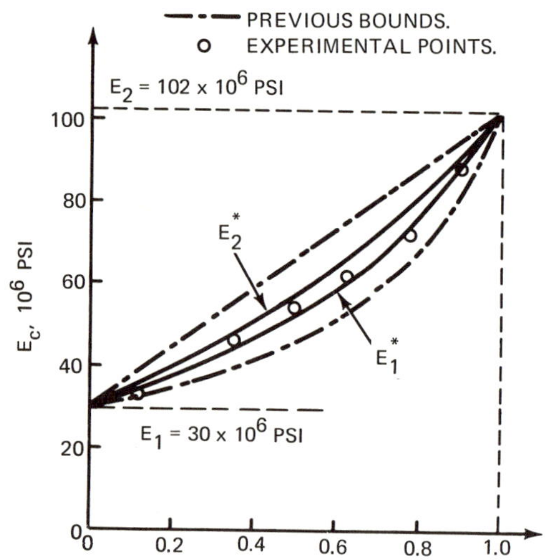

FIG. 2. Bounds for Young's modulus of tungsten carbide-cobalt alloy (from Ref. [8]).

for the shear modulus bounds of Eq. (23).

Hill [12] obtains equivalent bounds on the composite bulk modulus without resort to new variational principles. The use of the concepts of average stress and strain in the phases in combination with the extremum principles of complementary and strain energies are shown to be sufficient to obtain Eq. (22) for $n = 2$. Hill [12] also observed that for equal shear rigidities ($G_1 = G_2 = G_c$), these bounds coincide, providing an exact solution:

$$\frac{K_c}{K_R} = \left(1 + \frac{4G_cK_V}{3K_1K_2}\right) \Big/ \left(1 + \frac{4G_cK_R}{3K_1K_2}\right) \tag{25}$$

where K_V and K_R are the Voigt and Reuss bounds (for $n = 2$). Poisson's ratio and Young's modulus for such a composite become

$$\nu_c = \frac{V_1\nu_1 + V_2\nu_2 - \nu_1\nu_2}{1 - V_1\nu_2 - V_2\nu_1} \tag{26}$$

$$E_c = \frac{V_1E_1 + V_2E_2 - (E_1E_2/4G_c)}{1 - [(V_1E_2 + V_2E_1)/4G_c]} \tag{27}$$

while the shear modulus for the phases and the composite is just G_c.

Bounding the shear modulus in the same fashion proved to be impractical [12]. The bounds for the shear modulus, Eq. (23), coincide, not for equal phase bulk moduli, but rather when

$$G_1(1 + 2F_1)/F_1 = G_2(1 + 2F_2)/F_2 \tag{28}$$

The inapplicability of the Hashin-Shtrikman bounds when $G_1 - G_2$ and $K_1 - K_2$ are of opposite signs was eliminated by Walpole [13], who utilized the idea of polarization tensors in the classical variational theorems rather than the new theorems formulated by Hashin-Shtrikman [6]. Thus for a two-phase material, the moduli bounds were adjusted to

$$\frac{V_1}{1 + [(G_1 - G_2)V_2/(G_2 + G_\ell^*)]} \le \frac{G_c - G_2}{G_1 - G_2}$$

$$\le \frac{V_1}{1 + [(G_1 - G_2)V_2/(G_2 + G_g^*)]} \tag{29}$$

$$\frac{V_1}{1 + [(K_1 - K_2)V_2/(K_2 + K_\ell^*)]} \le \frac{K_c - K_2}{K_1 - K_2}$$

$$\le \frac{V_1}{1 + [(K_1 - K_2)V_2/(K_2 + K_g^*)]} \tag{30}$$

where, for $(G_1 - G_2)(K_1 - K_2) \ge 0$,

$$K_\ell^* = (4/3)G_2, \quad K_g^* = (4/3)G_1$$

$$G_\ell^* = \frac{3}{2}\left(\frac{1}{G_2} + \frac{10}{9K_2 + 8G_2}\right)^{-1}$$

$$G_g^* = \frac{3}{2}\left(\frac{1}{G_1} + \frac{10}{9K_1 + 8G_1}\right)^{-1} \tag{31}$$

and for $(G_1 - G_2)(K_1 - K_2) \le 0$,

$$K_\ell^* = (4/3)G_1, \quad K_g^* = (4/3)G_2$$

$$G_\ell^* = \frac{3}{2}\left(\frac{1}{G_2} + \frac{10}{9K_1 + 8G_2}\right)^{-1}$$

$$G_g^* = \frac{3}{2}\left(\frac{1}{G_1} + \frac{10}{9K_2 + 8G_1}\right)^{-1} \tag{32}$$

In spite of the fact that these bounds represent a substantial improvement over the Voigt-Reuss bounds, they are still far apart when a phase approaches complete rigidity or weakness. Also, for the bulk modulus, at least, no improvement is possible without utilizing additional information about geometries or packing. It is therefore not surprising that quite a bit of effort has been expended in finding improved bounds or moduli for more specifically defined composites. The calculation of bounds on the moduli for particulate (spherical) and fibrous composites will be studied in the following sections.

In contrast to the bounds on elastic moduli, little effort seems to have been spent on obtaining usable bounds on composite thermal properties. In the main, the analysis for bounds on thermal properties has been limited to bounds similar to those found upon elastic properties by Paul [3]. Thornburgh and Pears [14] obtained the following results for the thermal conductivities κ_{ci} of a composite with arbitrary phases:

$$\frac{1}{(V_1/\kappa_{1i}) + (V_2/\kappa_{2i})} \leq \kappa_{ci} \leq V_1\kappa_{1i} + V_2\kappa_{2i} \qquad (33)$$

The subscript i denotes the conductivity in the ith direction. These results are based on the assumption that the reinforcement and the matrix are arranged either in series or in parallel, and that the resulting thermal model is analogous to series and parallel connected electrical circuits. With only these types of bounds available for thermal properties, the methods and results of Chapter 3 seem to provide the more readily usable data.

IV. IMPROVED BOUNDS FOR COMPOSITE PROPERTIES--SPHERICAL REINFORCING PARTICLES

In the previous chapter, various packing arrangements were introduced to represent the composites. For the current reinforcement under consideration, the least restrictive model would be that containing concentric spheres. With this model, a packing arrangement is not actually defined, and while the phase volume fraction is fixed, the actual sphere radii are not. Thus variously sized approximately spherical particles, reinforcing a single matrix material, can all be accomodated by this model (Fig. 3, see also Section IV of Chapter 3).

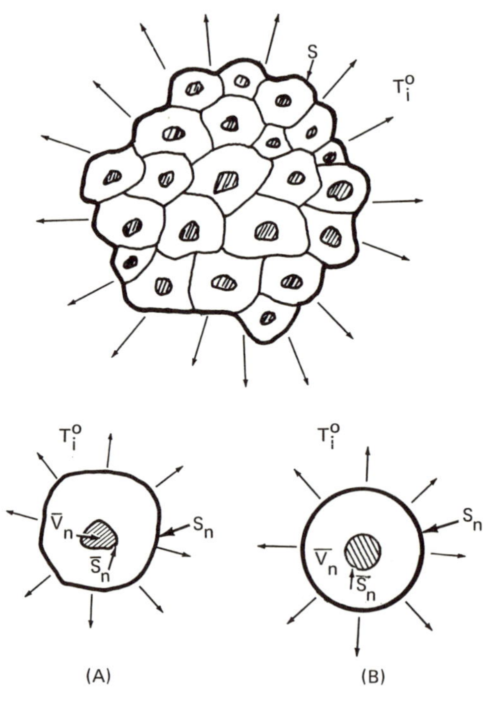

FIG. 3. Heterogeneous material modeled by use of concentric spheres (from Ref. [15]).

The spherical model was first analyzed by Hashin [15] using the classical variational principles of minimum potential and minimum complementary energy. The complementary and strain energies of the matrix material can be written as

$$U_0^{\sigma} = \frac{1}{2}\int_V \left\{\left[(\sigma^0)^2/9K_m\right] + \left[S_{ij}^0 S_{ij}^0/2G_m\right]\right\} dV \tag{34a}$$

$$U_0^{\varepsilon} = \frac{1}{2}\int_V \left[K_m(\varepsilon^0)^2 + 2G_m e_{ij}^0 e_{ij}^0\right] dV \tag{34b}$$

where σ^0 and ε^0 denote the isotropic parts of the stress and strain tensors, respectively, and S_{ij}^0 and e_{ij}^0 the deviatoric parts:

$$\sigma^0 = \sigma_{kk}^0, \quad \varepsilon^0 = \varepsilon_{kk}^0$$

$$S_{ij}^0 = \sigma_{ij}^0 - \frac{1}{3}\sigma^0\delta_{ij}, \quad e_{ij}^0 = \varepsilon_{ij}^0 - \frac{1}{3}\varepsilon^0\delta_{ij}$$

When N inclusions are inserted, loaded by surface tractions, we have

$$U^{\sigma} = U_0^{\sigma} + \delta U^{\sigma} = U_0^{\sigma} + \sum_{n=1}^{N} \delta U_n^{\sigma} \tag{35a}$$

where

$$\delta U_n^{\sigma} = \frac{1}{2}\int_{\overline{S}_n} \left(T_i^{\,0}\overline{u}_i^{\,n} - \overline{T}_i^{\,n}u_i^{\,0}\right) dS \tag{35b}$$

The quantities with index zero refer to the same body, loaded by the same surface tractions when no inclusions exist; $\overline{S}_n$ is the surface of nth inclusion and $\overline{u}_i^{\,n}$ and $\overline{T}_i^{\,n}$ are the components of displacement and traction vectors on $\overline{S}_n$. If, instead of tractions, displacements are prescribed, we have

$$U^{\varepsilon} = U_0^{\ \varepsilon} + \delta U^{\varepsilon} = U_0^{\ \varepsilon} + \sum_{n=1}^{N} \delta U_n^{\ \varepsilon} \tag{36a}$$

where

$$\delta U_n^{\ \varepsilon} = -\frac{1}{2}\int_{\overline{S}_n} \left(T_i^{\,0}\overline{u}_i^{\,n} - \overline{T}_i^{\,n}\, u_i^{\,0} \right) \tag{36b}$$

The energy of a quasihomogeneous, quasiisotropic composite can also be represented by Eq. (34) by merely changing the moduli to composite moduli K_c and G_c. If an admissible stress field is chosen for the composite, then we have from the principle of minimum complementary energy

$$U^{\sigma} \leq U_0^{\ \sigma} + \delta\tilde{U}^{\sigma} \tag{37}$$

Choosing an admissible displacement field and using the principle of minimum potential energy, we have

$$U^{\varepsilon} \leq U_0^{\ \varepsilon} + \delta\tilde{U}^{\varepsilon} \tag{38}$$

If the bounds on bulk modulus are required, the admissible stress or displacement states are made isotropic. Thus

$$T_i^{\,0} = \frac{1}{3}\sigma^0 n_i \tag{39a}$$

or

$$u_i^{\,0} = \frac{1}{3}\varepsilon^0 x_i \tag{39b}$$

where n_i are the components of the outward normal and x_i are Cartesian coordinates with respect to a system of axes in the heterogeneous body. Utilizing Eqs. (34)-(39), the bounds can be obtained in the form [15]

$$\frac{K_m}{1 + \left\{ \delta\tilde{U}^\sigma / [(\sigma^0)^2 V / 18K_m] \right\}} \leq K_c \leq K_m \left(1 + \frac{\delta\tilde{U}^\varepsilon}{\frac{1}{2} K_m (\varepsilon^0)^2 V} \right) \tag{40}$$

where V is the volume. Similarly, using the following admissible stress and displacement states for the deviatoric portions of the tensors,

$$T_1^0 = \tau n_2, \quad T_2^0 = \tau n_1, \quad T_3^0 = 0 \tag{41a}$$

or

$$u_1^0 = \frac{1}{2}\gamma x_2, \quad u_2^0 = \frac{1}{2}\gamma x_1, \quad u_3^0 = 0 \tag{41b}$$

bounds on the shear modulus are obtained as

$$\frac{G_m}{1 + \left[\delta\tilde{U}^\sigma / (\tau^2 V / 2G_m) \right]} \leq G_c \leq G_m \left(1 + \frac{\delta\tilde{U}^\varepsilon}{\frac{1}{2} G_m \gamma^2 V} \right) \tag{42}$$

In order to evaluate $\delta\tilde{U}^\sigma$ and $\delta\tilde{U}^\varepsilon$, it is necessary to assume an idealized geometry and select the appropriate stress and displacement fields. Hashin represents each composite inclusion as a sphere surrounded by a matrix sphere (Fig. 3c) such that the volumes are preserved. While this model specifies the ratio of the phase volumes, the actual sphere radii may vary.

To obtain the composite bulk modulus, the admissible stress state is chosen as hydrostatic, Eq. (39a), while the displacement state is taken as Eq. (39b). Thus, solving a simple elasticity problem [16], the expressions for $\delta\tilde{U}^\sigma$ and $\delta\tilde{U}^c$ may be derived. Substituting these into Eq. (40)

leads to the following bounds for the bulk modulus K_c (two-phase material):

$$K_1^* \leq K_c \leq K_2^*$$

$$K_1^* = \frac{K_m}{1 + \dfrac{(K_m - K_p)(4G_m + 3K_m)V_p}{K_m(4G_m + 3K_p) - 4G_m(K_m - K_p)V_p}}$$

$$K_2^* = K_m + (K_p - K_m)\frac{(4G_m + 3K_m)V_p}{4G_m + 3K_p + 3(K_m - K_p)V_p} \tag{43}$$

where the subscript p stands for particle inclusion and V_p denotes the volume fraction of the inclusion. It should be noted here that $K_1^* = K_2^*$, and thus the bulk modulus is exact for this idealized geometry. Furthermore, if in Eq. (43), we set $m = 1$, and $p = 2$, the resulting equation is (22) for K_1^* with $n = 2$. Similarly, if we set $m = 2$ and $p = 1$, the result is Eq. (22) for K_2^* with $n = 2$. Hence the earlier statement about the bounds in Eq. (22) being the closest possible for arbitrary phases is valid. When there is more than one kind of inclusion, Eq. (43) can be generalized by summing over each K_p^i and using the appropriate volume concentration V_p^i. In this more general case, K_1^* will no longer be equal to K_2^*.

For bounds on the shear modulus, the procedure is essentially the same except that the elasticity problem is much more difficult to solve. The bounds (for a two-phase material) are found by Hashin [15] as

$$G_1^* = \frac{G_m}{1 + [1 - (G_p)/G_m]y_1^{\sigma}\ V_p} \tag{44a}$$

$$G_2^* = G_m\left[1 + \left(\frac{G_p}{G_m} - 1\right)y_1^{\varepsilon}\ V_p\right] \tag{44b}$$

where y_1^{σ} and y_1^{ε} are given by the following simultaneous equations:

$$y_1^{\sigma}\left\{\frac{2[1 - (G_p/G_m)]}{5(1 - \nu_m)}\left(V_p^{7/3} - V_p^{5/3}\right)\right\} + y_2^{\sigma}\left\{[(7 - 10\nu_p) - (7 - 10\nu_m)I]4V_p^{7/3} - (7 + 5\nu_m)I\right\} = 0$$

$$y_1^{\sigma}\left\{\frac{G_p}{G_m} + \frac{7 - 5\nu_m}{15(1 - \nu_m)}\left(1 - \frac{G_p}{G_m}\right)(1 - V_p)\right\} + 21\, y_2^{\sigma}\, I\left(\frac{1}{V_p^{2/3}} - 1\right) = 1$$

$$y_1^{\varepsilon}\left\{\frac{2[1 - (G_p/G_m)]}{5(1 - \nu_m)}\left(V_p^{7/3} - V_p^{5/3}\right)\right\} + y_2^{\varepsilon}\left\{[(7 - 10\nu_p) - (7 - 10\nu_m)I]4V_p^{7/3} + 4(7 - 10\nu_m)I\right\} = 0 \tag{45}$$

$$y_1^{\varepsilon}\left\{\frac{G_p}{G_m} + \frac{7 - 5\nu_m}{15(1 - \nu_m)}\left(1 - \frac{G_p}{G_m}\right) + \frac{2(4 - 5\nu_m)}{15(1 - \nu_m)}\left(1 - \frac{G_p}{G_m}\right)V_p\right\} + 21\, y_2^{\varepsilon}\, I\left(\frac{1}{V_p^{2/3}} - 1\right) = 1$$

where

$$I = \frac{4(7 - 10\nu_p) + (G_p/G_m)(7 + 5\nu_p)}{35(1 - \nu_m)}$$

The solution may be generalized for n phases in a manner similar to that for the bulk modulus. The bounds on Young's modulus or Poisson's ratio can now easily be found from Eq. (24) and

$$\nu_c = (3K_c - 2G_c)/[2(3K_c + G_c)] \tag{46}$$

The bounds for the tungsten carbide-cobalt alloy of Ref. [4] for shear modulus and Poisson's ratio are presented in Figs. 4 and 5. Experimental comparison with this alloy for Young's modulus is made in Fig. 6. Since the spherical model bounds are the same for the bulk modulus and vary only slightly for the shear modulus, when compared to the arbitrary phase analysis, the results of Figs. 2 and 6 vary only slightly.

Lee and Westmann [17] extended Hashin's analysis to the case when the matrix material is reinforced with hollow spherical inclusions. Proceeding in the same manner as before, the only difference being that the inclusion itself contains a spherical cavity with $K = G = 0$, the bulk modulus bounds are again found to coincide.

$$\frac{K_c}{K_m} = \left\{ \frac{4G_p}{3K_m}\left[1 - \left(\frac{r_1}{r_f}\right)^3\right]\left[1 + \left(\frac{r_f}{r_m}\right)^3 \frac{4G_m}{3K_m}\right] + \left[\frac{4G_p}{3K_p} + \left(\frac{r_1}{r_f}\right)^3\right]\left[1 - \left(\frac{r_f}{r_m}\right)^3\right]\frac{4G_m}{3K_m} \right\}$$
$$\div \left\{ \frac{4G_p}{3K_m}\left[1 - \left(\frac{r_1}{r_f}\right)^3\right]\left[1 - \left(\frac{r_f}{r_m}\right)^3\right] + \left[\frac{4G_p}{3K_p} + \left(\frac{r_1}{r_f}\right)^3\right]\left[\frac{4G_m}{3K_m} + \left(\frac{r_f}{r_m}\right)^3\right] \right\} \tag{47}$$

where the various radii (r_1, r_f, r_m) are shown in Fig. 7.

The solution for bounds on the shear modulus, although in the same form as discussed earlier, led to 12 simultaneous equations each, which necessitated a computer solution. A parametric study of the results is presented in Figs. 8 and 9 for the shear modulus, since a closed-form solution is

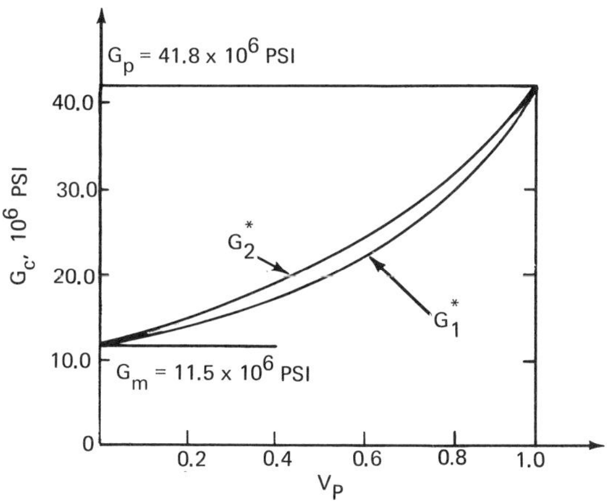

FIG. 4. Bounds for shear modulus (from Ref. [15]).

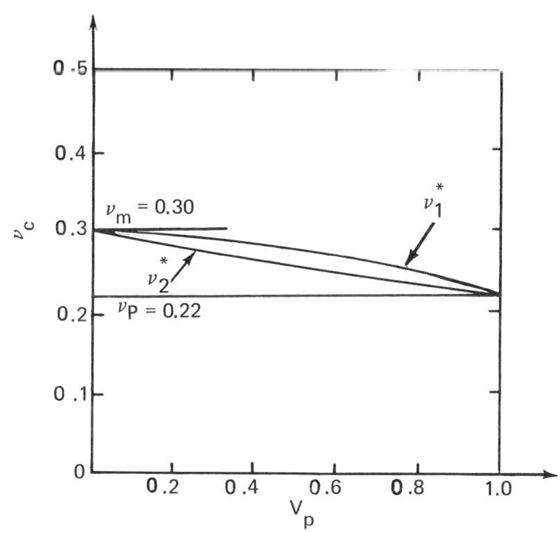

FIG. 5. Bounds for Poisson's ratio (from Ref. [15]).

unavailable. Figure 8 illustrates the effects of wall thickness of the inclusion upon the elastic properties of the composite and Fig. 9 indicates the effect of varying the material properties of the inclusion while holding the wall thickness fixed.

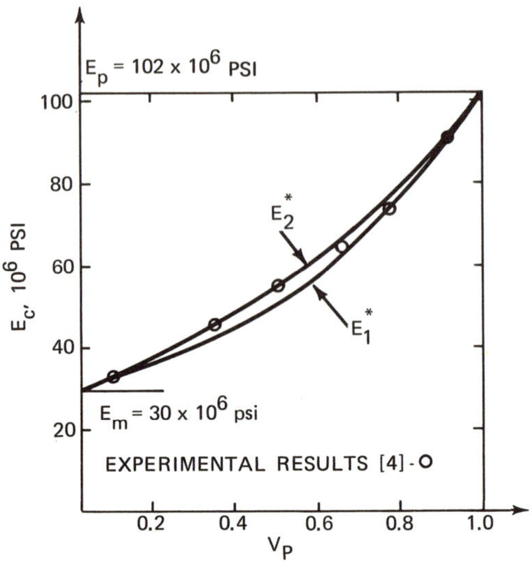

FIG. 6. Bounds for Young's modulus (from Ref. [15]).

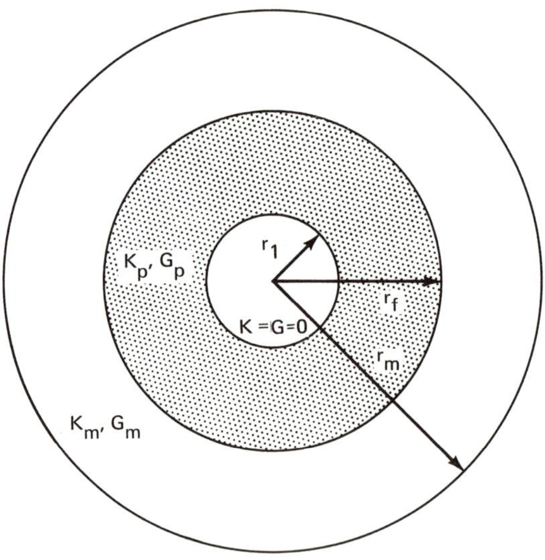

FIG. 7. Hollow-sphere reinforcement model.

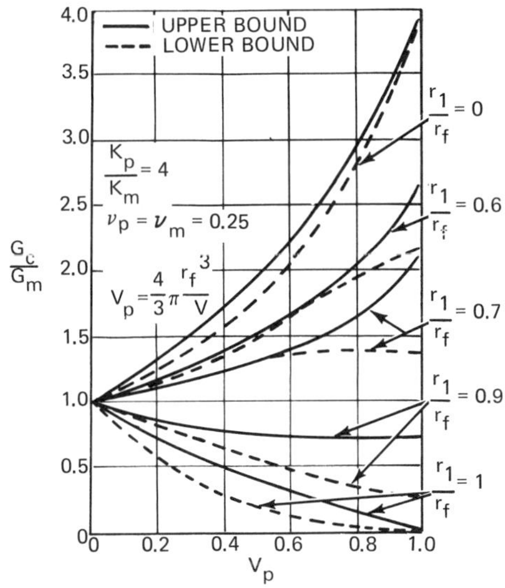

FIG. 8. Shear modulus for varying wall thickness (from Ref. [17]).

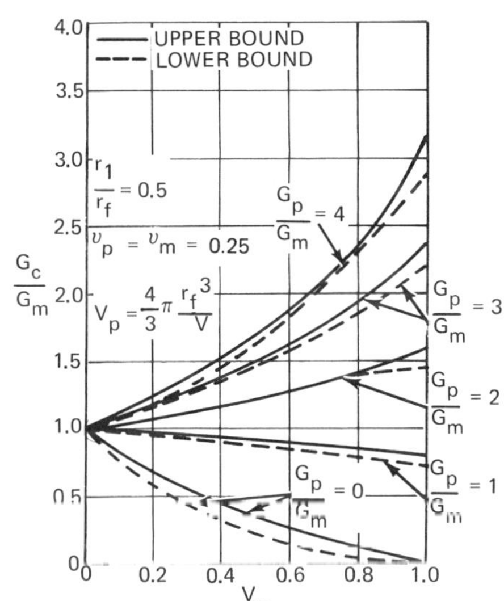

FIG. 9. Shear modulus for different moduli ratios (from Ref. [17]).

Before going to other inclusion geometries, it is worthwhile to examine the various bounds for the particulate composites. While the bounds obtained in this section are much closer than those originally proposed by Paul [3], there exists a drawback in the spherical model which is the same as that noted in Chapter 3, i.e., the perfect geometry assumption. Besides the fact that the inclusions are not usually perfect spheres, the assumption of zero contiguity is made. Contiguity can be defined as the average fraction of surface area over which inclusions are in contact. This property cannot be controlled in the making of alloy or composite specimens and thus if it were possible to hold everything else constant, it could be considered as a measure of the experimental data scatter. Such data scatter is shown in Fig. 10, in which experimental results for a tungsten-copper composite [18] are compared with Paul's [3] bounds. It is interesting to compare this figure with Figs. 1, 2 and 6. In Figs. 1, 2, and 6, the relative phase moduli ratios were

$$K_2/K_1 = 2.4, \quad G_2/G_1 = 3.6, \quad E_2/E_1 = 3.4$$

while for Fig. 10, the ratios are

$$K_2/K_1 = 2.08, \quad G_2/G_1 = 2.89, \quad E_2/E_1 = 2.77$$

Since in each case the tungsten-copper modulus ratios (Fig. 10) are smaller than those for tungsten carbide-cobalt (Figs. 1, 2, and 6), it is reasonable to expect that the Hashin [15] bounds would be closer to each other for this composite. From Fig. 10, however, it is obvious that such close bounds cannot possibly hope to enclose the experimental point scatter. Equally distressing is the observation that even the Hashin-Shtrikman [8] bounds for arbitrary phase geometry may not enclose all the experimental points. Indeed the best bounds here, if enclosing all experimental points is to be a measure, are the original

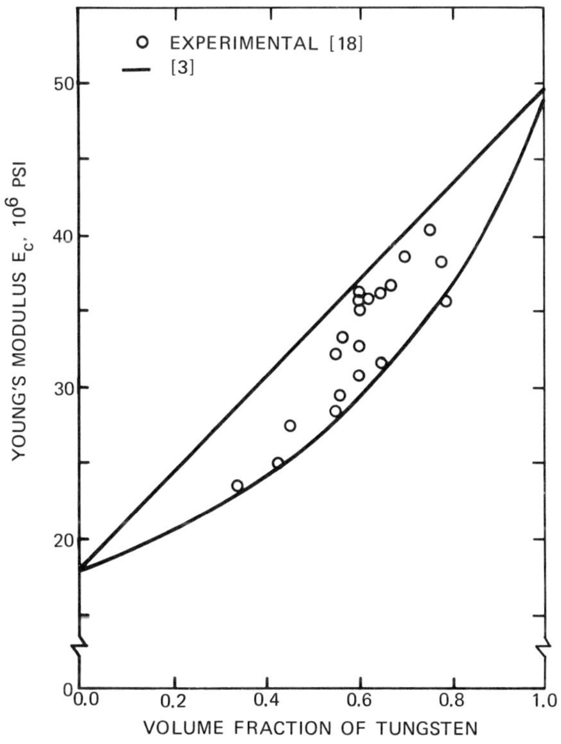

FIG. 10. Comparison of theoretical results with experimental data for tungsten-copper composite. Curves are based on Young's modulus 50×10^6 psi (18×10^6 psi) and Poisson's ratio 0.30 (0.35) for tungsten (copper) (from Ref. [19]).

ones presented by Paul [3]. For the specific alloy in question, the problem may be the sintering manufacturing technique which leads to porosity, rather than the contiguity problem. If porosity could be appropriately accounted for in the Hashin and Hashin-Shtrikman bounds by introducing void "enclosures" into the numerical calculations, these bounds may spread enough to encompass all the experimental points.

V. IMPROVED BOUNDS FOR COMPOSITE PROPERTIES--FIBER-REINFORCED COMPOSITES

In Chapter 3, it was remarked that a unidirectionally fiber-reinforced composite may be regarded as transversely isotropic. Such a composite material is characterized by five independent elastic constants (E_1, E_2, ν_{12}, G_{12}, and ν_{23}). Presently, we will find it convenient to regard E_1, K_{23}, G_{12}, G_{23}, and C_{11} as the five independent moduli. Here, K_{23} denotes the plane strain ($\varepsilon_1 = 0$) bulk modulus* and C_{11} is the uniaxial strain ($\varepsilon_2 = \varepsilon_3 = 0$) modulus,

$$C_{11} = \sigma_1/\varepsilon_1 = E_1(1 - \nu_{23})/(1 - \nu_{23} - 2\nu_{12}\nu_{21})$$

Note that E_2, ν_{12}, and ν_{23} are related to K_{23}, C_{11}, E_1, and G_{23} through the relations

$$E_2 = 2K_{23}(1 - \nu_{23})/[1 + 4(K_{23}\nu_{12}^2/E_1)]$$

$$\nu_{12}^2 = \frac{1}{4}(C_{11} - E_1)/K_{23}, \quad \nu_{23} = (E_2/2G_{23}) - 1 \tag{48}$$

Bounds for elastic properties of unidirectionally fiber-reinforced composites were derived by Hashin and Rosen [20] and Rosen et al. [21]. The reinforcing fibers are assumed to be of circular cross section but may be hollow or solid. Two cases were considered, (1) fibers of identical cross sections forming a hexagonal array in the transverse plane, and (2) fibers of different diameters but with the same ratio of inner to outer diameter, arranged in a random order in the transverse plane. For the hexagonal array, the representative volume element is one central fiber embedded in a

*See Chapter 3, Section III.C.

hexagonal prism (Fig. 11). The fiber is surrounded by a circular cylinder of radius r_m $\left(\pi r_m{}^2 \ell = V_1\right)$. This circular cylinder is in turn enclosed in a matrix shell of volume V_2. For the random array, the representative volume element is an average fiber of radius r_f enclosed in a circular cylinder. The radius of matrix cylinder r_m is selected such that the matrix volume fraction is preserved. In fact, the results for the random array follow from those for the hexagonal array in the limit as V_2 approaches zero. It is to be noted that bounds for the random array are not exact, due to the fact that cross sections of arbitrary shape are approximated by average circular cylinders.

The bounding analysis is accomplished by utilizing the variational theorems of minimum potential and minimum complementary energy in a manner equivalent to that discussed for the spherical model [15]. To bound the moduli, it is necessary to perform specific elasticity analyses as for

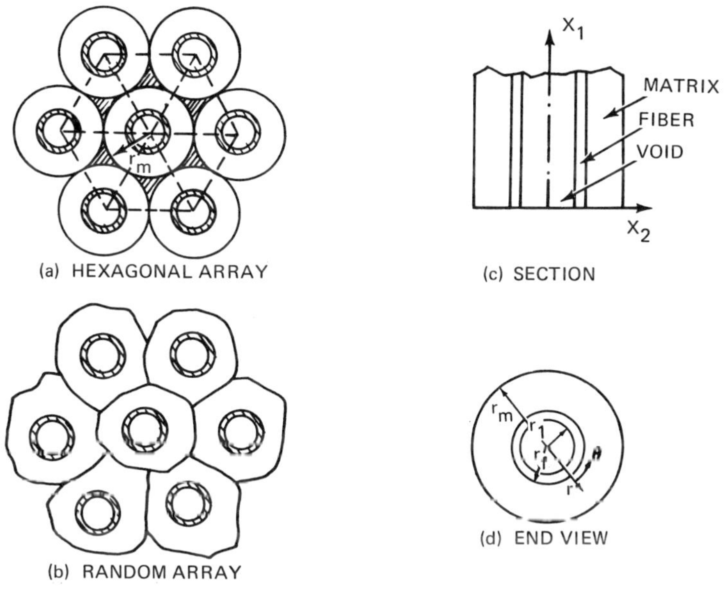

FIG. 11. Composite cylinder (from Ref. [20]).

particulate composites. The bounding of the plane-strain bulk modulus K_{23} is accomplished by first applying the plane-strain system

$$\varepsilon^0_{22} = \varepsilon^0_{33} = \varepsilon^0 \tag{49a}$$

all other components being zero, and then the stress sytem

$$\sigma^0_{22} = \sigma^0_{33} = \sigma^0 \tag{49b}$$

all other components being zero, to the composite cylinders, to obtain, respectively, the upper and lower bounds. For the hexagonal array, the bounds are

$$\frac{k_m}{(V_1/m_K) + V_2} \leq K_{23}(h) \leq k_m(m_K V_1 + V_2)$$

$$\begin{aligned} m_K = & \Big[(k_f/k_m)(1 - \xi^2)[1 + (1 - 2\nu_m)\zeta^2] \\ & + \{1 + [\xi^2/(1 - 2\nu_f)]\}(1 - \zeta^2)(1 - 2\nu_m)\Big] \\ & \div \Big[(k_f/k_m)(1 - \xi^2)(1 - \zeta^2) \\ & + \{1 + [\xi^2/(1 - 2\nu_f)]\}[\zeta^2 + (1 - 2\nu_m)]\Big] \end{aligned} \tag{50}$$

$$V_1 = (\pi/2\sqrt{3}) \approx 0.907, \quad V_2 = 1 - V_1$$

$$k_f = E_f/[2(1 - \nu_f - 2\nu_f^2)], \quad k_m = E_m/[2(1 - \nu_m - 2\nu_m^2)]$$

$$\xi^2 = (r_1/r_f)^2, \quad \zeta^2 = (r_f/r_m)^2$$

For the random array, the bounds coincide and are given by

$$K_{23}(r) = k_m m_k \tag{51}$$

Note that for the random array, ζ^2 equals the gross fiber volume fraction V_f. The results for solid fibers can be obtained by setting $r_1 = 0$. It is easy to verify that Eq. (51) reduces to Eq. (48) of Chapter 3 for $\xi = 0$.

The shear modulus G_{23} is bounded by first applying the displacement state

$$u_1^0 = 0, \quad u_2^0 = \frac{1}{2}\gamma^0 x_3, \quad u_3^0 = \frac{1}{2}\gamma^0 x_2 \tag{52a}$$

and then the stress vectors

$$T_1^0 = 0, \quad T_2^0 = \tau^0 n_3, \quad T_3^0 = \tau^0 n_2 \tag{52b}$$

to the composite cylinder models. The solution for these problems is difficult, and the bounding equations have to be evaluated numerically. This is true for both the hexagonal and random arrays. The bounds do not in general coincide for either of the two arrays.

The shear modulus G_{12} is bounded by first applying the pure shear strain

$$\varepsilon_{12}^0 = \varepsilon_{21}^0 = \frac{1}{2}\gamma^0 \tag{53a}$$

all other strains being zero, and then the stress state

$$\sigma_{12}^0 = \sigma_{21}^0 = \tau^0 \tag{53b}$$

all other components being zero, to the composite cylinder models. The bounds for the hexagonal array are found to be

$$G_m/[(V_1/m_G) + V_2] \leq G_{12}(h) \leq G_m(m_G V_1 + V_2) \tag{54}$$

where

$$m_G = \frac{(G_f/G_m)(1 - \xi^2)(1 + \zeta^2) + (1 + \xi^2)(1 - \zeta^2)}{(G_f/G_m)(1 - \xi^2)(1 - \zeta^2) + (1 + \xi^2)(1 + \zeta^2)}$$

If the array is random, the bounds concide and are given by

$$G_{12(r)} = G_m \frac{(G_f/G_m)(1 - \xi^2)(1 + V_f) + (1 + \xi^2)(1 - V_f)}{(G_f/G_m)(1 - \xi^2)(1 - V_f) + (1 + \xi^2)(1 + V_f)} \tag{55}$$

The results for solid fibers may be recovered by setting $\xi = 0$. It is easily seen that Eq. (55) reduces to Eq. (59) of Chapter 3 for $\xi = 0$.

In bounding the longitudinal Young's modulus $E_{1(r)}$ and Poisson's ratio $\nu_{12(r)}$ the upper bound is obtained by applying the strain system

$$\begin{aligned} &\varepsilon_{11}^0 = \varepsilon^0, \quad \varepsilon_{22}^0 = \varepsilon_{33}^0 = -\nu_{12}\varepsilon^0 \\ &\varepsilon_{12}^0 = \varepsilon_{23}^0 = \varepsilon_{31}^0 = 0 \end{aligned} \tag{56}$$

to the specimen. The lower bound is obtained from the application of $\sigma_{11}^0 = \sigma^0$ on the faces and $T_2^0 = T_3^0 = 0$ on the lateral surface. For the random array, the bounds coincide and are given by

$$\begin{aligned} &E_{1(r)} = E_m m_E \\ &\nu_{12(r)} = \frac{V_f E_f L_1 (1 - \xi^2) + V_m E_m L_2 \nu_m}{V_f E_f L_3 (1 - \xi^2) + V_m E_m L_2} \end{aligned} \tag{57}$$

where

$$m_E = \left[(1 - \xi^2)V_f \frac{E_f}{E_m} + V_m\right] \frac{E_m(D_1 - D_3F_1) + E_f(D_2 - D_4F_2)}{E_m(D_1 - D_3) + E_f(D_2 - D_4)}$$

$$D_1 = \frac{1 + \xi^2}{1 - \xi^2} - \nu_f, \quad D_2 = \frac{1 + V_f}{V_m} + \nu_m$$

$$D_3 = \frac{2\nu_f^2}{1 - \xi^2}, \quad D_4 = 2\nu_m^2 \frac{V_f}{V_m}$$

$$F_1 = \frac{\nu_m V_f E_f(1 - \xi^2) + \nu_f V_m E_m}{\nu_f\left(V_f E_f(1 - \xi^2) + V_m E_m\right)}, \quad F_2 = \frac{\nu_f}{\nu_m} F_1$$

$$L_1 = 2\nu_f\left(1 - \nu_m^2\right)V_f + V_m(1 + \nu_m)\nu_m$$

$$L_2 = V_f\left[(1 + \nu_f)\xi^2 + 1 - \nu_f - 2\nu_f^2\right] \tag{58}$$

$$L_3 = 2\left(1 - \nu_m^2\right)V_f + (1 + \nu_m)V_m, \quad V_m = 1 - V_f$$

If $\nu_m = \nu_f$, then Eqs. (57) and (58) reduce to

$$E_{1(r)} = V_f E_f(1 - \xi^2) + V_m E_m \quad \text{and} \quad \nu_{12(r)} = \nu_m = \nu_f$$

We also note here that Eqs. (57) become identical with Eqs. (34) and (38) of Chapter 3 for $\xi = 0$. For the hexagonal array, the bounds for $E_{1(h)}$ are

$$E_m/\left[\left(V_1/m_E'\right) + V_2\right] \leq E_{1(h)} \leq E_m\left(m_E' V_1 + pV_2\right) \tag{59}$$

where

$$p = \left[1 - \nu_m - 4\nu_m\nu_{12(r)}' + 2\left(\nu_{12(r)}'\right)^2\right]/\left(1 - \nu_m - 2\nu_m{}^2\right) \tag{60}$$

The primes on m_E and $\nu_{12(r)}$ in Eqs. (59) and (60) indicate that these quantities have to be evaluated from Eqs. (58) by replacing V_f and V_m by ζ^2 and $1 - \zeta^2$, respectively. For $\nu_f = \nu_m$, the bounds upon $E_{1(h)}$ and $\nu_{12(h)}$ coincide. If $\nu_f \neq \nu_m$, the bounds upon $\nu_{12(h)}$ must be obtained indirectly from bounds on $K_{23(h)}$, $E_{1(h)}$, and $C_{11(h)}$ (given below), thus resulting in a much wider spacing of the bounds.

Finally the bounds for the modulus C_{11} may be obtained by solving the problems of the specimen under uniform macroscopic stress or strain in the x_1 direction and preventing lateral deformation in the x_2-x_3 plane by a rigid enclosure. Thus for the random array, coincident bounds can be found as

$$C_{11}(r) = E_1(r) + 4(\nu_{12}(r))^2 K_{23}(r) \tag{61}$$

where the necessary quantities have been previously defined. For the hexagonal array, the bounds are

$$\frac{1}{[V_1/C'_{11}(r)] + (qV_2/E_m)} \le C_{11}(h) \le C'_{11}(r)V_1 + \frac{E_m(1 - \nu_m)}{(1 + \nu_m)(1 - 2\nu_m)} V_2 \tag{62}$$

where

$$q = 1 - 4\nu_m\eta + 2(1 - \nu_m)\eta^2$$

$$\eta = 2\nu'_{12}(r)K'_{23}(r)/C'_{11}(r)$$

and the primed quantities may be obtained from Eqs. (51), (57), and (61) by substituting ζ^2 for V_f and $(1 - \zeta^2)$ for V_m.

The Hashin-Rosen [20] results are graphically presented in Figs. 12-16, and a comparison with experiment and other predictions is given in Fig. 17. Figures 12-16 are for glass composites, where $\xi = 0$ signifies solid fibers. Figure 17 compares the elasticity solution of Whitney and Riley [22] (see Chapter 3) and the upper bound solution of Hashin-Rosen ($E_{2(r)}$) with experiment for a boron epoxy composite [22]. It is interesting to note that the

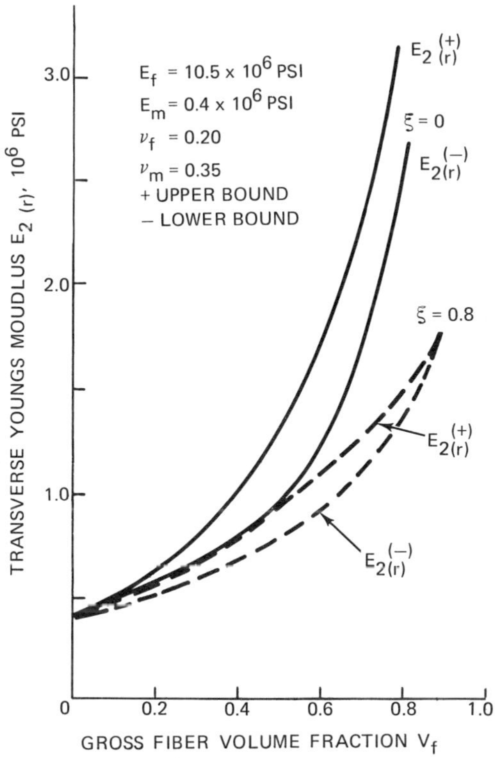

FIG. 12. Bounds on the effective transverse Young's modulus for solid and hollow glass fibers in a plastic matrix (parallel fibers, random array) (from Ref. [21]).

experimental results for the transverse Young's modulus are bounded by the Whitney-Riley formula and the Hashin-Rosen upper bound. They are consistently slightly higher than the Hashin-Rosen upper bound. The Hashin-Rosen predictions for the shear modulus $G_{12(r)}$ are compared with experimental data in Fig. 10. It can be seen that those predictions do not correlate as well as those for Young's modulus with experimental observations.

In previous discussions, we assumed that the reinforcements did not touch (zero contiguity factor). Certain

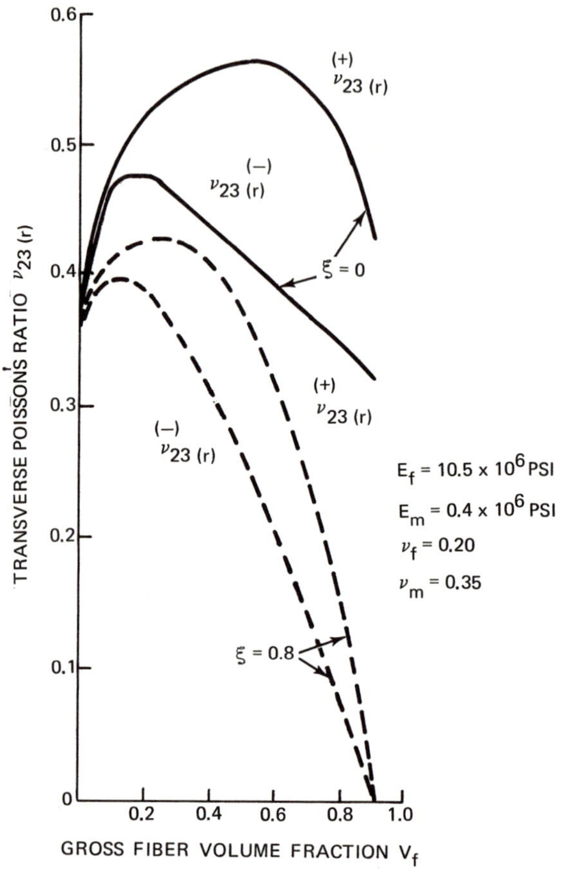

FIG. 13. Bounds on the effective transverse Poisson's ratio for solid and hollow glass fibers in a plastic matrix (parallel fibers, random array) (from Ref. [21]).

empirical corrections for this, and also for the fact that the fibers may be misaligned (not perfectly straight or parallel), were proposed by Tsai [23]. Thus, for example, the law-of-mixtures predictions for the longitudinal modulus is multiplied by a factor $c \leq 1$, to account for possible fiber misalignment. Other moduli are adjusted in a similar manner. The main problem in utilizing the c

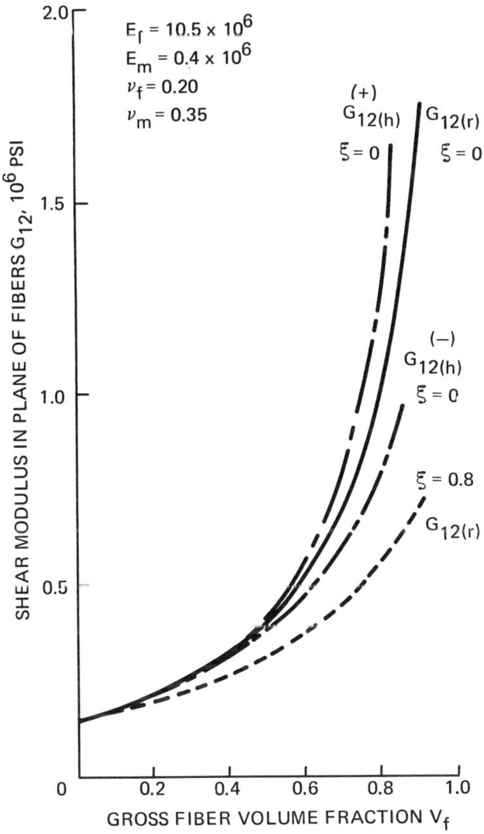

FIG. 14. Effective shear modulus in the plane of the fibers for solid and hollow glass fibers in a plastic matrix (parallel fibers, random and hexagonal array) (from Ref. [21]).

factor is the lack of knowledge as to what it may be for any given composite specimen.

As we noted earlier, one of the problems with bounds upon material properties can be that experimental points do not all fall within the enclosed area. This may be due to the geometric assumptions, such as a specific regular packing arrangement, and zero contiguity. Since most methods of predicting composite material properties have

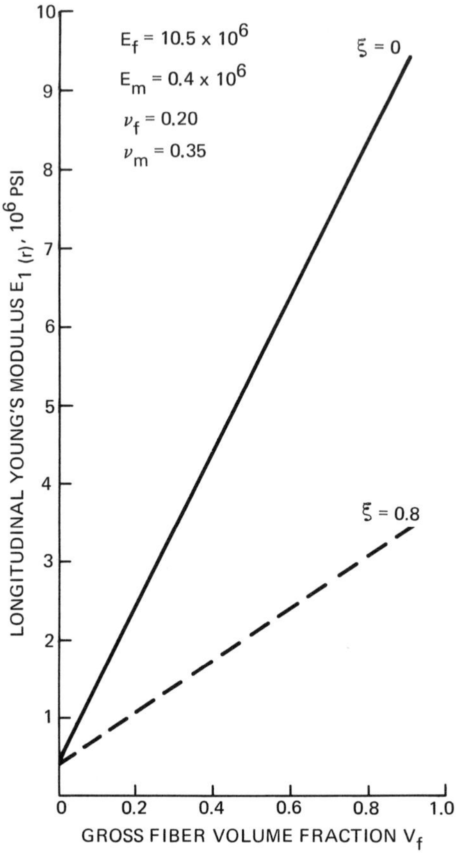

FIG. 15. Effective longitudinal Young's modulus for solid and hollow glass fibers in a plastic matrix (parallel fibers, random array) (from Ref. [21]).

been discussed, i.e., mechanics-of-materials theory, elasticity and numerical analyses, and finally bounding by minimal principles, it is interesting now to compare all these predictions with the available experimental data. An attempt at this is made in Figs. 19-27.

Figures 19-23 and 25-26 present a detailed study of glass-epoxy composites, while Figs. 24 and 27 present some

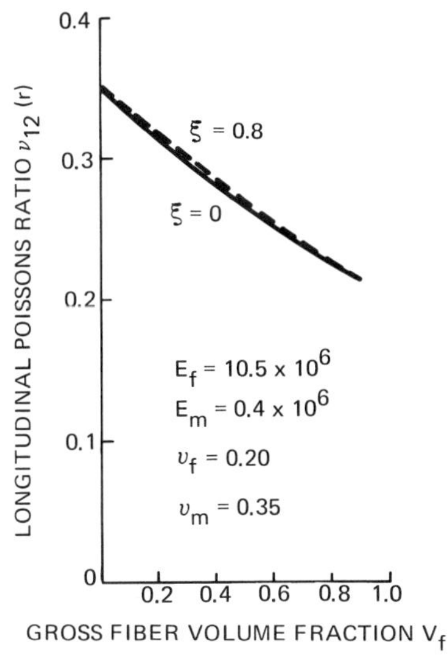

FIG. 16. Effective longitudinal Poisson's ratio for solid and hollow glass fibers in a plastic matrix (parallel fibers, random array) (from Ref. [21]).

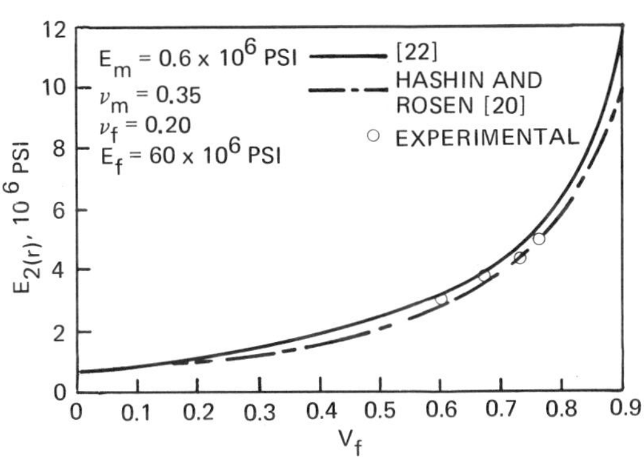

FIG. 17. Transverse modulus for boron-epoxy composite as a function of fiber volume fraction (from Ref. [22]).

details for boron- and graphite-fiber-reinforced composites, respectively. Since most theories predict a law-of-mixtures approximation for some moduli, Fig. 19 presents a comparison

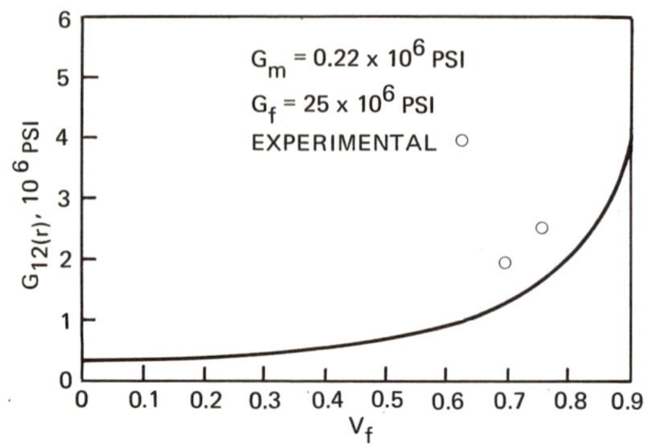

FIG. 18. Shear modulus for boron-epoxy composite as a function of fiber volume fraction (from Ref. [22]).

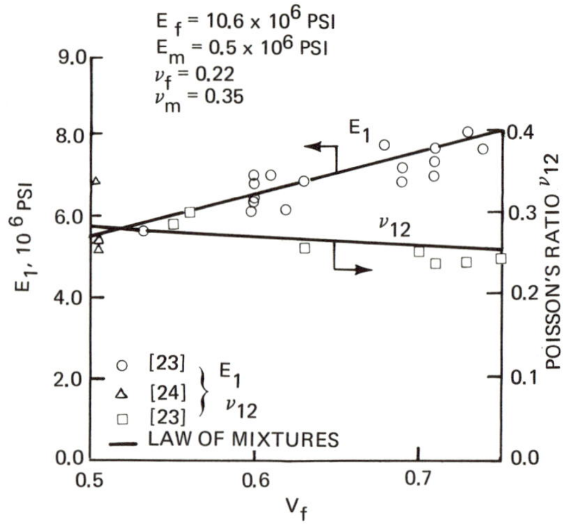

FIG. 19. Comparison of E_1 and ν_{12} (E-glass-epoxy) (from Ref. [25]).

of this prediction with experimental results. It can be seen that if this is viewed as an upper bound prediction on the Young's modulus E_1 [3], experimental scatter still provides some points above the bound. Figure 20 compares various theoretical predictions of the transverse Poisson's ratio. It can be seen that the Hashin-Dow-Rosen [20, 21,

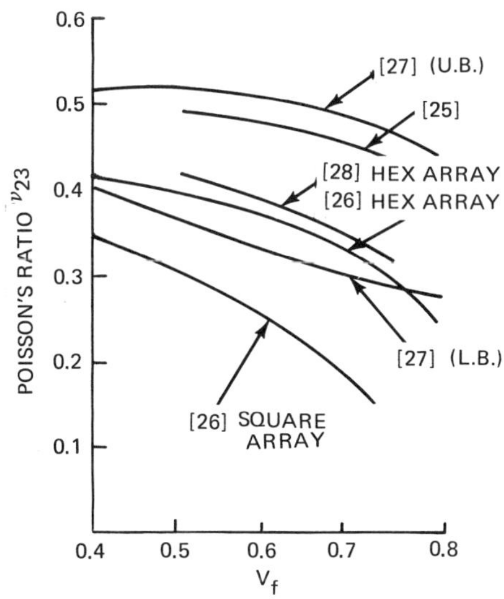

FIG. 20. Comparison of ν_{23} (E-glass-epoxy) (from Ref. [29]).

27] bounds appear to bound all the theories except for a finite-element analysis of a square-packed array [26].

Figure 21 shows experimental and theoretical predictions for the transverse Young's modulus of a glass-epoxy composite. In Fig. 21, the Hashin-Dow-Rosen [20, 21, 27] bound seems to be a good lower bound, but the best upper bound seems a combination of the Ekvall [30] mechanics-of-materials model (see Chapter 3) for low fiber volume ratios, and the Adams-Doner-Thomas [31] (the finite-difference, numerical relaxation model discussed in Chapter 3) solution at high fiber volume ratios. The wide scatter in experimental results is even more evident in this figure. Figure 22 presents the same type of results for the shear moduous G_{12}. The optimum lower bound, in this case, appears to be the Adams-Doner-Thomas calculations, although the Hashin-Dow-Rosen random array prediction gives the lower bound for all theories and experimental values. The

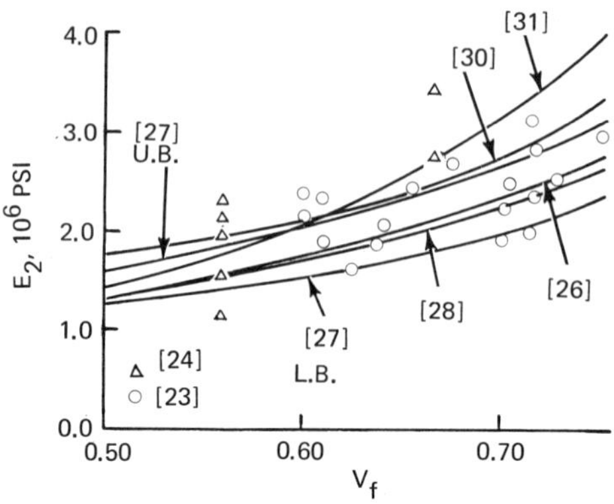

FIG. 21. Comparison of E_2 (E-glass-epoxy) (from Ref. [25]).

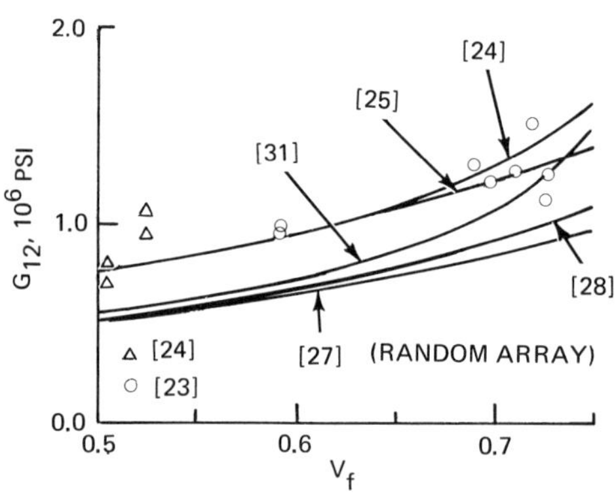

FIG. 22. Comparison of G_{12} (E-glass-epoxy) (from Ref. [25]).

upper bound again is the Greszczuk-Ekvall type of model. (The Greszczuk model was discussed in Chapter 3 in reference to the analytic prediction of composite thermal properties.)

Figures 23 and 24 demonstrate the range of the various theories in predicting the transverse Young's modulus for S-glass-epoxy and boron-epoxy composites, respectively.

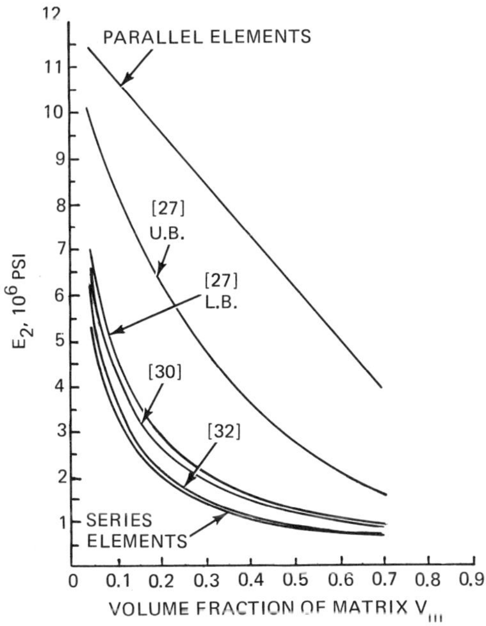

FIG. 23. Comparison of E_2 (S-glass-epoxy) (from Ref. [33]).

The gap between the Reuss (series) and Voigt (parallel) bounds makes it immediately obvious why these bounds are impractical to use. The closest bounds are the Hashin-Dow-Rosen bounds. The fact that these may be too close to encompass the experimental scatter has been demonstrated earlier. From the previous figures, it can be seen that the upper bound needs adjustment still upward.

Figures 25-27 present the predicted thermal properties of glass-epoxy and graphite-epoxy composites. It is seen that the predictions for the thermal coefficients of expansion α_1 and α_2 and the heat conductivity are all rather close to one another. The Levin bounds [35], however, do not enclose the other theoretical predictions. These bounds were calculated in a manner similar to the bounds on elastic properties calculated by Hill [12,41], and thus

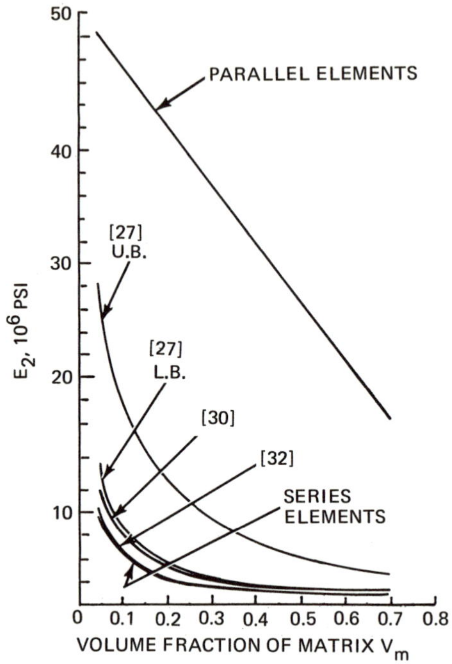

FIG. 24. Comparison of E_2 (boron-epoxy) (from Ref. [33]).

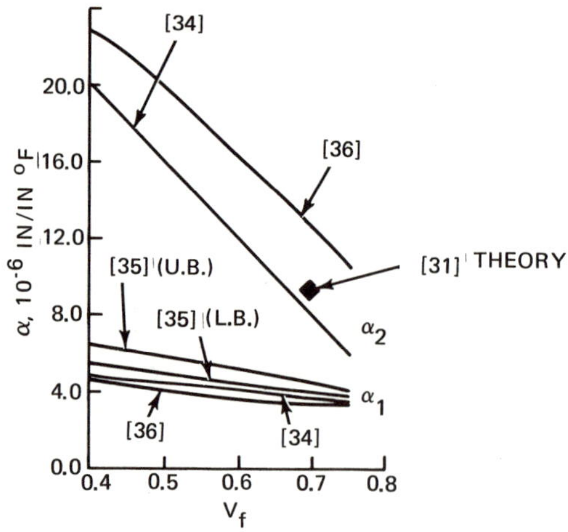

FIG. 25. Comparison of α_1 and α_2 (E-glass-epoxy) (from Ref. [29]).

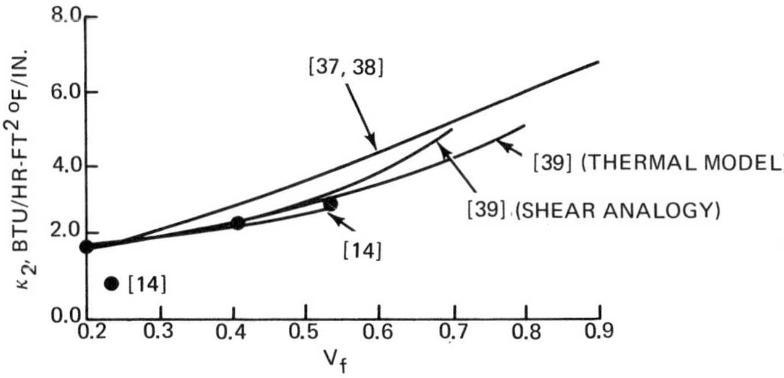

FIG. 26. Heat conductivities comparison (E-glass-epoxy) (from Ref. [29]).

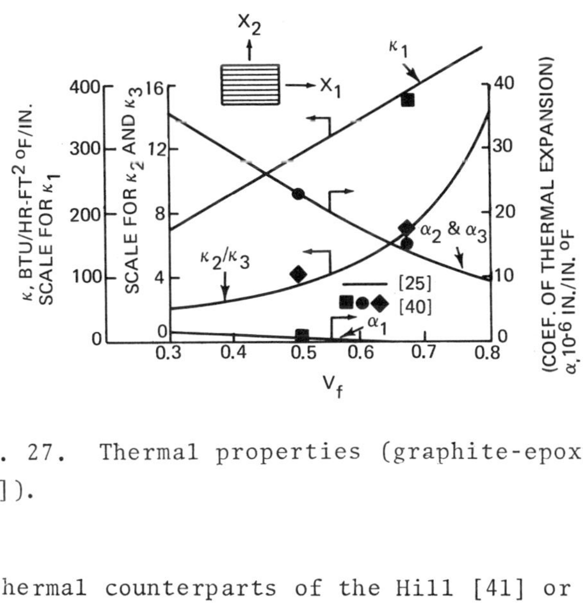

FIG. 27. Thermal properties (graphite-epoxy) (from Ref. [29]).

are the thermal counterparts of the Hill [41] or Hashin-Rosen [20] bounds for composite elastic properties. The Chamis formulas [25] show good correlation with experiment for the graphite-reinforced composite. These formulas are semiempirical and therefore good correlation for a given set of experimental data must not be construed to imply their general applicability.

The above-discussed results convey the statistical

nature of the composite material properties by the scatter evidenced in the experimental data. Thus, perhaps the recommendation in Ref. [20] that some statistical composite geometry function be used to improve the property bounds is one which needs further study.

VI. INELASTIC BEHAVIOR

There has been little research effort in this area, partly because a completely satisfactory elastic analysis is still not available. However, some work in this area has been done by Hill [41,42] and Hashin [43]. The finite-element analyses that have been attempted were mentioned in Chapter 3.

In Ref. [41], Hill finds composite bounds upon the elastic properties essentially equivalent to the Hashin-Rosen [20] bounds. In extending this work to account for inelastic behavior [42], the assumption is made that during continued deformation, the stress-strain laws retain transverse isotropy and diagonal symmetry. Thus, one is allowed to introduce the concept of instantaneous moduli into the previously discussed bounding formulas. What is necessary in this development is only the definition of the instantaneous elastic moduli for matrix and reinforcement. Special cases may then be obtained for a composite with an elasto-plastic matrix. Hill [42] provides such information only for the longitudinal modulus.

Hashin [43] treats the composite cylinder model discussed previously in which the composite is now represented by a linear viscoelastic matrix and parallel elastic fibers. The procedure in the analysis is identical to that outlined earlier for the analysis of elastic property bounds. The composite model is subjected to various overall strain states and the corresponding viscoelasticity problems are solved. All the fiber composite constants are obtained either explicitly or in bounded form. Special simplifications for rigid fibers are obtained. Unfortunately, no correlation of the theoretical predictions with experimental data is available. Thus the usefulness of these models in predicting inelastic effects is as yet undetermined.

REFERENCES

1. W. Voigt, Lehrbuch der Kristellphysik, Teubner, Berlin, 1910.
2. A. Reuss, "Calculation of Flow Limits of Mixed Crystals on the Basis of the Plasticity of Single Crystals," Z. Angew. Math. Mech., 9, 49 (1929).
3. B. Paul, "Prediction of Elastic Constants of Multi phase materials," Trans. Met. Soc. AIME, 218, 36-41 (1960).
4. C. Nishimatsu and J. Gurland, "Experimental Survey of the Deformation of Hard-Ductile Two-Phase Alloy System, WC-Co," Technical Report No. 2, Division of Engineering, Brown University, September 1958.
5. R. Kieffer and P. Schwartzkopf, Hartstoffe und Hartmetalle, Springer, Vienna, 1953.
6. Z. Hashin and S. Shtrikman, "On Some Variational Principles in Anisotropic and Nonhomogeneous Elasticity," J. Mech. Phys. Solids. 10, 335-342 (1962).
7. Z. Hashin and S. Shtrikman, "A Variational Approach to the Theory of the Elastic Behavior of Polycrystals," J. Mech. Phys. Solids, 10, 343-352 (1962).
8. Z. Hashin and S. Shtrikman, "A Variational Approach to the Theory of the Elastic Behavior of Multiphase Materials," J. Mech. Phys. Solids, 11, 127-140 (1963).

9. Z. Hashin, "Theory of Mechanical Behavior of Heterogeneous Media," Appl. Mech. Rev., 17 (1), 1-9 (1964).

10. J. D. Eshelby, "The Force on an Elastic Singularity," Phil. Trans. Roy. Soc., Series A, 244, 87-112 (1951).

11. E. Kröner, Kontinuumstheorie der Versetzungen und Eigenspannungen, Springer-Verlag, Berlin, 1958.

12. R. Hill, "Elastic Properties of Reinforced Solids: Some Theoretical Principles," J. Mech. Phys. Solids, 11, 357-372 (1963).

13. L. Walpole, "On Bounds for the Overall Elastic Moduli of Inhomogeneous Systems--I," J. Mech. Phys. Solids, 14, 151-162 (1966).

14. J. D. Thornburgh and C. D. Pears, "Prediction of the Thermal Conductivity of Filled and Reinforced Plastics," paper no. 65/WA/HT-4, ASME Winter Annual Meeting, 1965.

15. Z. Hashin, "The Elastic Moduli of Heterogeneous Materials," J. Appl. Mech., 29, 143-150 (1962).

16. A. E. H. Love, A Treatise on the Mathematical Theory of Elasticity, Dover, New York, 1949, p. 142.

17. K. J. Lee and R. A. Westmann, "Elastic Properties of Hollow-Sphere-Reinforced Composites," J. Composite Materials, 4, 242-252 (1970).

18. R. Krock, "Effect of Composition and Dispersed-Phase Particle-Size Distribution on the Static Elastic Moduli of Tungsten-Copper Composite Materials," Proc. ASTM, 63, 605-612 (1963).

19. G. P. Sendeckyj, "Statistical Theory of Elastic Moduli of Particulate Composites," J. Composite Materials, 3, 177-181 (1969).

20. Z. Hashin and B. W. Rosen, "The Elastic Moduli of Fiber-Reinforced Materials," J. Appl. Mech., 31, 223-232 (1964).

21. B. W. Rosen, N. F. Dow, and Z. Hashin, "Mechanical Properties of Fibrous Composites," NASA CR-31, April 1964.

22. J. M. Whitney and M. B. Riley, "Elastic Properties of Fiber-Reinforced Composite Materials," AIAA J., 4, 1537-1542 (1966).

23. S. W. Tsai, "Structural Behavior of Composite Materials," NASA CR-71, July 1964.

24. L. B. Greszczuk, "Theoretical and Experimental Studies on Properties and Behavior of Filamentary Composites," Society of Plastics Industry 21st Annual Conference, Section 5-B, 1966.

25. C. C. Chamis, "Design-Oriented Analysis and Structural Synthesis of Multilayered Filamentary Composites," Ph.D. Thesis, Case Western Reserve Univ., September 1967.

26. R. L. Foye, "An Evaluation of Various Engineering Estimates of the Transverse Properties of Unidirectional Composites," Science of Advanced Materials and Process Engineering Series, Vol. 10, Western Periodicals Co., pp. G-31 to G-42, 1966.

27. N. F. Dow and B. W. Rosen, "Evaluation of Filament-Reinforced Composites for Aerospace Structural Applications," NASA CR-207, April 1965.

28. G. Pickett and M. W. Johnson, "Analytical Procedures for Predicting the Mechanical Properties of Fiber-Reinforced Composites," Air Force Materials Laboratory Report No. AFML-TR-65-220, Pt. 2, 1966.

29. C. C. Chamis and G. P. Sendeckyj, "Critique on Theories Predicting Thermoelastic Properties of Fibrous Composites," J. Composite Materials, 2, 332-358 (1968).

30. J. C. Ekvall, "Structural Behavior of Monofilament Composites," AIAA 6th Structures and Materials Conference, April 1965.

31. D. F. Adams, D. R. Doner, and R. L. Thomas, "Mechanical Behavior of Fiber-Reinforced Composites Materials," Air Force Materials Laboratory Report No. AFML-TR-67-96, May 1967.

32. J. C. Ekvall, "Elastic Properties of Orthotropic Monofilament Laminates," ASME Aviation Conference, Paper 61-AV-56, 1961.

33. L. T. Calcote and G. C. Grimes, "Investigation of Structural Design Concepts for Fibrous Aircraft Structures," Air Force Flight Dynamics Laboratory Report AFFDL-TR-67-29, Vol. II, 1967.

34. L. B. Greszczuk, "Thermoelastic Properties of Filamentary Composites," AIAA 6th Structures and Materials Conference, April 1965.

35. V. M. Levin, "Thermal Expansion Coefficients of Heterogeneous Materials," Mekhanika Tverdovo Tela, 2 (1), 88-94 (1967) [Eng. trans.: Mechanics of Solids, 2 (1), 58-61 (1967)].

36. A. A. Kilchinskii, "On One Model for Determining Thermoelastic Characteristics of Fiber-Reinforced Materials," Prikladnaia Mekhanika, 1 (12), 65 (1965).

37. L. P. Khoroshun, "Some Problems of the Correlation Theory of Structurally Inhomogeneous Elastic Media," Mekhanika Polimerov, 2 (3), 365-371 (1966) [Eng. trans: Polymer Mechanics, 2 (3), 229-232 (1966)].

38. L. P. Khoroshun, "Thermoelastic Properties of Stochastically Reinforced Materials," Prikladnaia Mekhanika, 2(9), 99-106 (1966) [Eng. trans.: Soviet Applied Mechanics, 2 (9), 62-66 (1966)].

39. S. G. Springer and S. W. Tsai, "Thermal Conductivities of Unidirectional Materials," J. Composite Materials, 1, 166-173 (1967).

40. O. L. Blakslee, A. A. Pallozi, W. A. Doig, G. B. Spence, and D. P. Hanley, "Fabrication, Testing, and Design Studies with 'Thornel' Graphite-Fiber, Epoxy Resin Composites," in Advances in Structural Composites, SAMPE 12th National Symposium, Sec. AC-6, 1967.

41. R. Hill, "Theory of Mechanical Properties of Fiber-Strengthened Materials: I. Elastic Behavior," J. Mech. Phys. Solids, 12, 199-212 (1964).

42. R. Hill, "Theory of Mechanical Properties of Fiber-Strengthened Materials: II. Inelastic Behavior," J. Mech. Phys. Solids, 12, 213-218 (1964).

43. Z. Hashin, "Viscoelastic Fiber-Reinforced Materials," AIAA J., 4, 1411-1417 (1966).

CHAPTER 5

MULTILAYER LAMINATES

I. INTRODUCTION

In the previous chapters, we were primarily concerned with the material properties of single layers of unidirectionally fiber-reinforced composites. However, most fabricated composites consist of several layers glued together at various orientations to each other. Examples of this can be found in the multiply fiberglass constructions and in the boron-fiber-reinforced composite parts in aircraft. In this chapter, we will consider the elastic properties of composites containing stacks of unidirectionally fiber-reinforced layers. A simple discussion of this problem based upon the stacking of layers is given by Ekvall [1].

II. STACKING ANALYSIS

Consider first a single layer containing fibers parallel to the direction x_1 (see Fig. 1). In Chapter 3 we had the following stress-strain relationships (valid under biaxial loading) referred to the principal axes x_1-x_2:

$$\sigma_1 = \sigma_{11} = [E_1/(1 - \nu_{12}\nu_{21})](\varepsilon_1 + \nu_{21}\varepsilon_2)$$

$$\sigma_2 = \sigma_{22} = [E_2/(1 - \nu_{21}\nu_{12})](\varepsilon_2 + \nu_{12}\varepsilon_1)$$

and

$$\sigma_6 = \tau_{12} = G_{12}\varepsilon_6 \qquad (1)$$

where

$$\begin{Bmatrix} \varepsilon_1 \\ \varepsilon_2 \\ \varepsilon_6 \end{Bmatrix} = \begin{Bmatrix} \varepsilon_{11} \\ \varepsilon_{22} \\ \gamma_{12} \end{Bmatrix}$$

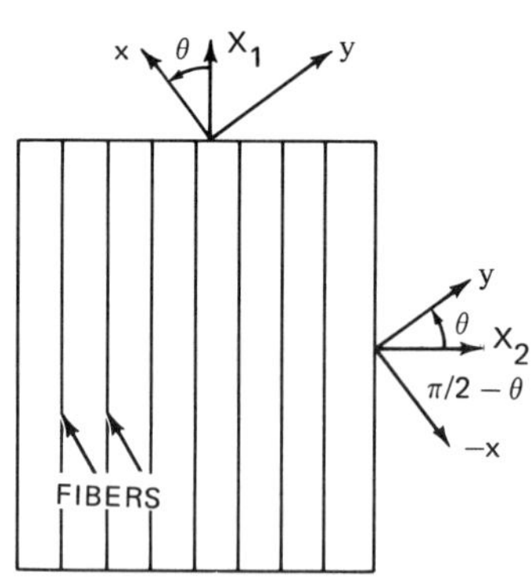

FIG. 1. System of axes.

Equations (1) can be more conveniently expressed in the matrix notation:

$$\{\sigma\} = [Q_1]\{\varepsilon\} \tag{2}$$

where

$$\{\sigma\} = \begin{Bmatrix} \sigma_1 \\ \sigma_2 \\ \sigma_6 \end{Bmatrix}, \quad \{\varepsilon\} = \begin{Bmatrix} \varepsilon_1 \\ \varepsilon_2 \\ \varepsilon_6 \end{Bmatrix}, \quad [Q_1] = \begin{bmatrix} E_1/\Lambda & E_2\nu_{12}/\Lambda & 0 \\ E_2\nu_{12}/\Lambda & E_2/\Lambda & 0 \\ 0 & 0 & G_{12} \end{bmatrix}$$

with $\Lambda = 1 - \nu_{12}\nu_{21}$ and $E_2\nu_{12} = E_1\nu_{21}$. Equations (2) may be inverted to express the strains in terms of the stresses, i.e.,

$$\{\varepsilon\} = [Q_1]^{-1}\{\sigma\} \tag{3}$$

Let us now consider another system of orthogonal axes, x-y, inclined at some angle θ to the principal axes x_1-x_2:

$$\begin{Bmatrix} x \\ y \end{Bmatrix} = \begin{bmatrix} \ell_1 & \ell_2 \\ m_1 & m_2 \end{bmatrix} \begin{Bmatrix} x_1 \\ x_2 \end{Bmatrix} \tag{4}$$

where ℓ_1, m_1, ℓ_2, and m_2 are the direction cosines given by

$$\begin{bmatrix} \ell_1 & m_1 \\ \ell_2 & m_2 \end{bmatrix} = \begin{bmatrix} \cos\theta & -\sin\theta \\ \sin\theta & \cos\theta \end{bmatrix} \tag{5}$$

The stresses and the strains, referred to the x-y system of axes,* are given by the matrix equations [2]

$$\{\sigma'\} = [Q_2]\{\sigma\} \tag{6}$$

and

$$\{\varepsilon'\} = [Q_3]\{\varepsilon\} \tag{7}$$

where

$$\{\sigma'\} = \begin{Bmatrix} \sigma_{xx} \\ \sigma_{yy} \\ \tau_{xy} \end{Bmatrix}, \quad \{\varepsilon'\} = \begin{Bmatrix} \varepsilon_{xx} \\ \varepsilon_{yy} \\ \gamma_{xy} \end{Bmatrix}$$

$$[Q_2] = \begin{bmatrix} \ell_1^2 & m_1^2 & 2\ell_1 m_1 \\ \ell_2^2 & m_2^2 & 2\ell_2 m_2 \\ \ell_1 \ell_2 & m_1 m_2 & (\ell_1 m_2 + \ell_2 m_1) \end{bmatrix}$$

$$[Q_3] = \begin{bmatrix} \ell_1^2 & m_1^2 & \ell_1 m_1 \\ \ell_2^2 & m_2^2 & \ell_2 m_2 \\ 2\ell_1 \ell_2 & 2m_1 m_2 & (\ell_1 m_2 + \ell_2 m_1) \end{bmatrix}$$

Using Eqs. (2), (6), and (7), we obtain the stress-strain relationship

$$\{\sigma'\} = [C]\{\varepsilon'\} = [Q_2][Q_1][Q_3]^{-1}\{\varepsilon'\} \tag{8}$$

*Note that the layer is orthotropic (four constants) with respect to the x_1-x_2 system of axes, and anisotropic (six constants) with respect to the x-y system of axes.

where

$$[C] = \begin{bmatrix} C_{11} & C_{12} & C_{16} \\ C_{21} & C_{22} & C_{26} \\ C_{61} & C_{62} & C_{66} \end{bmatrix}$$

with [3]

$$C_{11} = (1/\Lambda)[E_1 \cos^4\theta + E_2 \sin^4\theta + 2(\sin^2\theta)(\cos^2\theta) \cdot (E_2\nu_{12} + 2G_{12}\Lambda)]$$

$$C_{12} = C_{21} = (1/\Lambda)[E_2\nu_{12}(\cos^4\theta + \sin^4\theta) + (\sin^2\theta)(\cos^2\theta)(E_1 + E_2 - 4G_{12}\Lambda)]$$

$$C_{16} = C_{61} = (1/\Lambda)[(\sin\theta)(\cos^3\theta)(E_1 - E_2\nu_{12} - 2G_{12}\Lambda) - (\sin^3\theta)(\cos\theta)(E_2 - E_2\nu_{12} - 2G_{12}\Lambda)]$$

$$C_{22} = (1/\Lambda)[E_1 \sin^4\theta + E_2 \cos^4\theta + 2(\sin^2\theta)(\cos^2\theta) \cdot (E_2\nu_{12} + 2G_{12}\Lambda)]$$

$$C_{26} = C_{62} = (1/\Lambda)[(\sin^3\theta)(\cos\theta)(E_1 - E_2\nu_{12} - 2G_{12}\Lambda) - (\sin\theta)(\cos^3\theta)(E_2 - E_2\nu_{12} - 2G_{12}\Lambda)]$$

$$C_{66} = (1/\Lambda)[(\sin^2\theta)(\cos^2\theta)(E_1 + E_2 - 2E_2\nu_{12}) + G_{12}\Lambda(\cos^2\theta - \sin^2\theta)^2]$$

Alternatively, the strains may be expressed in terms of the stresses:

$$\{\varepsilon'\} = [S]\{\sigma'\} = [C]^{-1}\{\sigma'\} = [Q_3][Q_1]^{-1}[Q_2]^{-1}\{\sigma'\} \qquad (9)$$

where

$$[S] = [C]^{-1} = \begin{bmatrix} S_{11} & S_{12} & S_{16} \\ S_{21} & S_{22} & S_{26} \\ S_{61} & S_{62} & S_{66} \end{bmatrix}$$

with [1]

$$S_{11} = \frac{1}{E_x} = \frac{\cos^4\theta}{E_1} + \frac{\sin^4\theta}{E_2} + \left[\frac{1}{G_{12}} - \frac{2}{E_{12}}\right] \sin^2\theta \cos^2\theta$$

$$S_{12} = S_{21} = \frac{1}{E_{xy}} = -\frac{\cos^4\theta}{E_{12}} - \frac{\sin^4\theta}{E_{12}} + (\sin^2\theta)(\cos^2\theta)\left[\frac{1}{E_1} + \frac{1}{E_2} - \frac{1}{G_{12}}\right]$$

$$S_{16} = S_{61} = (\cos^3\theta)(\sin\theta)\left[\frac{2}{E_1} + \frac{2}{E_{12}} - \frac{1}{G_{12}}\right] - (\sin^3\theta)(\cos\theta)\left[\frac{2}{E_2} + \frac{2}{E_{12}} - \frac{1}{G_{12}}\right]$$

$$S_{22} = \frac{1}{E_y} = \frac{\sin^4\theta}{E_1} + \frac{\cos^4\theta}{E_2} + \left[\frac{1}{G_{12}} - \frac{2}{E_{12}}\right] (\sin^2\theta)(\cos^2\theta)$$

$$S_{26} = S_{62} = (\sin^3\theta)(\cos\theta)\left[\frac{2}{E_1} + \frac{2}{E_{12}} - \frac{1}{G_{12}}\right] - (\cos^3\theta)(\sin\theta)\left[\frac{2}{E_2} + \frac{2}{E_{12}} - \frac{1}{G_{12}}\right]$$

$$S_{66} = \frac{1}{G_{xy}} = \frac{\cos^4\theta}{G_{12}} + \frac{\sin^4\theta}{G_{12}} + 2(\sin^2\theta)(\cos^2\theta)\left[\frac{2}{E_1} + \frac{2}{E_2} + \frac{4}{E_{12}} - \frac{1}{G_{12}}\right]$$

and

$$E_{12} = E_2/\nu_{21} = E_1/\nu_{12}, \quad E_{xy} = E_y/\nu_{yx} = E_x/\nu_{xy}$$

The relationships for the S_{ij} are plotted in Fig. 2, for S-glass/epoxy resin composites as a function of orientation angle θ and fiber volume fraction V_f. The figures show that all the elastic moduli vary with the orientation angle, θ.

Consider a multilayer composite containing n layers with layer thicknesses h_i (i - 1, ..., n) and fiber reinforcement oriented at angles θ_i (i = 1, ..., n) to the x-y system of axis. We can write, for each layer,

$$\{\varepsilon'\}_i = [S]_i\{\sigma'\}_i \tag{10}$$

It is assumed in Ref. [1] that the layers are firmly bonded together to prevent slipping, such that the strains in the individual layers are equal to the composite strain $\{\varepsilon_c'\}$:

$$\{\varepsilon'\}_1 = \ldots = \{\varepsilon'\}_n = \{\varepsilon_c'\} \tag{11}$$

This is strictly true only for in-plane loadings. If the multilayer laminate is loaded in bending, it undergoes some change in curvature, and the cross-section strain may be written as

$$\{\varepsilon'\} = \{\varepsilon^{0\prime}\} + x_3\{\kappa\} \tag{12}$$

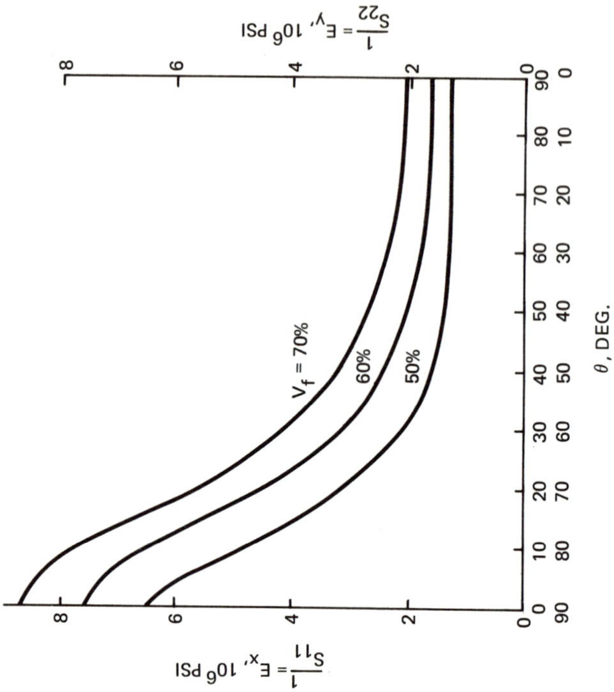
$\frac{1}{S_{22}} = E_y$, 10^6 PSI
$\frac{1}{S_{11}} = E_x$, 10^6 PSI
V_f = 70%
60%
50%
θ, DEG.

FIG. 2b.

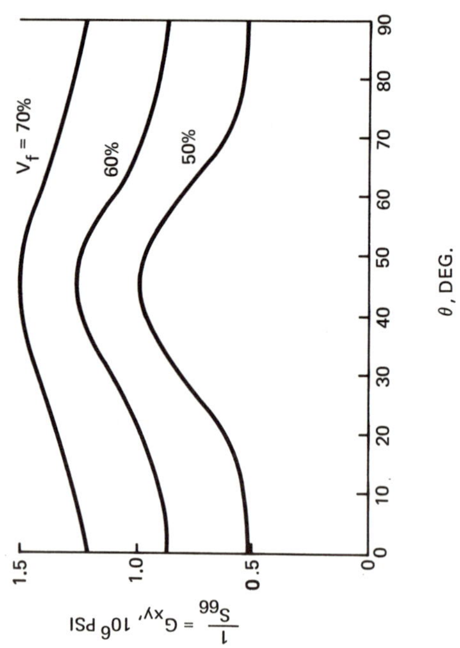
$\frac{1}{S_{66}} = G_{xy}$, 10^6 PSI
V_f = 70%
60%
50%
θ, DEG.

FIG. 2a.

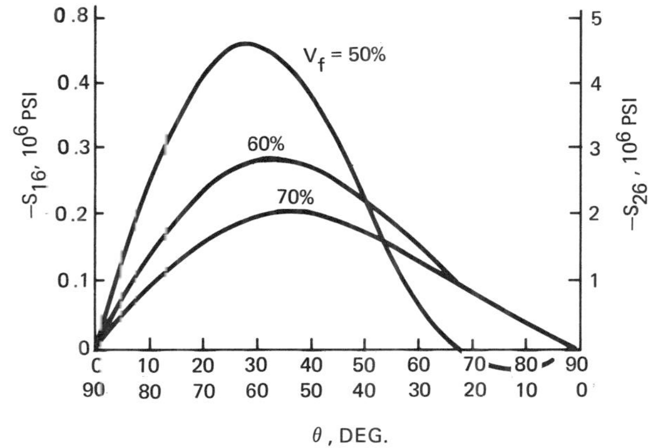

FIG. 2c.

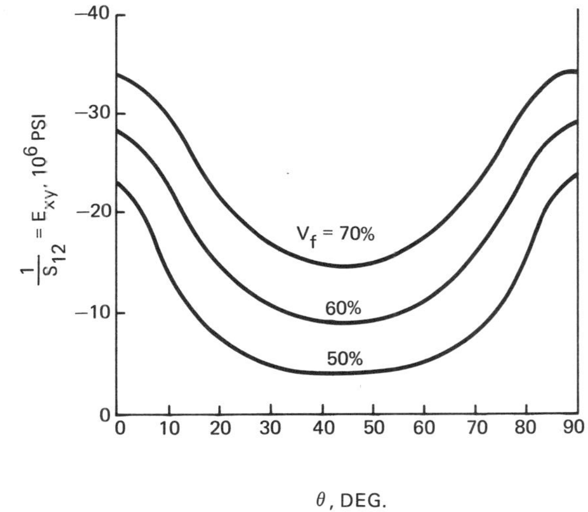

FIG. 2d.

FIG. 2a-2d. Relationships between geometric material properties for S Glass-Epoxy composites and V_f, θ (from Ref. [1]).

where κ is the curvature, and the superscript o refers to some reference surface from which x_3 measures the normal distance. The no-slip assumption, in this case, ensures strain continuity across the layer boundaries, but not necessarily the equality of strain in all the layers. Also, most laminated composites are used at a different temperature than the one at which they were fabricated. This induces thermal stresses between the layers. We will return to these questions in Section IV.

Assuming that each layer is homogeneous through its thickness, we can write, from the equilibrium of forces,

$$\sum_{i=1}^{n} \{\sigma'\}_i (h_i/h) = \{\sigma_c'\} \tag{13}$$

where h is the total thickness of the composite. Combining Eqs. (10), (11), and (13), we obtain

$$\left(\sum_{i=1}^{n} [S]_i^{-1} \frac{h_i}{h}\right)\{\varepsilon_c'\} = \left(\sum_{i=1}^{n} [C]_i \frac{h_i}{h}\right)\{\varepsilon_c'\} = \{\sigma_c'\} \tag{14}$$

Therefore, the stress-strain relations for the composite are

$$\{\sigma_c'\} = [\bar{A}]\{\varepsilon_c'\} = \left(\sum_{i=1}^{n} [C]_i \frac{h_i}{h}\right)\{\varepsilon_c'\} \tag{15}$$

and

$$\{\varepsilon_c'\} = [\bar{A}]^{-1}\{\sigma_c'\} = \left(\sum_{i=1}^{n} [C]_i \frac{h_i}{h}\right)^{-1}\{\sigma_c'\} \tag{16}$$

Thus, given the elastic properties of each separate layer, $[C]_i$, the elastic properties of the composite, $[\bar{A}]$, can be obtained from Eqs. (15) and (16)

In the following discussions, we will drop the primes in notation on the stresses and strains.

III. APPLICATIONS TO FILAMENT-WOUND STRUCTURES

Filament-wound pressure vessels generally display symmetry with respect to the principal axes x_1-x_2. Let us assume that x and y are, respectively, the meridional and tangential (hoop) axes in a layered shell. Then, the number of layers oriented at angle θ is equal to the number of layers oriented at angle $-\theta$. This implies that the resulting material is orthotropic and that the shear strain ε_{xy} is zero [3]. The strain assumptions of the membrane analysis of shells are compatible with the in-plane loading analysis of anisotropic layered plates discussed in the preceding section. Therefore, the overall stress resultants of membrane theory can be written as

$$N_x = \sigma_{cxx} h = \left[\bar{E}_x h / \left(1 - \bar{\nu}_{yx}\bar{\nu}_{xy}\right)\right]\left[\varepsilon_{xx} + \bar{\nu}_{yx}\varepsilon_{yy}\right]$$

$$N_y = \sigma_{cyy} h = \left[\bar{E}_y h / \left(1 - \bar{\nu}_{yx}\bar{\nu}_{xy}\right)\right]\left[\bar{\nu}_{xy}\varepsilon_{xx} + \varepsilon_{yy}\right] \qquad (17)$$

where

$$\bar{E}_x = \bar{A}_{11} - \left(\bar{A}_{12}^2/\bar{A}_{22}\right), \quad \bar{E}_y = \bar{A}_{22} - \left(\bar{A}_{12}^2/\bar{A}_{11}\right)$$

$$\bar{\nu}_{xy} - \bar{A}_{12}/\bar{A}_{22}, \quad \bar{\nu}_{yx} - \bar{A}_{12}/\bar{A}_{11}$$

The condition of balanced design in filament winding is achieved when the fibers support all the applied load.

In a netting analysis (see Chapter 2), the condition for balanced design can be written as*

$$N_y/N_x = \tan^2\theta \tag{18}$$

where θ is the local helix angle. If, however, the effect of the resin is taken into account, a new balanced design condition may be defined by setting σ_2 equal to zero [3]. Then, from Eq. (1),

$$\varepsilon_2 = -\nu_{12}\varepsilon_1 \tag{19}$$

---

*In a netting analysis, only fibers can support the load. Thus, in the x_1-x_2 system of axis, the stress vector is

$$\begin{Bmatrix} \sigma_1 \\ 0 \\ 0 \end{Bmatrix}$$

For a fiber inclined at an angle $\pm\theta$, the stress vector (in the x-y system is

$$\begin{Bmatrix} \sigma_1 \cos^2\theta \\ \sigma_1 \sin^2\theta \\ \pm\sigma_1 \sin\theta\cos\theta \end{Bmatrix}$$

Therefore, using the membrane theory, we have

$$N_x = \sigma_1\left(\cos^2\theta\right)h, \quad N_y = \sigma_1\left(\sin^2\theta\right)h$$

$$N_y/N_x = \tan^2\theta$$

We obtain, upon combining Eqs. (7) and (19)

$$\varepsilon_{xx}/\varepsilon_{yy} = -\left(\nu_{12}\tan^2\theta + 1\right)/\left(\nu_{12} + \tan^2\theta\right) \tag{20}$$

Substitution of Eq. (20) into Eq. (17) yields

$$\frac{N_y}{N_x} = \frac{\bar{E}_y}{\bar{E}_x}\,\frac{(\tan^2\theta)(1 - \bar{\nu}_{xy}\nu_{12}) + \nu_{12} - \bar{\nu}_{xy}}{(\tan^2\theta)(\bar{\nu}_{yx} - \nu_{12}) + \bar{\nu}_{yx}\nu_{12} - 1} \tag{21}$$

Note that in this case, as in the pure netting analysis, the filament winding pattern and the dome shape (defined by the membrane stress resultant ratio) are not independent. Specification of one automatically determines the other.

In filament-wound structures, juncture discontinuity stresses, such as those arising between end domes and cylinders, must also be considered. These stresses, as well as those in the area of the apex, will produce bending and violate the assumptions of membrane analysis. The bending in the juncture areas can be minimized by adjusting the dome meridional shape. Since the bending is caused by the discontinuity in the meridional radius of curvature r_1 at the juncture, a shape such as a Cassinian dome (Fig. 3) could be used. This shape is the ideal transition curve, since r_1 approaches infinity at the cylinder juncture. However, this curve does not satisfy Eq. (21)

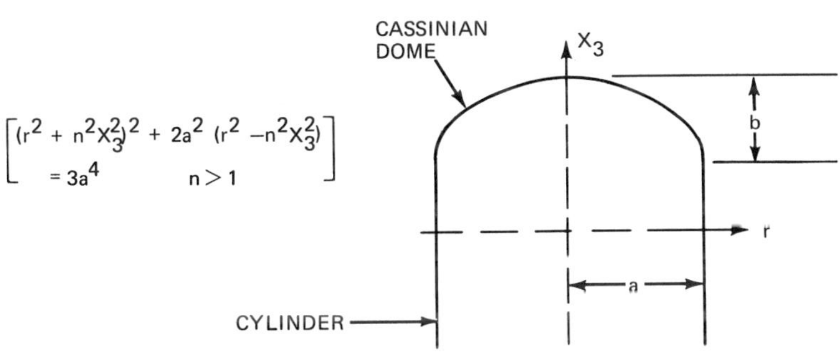

FIG. 3. Cassini curve.

anywhere, and the balanced design concept is lost. In practice, the juncture bending is minimized by adding a form factor Z (≤ 2) to Eq. (21):

$$\frac{N_y}{N_x} = \frac{\bar{E}_y}{\bar{E}_x} \frac{(\tan^2\theta)(1 - \bar{\nu}_{xy}\nu_{12}) + \nu_{12} - \bar{\nu}_{xy}}{(\tan^2\theta)(\bar{\nu}_{xy} - \nu_{12}) + \bar{\nu}_{yx}\nu_{12} - 1} + (2.0 - Z) \qquad (22)$$

The form factor Z is varied near the juncture to match the free edge deflections of the components and thus minimize the local bending [3]. Further from the juncture, Z is increased to 2.0, to maintain the balanced design condition. The effect upon the actual dome contour of choosing a particular Z_0 is usually quite small (see Fig. 4).

The use of membrane analysis to analyze filament-wound shells is due to its simplicity and the accuracy of the predictions. This simplicity is lost somewhat when a two-component juncture is to be considered, since free edge deflections of the components must be calculated. The form factor considerations become even more complex for multi-junctures (e.g., skirt attachment) and lead to an iteration-type solution [3]. Also, near an apex which is not both isotropic and spherical, the membrane theory breaks down at meridional angles smaller than 0.2 radians [4]. For a detailed consideration of these limitations of the membrane analysis, reference is made to [3] and [4].

IV. GENERALIZED THEORY OF MULTILAYER LAMINATES

In Section II, we derived the stress-strain relationships for the laminated composite under the equal strain assumption of Eq. (11). Presently, we will relax this assumption to allow also for the application of out-of-plane loads and bending moments to the laminated composite. It should also be noted that most composites are constructed at elevated temperatures. This causes a differential expansion (or contraction) between adjoining layers when the

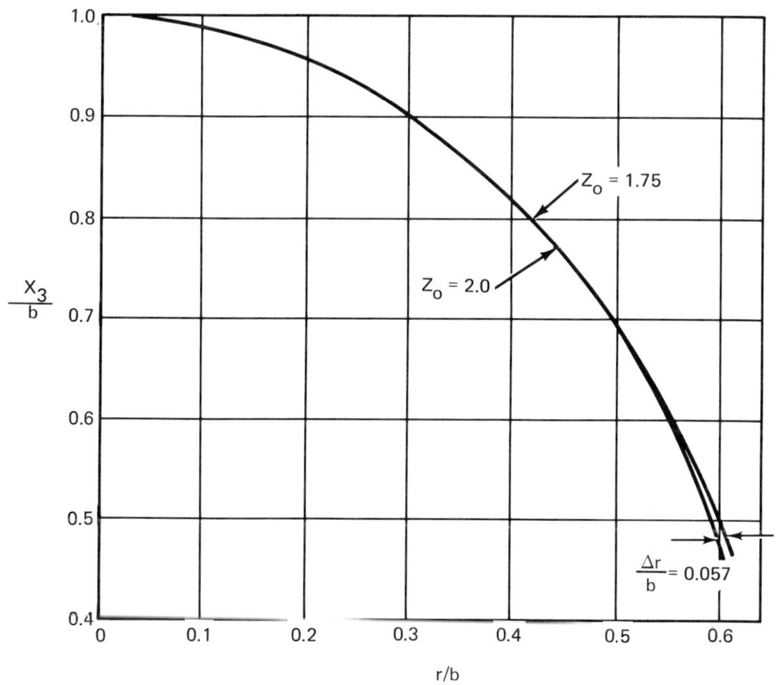

FIG. 4. Effect of Z_0 on dome contour (from Ref. [3]).

service temperature is different from the fabrication temperature. In the following, we will discuss the general thermoelastic analysis of laminated composites [5], [6].

As before, we will assume that each constituent layer is homogeneous and anisotropic* and is in a state of two-dimensional stress. For an orthotropic plate, the thermal expansion tensor α_{ij} $(i,j = 1,2,3)$ has only two non-zero components, namely α_{11} and α_{22}. However, in the x-y system of axes, there are three nonzero components,

*With respect to the x-y system of axis (orthotropic with respect to the principal axes x_1-x_2).

$$\begin{Bmatrix} \alpha_1 \\ \alpha_2 \\ \alpha_6 \end{Bmatrix} = \begin{Bmatrix} \alpha_{xx} \\ \alpha_{yy} \\ \alpha_{xy} \end{Bmatrix} = \begin{bmatrix} \ell_1^2 & m_1^2 & 2\ell_1 m_1 \\ \ell_2^2 & m_2^2 & 2\ell_2 m_2 \\ \ell_1 \ell_2 & m_1 m_2 & (\ell_1 m_2 + \ell_2 m_1) \end{bmatrix} \begin{Bmatrix} \alpha_{11} \\ \alpha_{22} \\ 0 \end{Bmatrix} \tag{23}$$

The stress-strain relationships for the kth anisotropic layer may be written as

$$\varepsilon_i^{(k)} = S_{ij}^{(k)} \sigma_j^{(k)} + \alpha_i^{(k)} T \tag{24}$$

and

$$\sigma_i^{(k)} = C_{ij}^{(k)} \left(\varepsilon_j^{(k)} - \alpha_j^{(k)} T \right) \tag{25}$$

where $i, j = 1,2,6$ and

$$\begin{Bmatrix} \sigma_1 \\ \sigma_2 \\ \sigma_6 \end{Bmatrix} = \begin{Bmatrix} \sigma_{xx} \\ \sigma_{yy} \\ \tau_{xy} \end{Bmatrix}, \begin{Bmatrix} \varepsilon_1 \\ \varepsilon_2 \\ \varepsilon_6 \end{Bmatrix} = \begin{Bmatrix} \varepsilon_{xx} \\ \varepsilon_{yy} \\ \gamma_{xy} \end{Bmatrix}$$

Equation (24) implies that the total strain is the sum of the mechanical strain and the free thermal strain. The stiffness tensor $C_{ij}^{(k)}$ and the compliance tensor, $S_{ij}^{(k)}$ for the kth layer are given by Eqs. (8) and (9).

In deriving the stress-strain relationships for a laminated anisotropic composite, we will assume that

(a) There is no deformation of the cross section in a direction normal to the plane of the plate (no thickness stretch).

(b) Normals to the cross section remain straight and normal after deformation.

In this case, the strain across the composite thickness is

$$\varepsilon_i = \varepsilon_i^{\,0} + x_3\kappa_i, \quad i = 1,2,6 \tag{12'}$$

From the classical plate theory [7], the stress resultants for the total multilayer laminate are defined as

$$\overline{N}_i = N_i + N_i^T = A_{ij}\varepsilon_j^0 + B_{ij}\kappa_j$$

$$\overline{M}_i = M_i + M_i^T = B_{ij}\varepsilon_j^0 + D_{ij}\kappa_j \tag{26}$$

where

$$\left(N_i,\ M_i\right) = \int_{-h/2}^{h/2} \sigma_i(1,\ x_3)\,dx_3 \tag{27}$$

$$\left(N_i^T,\ M_i^T\right) = \int_{-h/2}^{h/2} C_{ij}\alpha_j T(1,\ x_3)\,dx_3 \tag{28}$$

$$\left(A_{ij},\ B_{ij},\ D_{ij}\right) = \int_{-h/2}^{h/2} C_{ij}(1,\ x_3,\ x_3^2)\,dx_3 \tag{29}$$

Note that Eqs. (26)-(29) provide the link between the thermal and the nonthermal analyses by treating the thermal effects as equivalent mechanical loads N_i^T and M_i^T.

Combining Eqs. (26), the stress at any location across the thickness of the laminated structure may be written in the matrix form:

$$\begin{Bmatrix} \overline{N} \\ M \end{Bmatrix} = \begin{bmatrix} A & B \\ B & D \end{bmatrix} \begin{Bmatrix} \varepsilon^0 \\ \kappa \end{Bmatrix} \tag{30}$$

Since the matrices [A], [B], and [D] are symmetric, the composite matrix given by Eq. (30) is also symmetric.

Solving Eq. (26a) for $\{\varepsilon^0\}$, we obtain

$$\{\varepsilon^0\} = [A]^{-1}\{\bar{N}\} - [A]^{-1}[B]\{\kappa\} \tag{31}$$

Substitution of Eq. (31) into Eq. (26b) yields

$$\{\bar{M}\} = [B][A]^{-1}\{\bar{N}\} + \Big([D] - [B][A]^{-1}[B]\Big)\{\kappa\} \tag{32}$$

The set of equations (31) and (32) may be rewritten as

$$\begin{Bmatrix} \varepsilon^0 \\ \bar{M} \end{Bmatrix} = \begin{bmatrix} A^* & B^* \\ H^* & D^* \end{bmatrix} \begin{Bmatrix} \bar{N} \\ \kappa \end{Bmatrix} \tag{33}$$

where

$$[A^*] = [A]^{-1}, \quad [B^*] = -[A]^{-1}[B]$$

$$[H^*] = [B][A]^{-1}, \quad [D^*] = [D] - [B][A]^{-1}[B]$$

Note that the composite matrix given by Eq. (33) is not symmetric, since $[B^*] \neq [H^*]$. Solving Eq. (32) for $\{\kappa\}$, we have

$$\{\kappa\} = [D^*]^{-1}\{\bar{M}\} - [D^*]^{-1}[H^*]\{\bar{N}\} \tag{34}$$

Substitution of Eq. (34) into Eq. (31) yields

$$\{\varepsilon^0\} = \Big([A^*] - [B^*][D^*]^{-1}[H^*]\Big)\{\bar{N}\} + [B^*][D^*]^{-1}\{\bar{M}\} \tag{35}$$

Combining Eqs. (34) and (35), there follows

$$\begin{Bmatrix} \varepsilon^0 \\ \kappa \end{Bmatrix} = \begin{bmatrix} A' & B' \\ B' & D' \end{bmatrix} \begin{Bmatrix} \overline{N} \\ \overline{M} \end{Bmatrix} \tag{36}$$

where

$$[A'] = [A^*] - [B^*][D^*]^{-1}[H^*]$$

$$[B'] = [B^*][D^*]^{-1} = -[D^*]^{-1}[H^*]$$

$$[D'] = [D^*]^{-1}$$

The composite "prime" matrix is symmetric, since it is the inverse of a symmetric matrix, Eq. (30). Combining Eqs. (12), (25), and (36), the stress in the kth layer is found to be

$$\left\{\sigma^{(k)}\right\} = \left[C^{(k)}\right] \left\{ \left([A'] + x_3^{(k)}[B']\right)\{\overline{N}\} + \left([B'] + x_3^{(k)}[D']\right)\{\overline{M}\} - \left\{\alpha^{(k)}\right\} T \right\} \tag{37}$$

Equation (37) is the general relationship expressing laminate stresses as functions of stress resultants, bending moments, and the composite temperature distribution.

The matrices A_{ij}, B_{ij}, and D_{ij} are obtained by integrating C_{ij} along the x_3-axis. This is a scalar operation, and does not, therefore, alter the tensorial character of C_{ij}. Hence, in general, the composite matrix equation (30) or its equivalent equation (36), has 18 independent elements. This number can be reduced for certain situations, as follows:

(a) If the plate is homogeneous, then the C_{ij} are not functions of x_3. Hence, from Eq. (29), we have

$$A_{ij} = (12/h^2)D_{ij}, \quad B_{ij} = 0 \tag{38}$$

(b) If all the layers in the plate are isotropic, then for each layer, there are only two independent constants:

$$C_{11} = C_{22}, \quad C_{16} = C_{26} = 0$$

$$C_{66} = (1/2)(C_{11} - C_{12}) \tag{39}$$

Thus, each of the matrices A_{ij}, B_{ij}, and D_{ij}, has two independent constants, and the total number for this case is six.

(c) If the plate is homogeneous and isotropic, then obviously the number of independent constants is two.

(d) If the C_{ij} are even functions of x_3, then from Eq. (29), we have

$$B_{ij} \equiv 0 \tag{40}$$

This reduces the number of independent constants from 18 to 12.

The preceding discussion demonstrates that the number of independent constants is related to the symmetry properties of C_{ij} and to the type of lamination used. We will now consider two special cases of lamination: cross-ply and angle-ply. These types of construction are often encountered in practical applications.

V. CROSS-PLY COMPOSITES

The cross-ply lamination consists of n orthotropic layers stacked with an alternating orientation of 90° between the layers. We assume that the x and the y axes are parallel to the fiber reinforcement in the odd- and the even-numbered layers, respectively. All the odd layers have the same thickness. The even layers also have the same thickness, but not necessarily equal to that of the odd layers. The components of C_{ij} for the odd layers then are C_{11}, C_{22}, C_{12}, and C_{66}, with $C_{16} = C_{26} = 0$. The components of C_{ij} for the even layers are identical to those for the odd layers, except that C_{11} and C_{22} are interchanged.

Assuming now, that each unit layer is homogeneous, the integral equation (29) may be written as

$$A_{ij} = \sum_{k=1}^{n} C_{ij}^{(k)} (h_{k+1} - h_k)$$

$$B_{ij} = (1/2) \sum_{k=1}^{n} C_{ij}^{(k)} \left(h_{k+1}^2 - h_k^2\right)$$

$$D_{ij} = (1/3) \sum_{k=1}^{n} C_{ij}^{(k)} \left(h_{k+1}^3 - h_k^3\right) \tag{41}$$

The summations of Eqs. (41) may be expressed in closed form for cross-ply composites [5] by using the properties of series. For example,

$$A_{11} = \sum_{k=1}^{n} C_{11}^{(k)} (h_{k+1} - h_k) = C_{11} h_{odd} + C_{22} h_{even}$$

$$+ C_{11} h_{odd} + C_{22} h_{even} + \cdots$$

Define

$$h = \sum h_{odd} + \sum h_{even}, \quad m = \left(\sum h_{odd}\right) / \left(\sum h_{even}\right)$$

Therefore

$$A_{11} = \left(hC_{11}\sum h_{odd} + hC_{22}\sum h_{even}\right)/h$$

$$= \left(hC_{11}\sum h_{odd} + hC_{22}\sum h_{even}\right)/\left(\sum h_{odd} + \sum h_{even}\right)$$

$$= (hmC_{11} + hC_{22})/(m + 1)$$

The components of the matrices [A], [B], and [D] are as follows:

(a) For an odd number of layers n:

$$\begin{aligned}
A_{11} &= [1/(1 + m)](m + F)hC_{11} \\
A_{22} &= [1/(1 + m)](1 + mF)hC_{11} \\
A_{12} &= hC_{12}, \quad A_{66} = hC_{66}, \quad A_{16} = A_{26} = 0 \\
B_{ij} &= 0, \quad i, j = 1, 2, 6 \\
D_{11} &= [(F - 1)P + 1](h^3/12)C_{11} \\
D_{22} &= [(1 - F)P + F](h^3/12)C_{11} \\
D_{12} &= (h^3/12)C_{12}, \quad D_{66} = (h^3/12)C_{66}, \quad D_{16} = D_{26} = 0
\end{aligned} \tag{42}$$

where

$$F = \frac{C_{22}}{C_{11}} = \frac{E_2}{E_1}$$

$$P = \left(\frac{1}{1 + m}\right)^3 + \frac{m(n - 3)[m(n - 1) + 2(n + 1)]}{(n^2 - 1)(1 + m)^3}$$

(b) For an even total number of layers n, the same

components exist as for case (a) except for the following changes:

$$B_{11} = -B_{22} = \{m(F - 1)/[n(1 + m)^2]\} h^2 C_{11}$$

$$D_{11} = [(F - 1)Q + 1](h^3/12)C_{11} \qquad (43)$$

$$D_{22} = [(1 - F)Q + F](h^3/12)C_{11}$$

where

$$Q = \frac{1}{1 + m} - \frac{8m(m - 1)}{n^2(1 + m)^3}$$

The matrices A_{ij} and D_{ij} are found to be orthotropic, and B_{ij} is identically zero for cross-ply construction except for B_{11} ($= -B_{22}$) when n is even. The equations for A_{11} and A_{22} are plotted for various values of F in Fig. 5. The case $F = 0$ (Fig. 5) corresponds to a netting analysis. The equations for B_{11}, D_{11}, and D_{22} are plotted in Figs. 6 and 7. The physical significance of B_{11} can be interpreted as a measure of the shifting of the neutral plane. This shifting is maximum for $n = 2$. For large n, B_{11} approaches zero. Note that for the [B] and [D] matrices, the case $F = 0$ does not correspond to a netting analysis. As far as a pure netting analysis is concerned, both the [B] and [D] matrices must be identically zero. This follows directly from the assumption that only the fibers are able to carry load (in a direction parallel to their axes).

VI. ANGLE-PLY COMPOSITES

The angle-ply composite consists of orthotropic layers (or plies) with an arbitrary alternating angle of orienta-

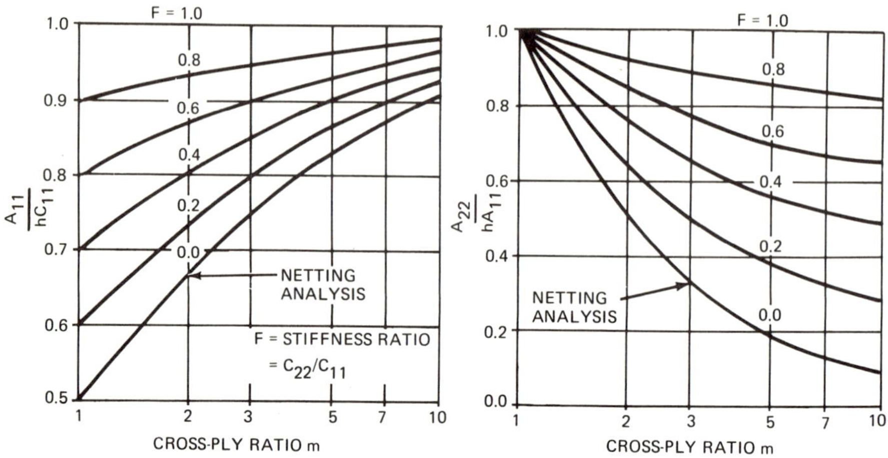

FIG. 5. Stiffness components A_{11} and A_{22} (from Ref. [5]).

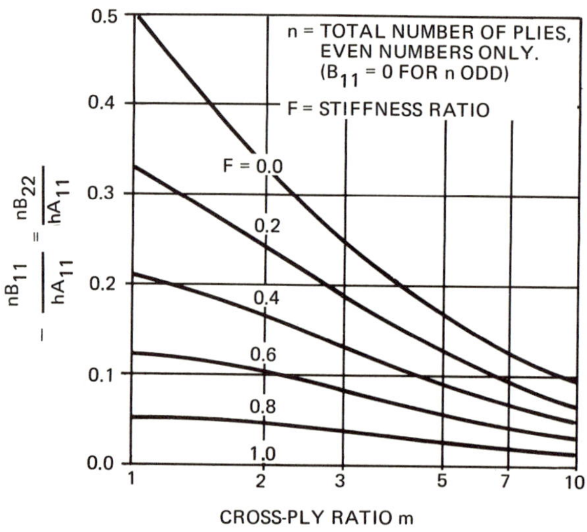

FIG. 6. Coupling term B_{11} (from Ref. [5]).

tion between the layers. The odd-numbered plies are at an angle $-\theta$ from the x axis and the even-numbered plies are

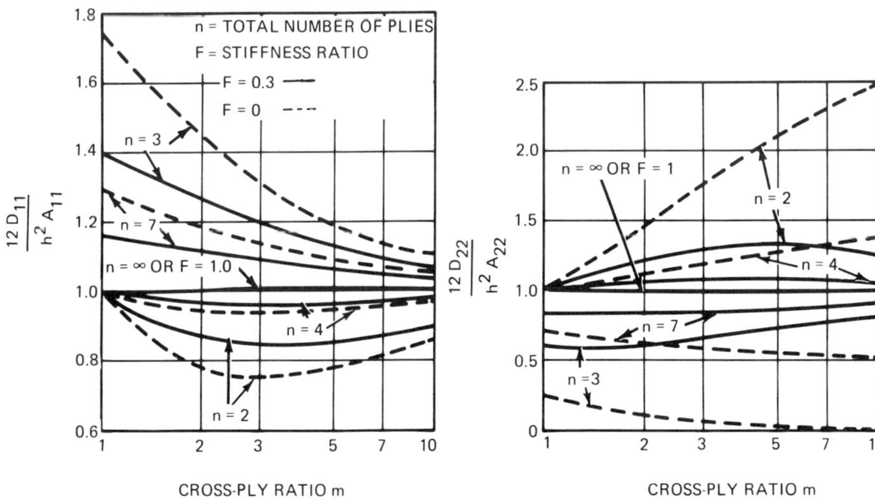

FIG. 7. Flexural rigidities D_{11} and D_{22} (from Ref. [5]).

at an angle θ (see Fig. 8). It is assumed that all the layers have the same thickness. The general summation equations (41) are also valid for angle-ply composites. By expanding the series again, simple closed-form equations can be obtained for the components of the [A], [B], and [D] matrices [5]. The stiffness matrix C_{ij} derived in Section I, corresponds to the even-numbered layer orientation.

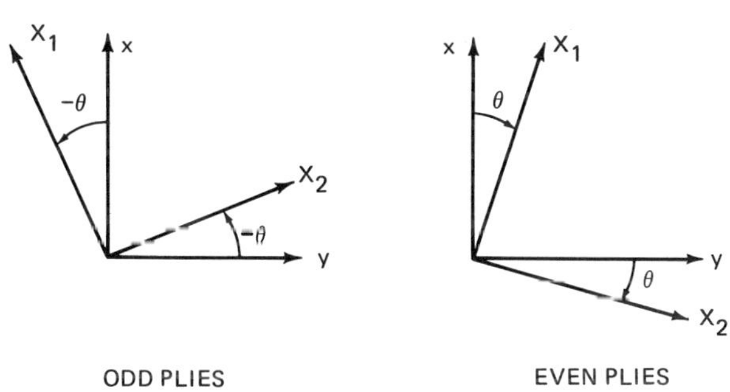

FIG. 8. Orientation of odd and even plies.

The C_{ij} for the odd-numbered layers are the same as for the even-numbered layers except that the components C_{16} and C_{26} differ in sign. The equations for the various components of the [A], [B], and [D] matrices are given as follows:

(a) For an odd number of layers n,

$$(A_{11}, A_{22}, A_{12}, A_{66}) = h(C_{11}, C_{22}, C_{12}, C_{66})$$

$$(A_{16}, A_{26}) = (h/n)(C_{16}, C_{26})$$

$$B_{ij} = 0, \quad i, j = 1, 2, 6 \tag{44}$$

$$(D_{11}, D_{22}, D_{12}, D_{66}) = (h^3/12)(C_{11}, C_{22}, C_{12}, C_{66})$$

$$(D_{16}, D_{26}) = (h^3/12)[(3n^2 - 2)/n^3](C_{16}, C_{26})$$

(b) For an even number of layers n, the same components exist as for the odd number of layers except for the following changes:

$$A_{16} = A_{26} = 0, \quad (B_{16}, B_{26}) = -(h^2/n)(C_{16}, C_{26})$$

$$D_{16} = D_{26} = 0 \tag{45}$$

The A_{ij} are plotted as a function of the lamination angle in Fig. 9 for a representative angle-ply composite. For n even, A_{16} and D_{16} are zero. For n odd, A_{16} and D_{16} are inversely proportional to n, and hence have a maximum for $n = 3$. The matrices A_{ij} and D_{ij} are orthotropic for n even and anisotropic for n odd. For n even, B_{16} and B_{26} are nonzero and are inversely proportional to n. The parametric variation of B_{16} and D_{16} is shown in Fig. 10.

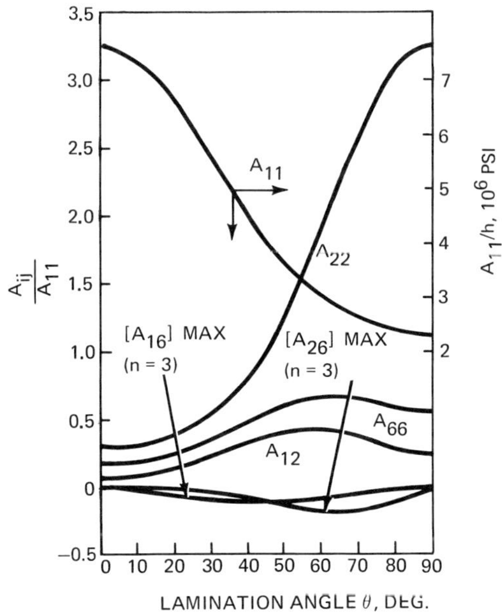

FIG. 9. A_{11} and dimensionless A_{ij} for representative angle-ply (reproduced from Ref. [5]).

VII. EXPERIMENTAL RESULTS

Tsai [5] presents some experimental data to verify the analytical results discussed above. The test specimen layers were made up of unidirectional glass fibers pre-impregnated with an epoxy resin. The resin material was supplied by U. S. Polymeric Company, with a designation of E-787-NUF. The resin content was within 13-20%. The laminated specimens consisted of two or three layers. The test results were obtained by measuring the surface strains of the loaded specimens. The measured components of the [A′], [B′], and [D′] matrices agreed reasonably well with the theoretically predicted values for both cross-ply and angle-ply laminations. We show a comparison between the theoretically predicted and experimentally measured values

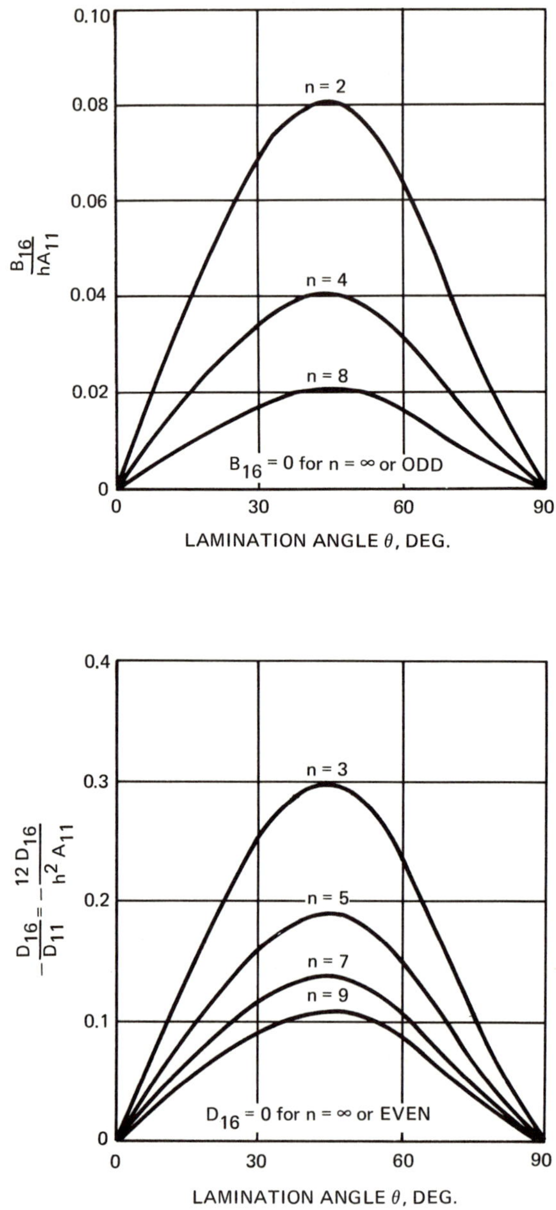

Fig. 10. B_{16} and D_{16} for representative angle ply (from Ref. [5]).

for the cross-ply composites in Fig. 11. The agreement is observed to be quite good. Thus, given the elastic properties (E_1, E_2, ν_{12}, G_{12}) of single layers, the elastic constants of the laminated structure can be predicted accurately.

The laminate analysis developed here may also be employed in an approximate manner to calculate the elastic properties of randomly oriented and two- and three-dimensional fibrous composites [8], [9]. As an example, let us consider a two-dimensional random array of fibers. We will assume that the thickness of the sheet of fibers is considerably less than the average length of fibers. Such a material may then be modeled as a laminated quasiisotropic (0°, 90°, ± 45°) system [8]. Halpin and Pagano [8] demonstrate that the stiffness of a random short fiber composite

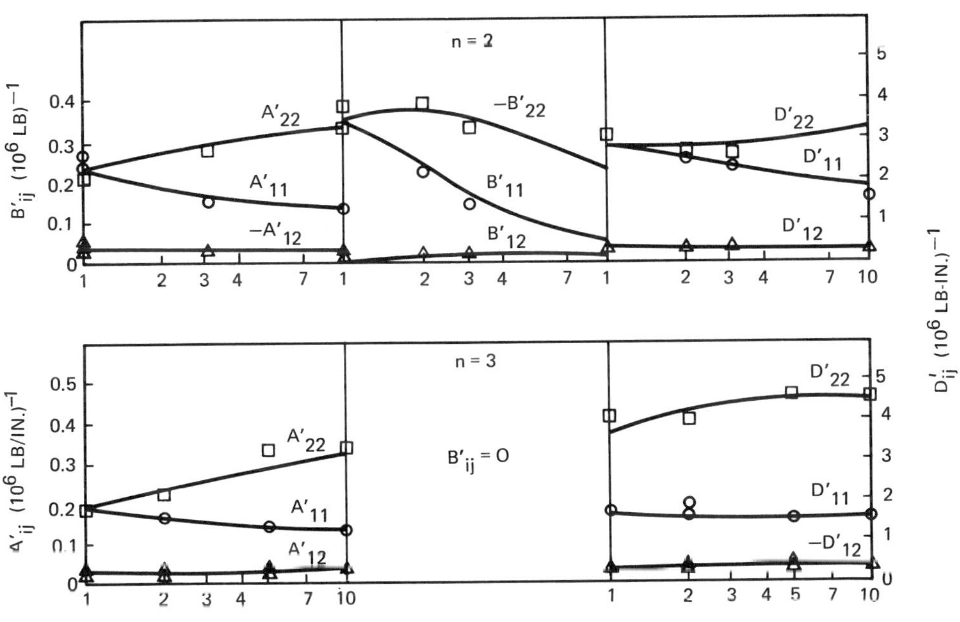

FIG. 11. Cross-ply composites (reproduced from Ref. [5]).

material of a certain fiber volume fraction V_f and average fiber length to average fiber diameter ratio ℓ/d_f is equivalent to a laminated quasiisotropic material of the same V_f and ℓ/d_f ratio. We shall not, however, pursue this question any further here. See Refs. [8], [9] for more details.

This concludes our discussion of the elastic material properties of the composite. In this part of the book (Chapters 2-5), we have discussed the various models employed in analyzing material properties of particle and fiber reinforced composites (Chapters 2-4). The behavior of laminated composites was considered finally in this chapter. We turn now to the more complicated problem of predicting composite strength. This forms the subject matter of the second part (Chapters 6-11) of the book.

REFERENCES

1. J. C. Ekvall, "Structural Behavior of Mono-Filament Composites," Paper presented at the 6th AIAA Structures and Materials Conference, April 1965.
2. I. S. Sokolnikoff, "Mathematical Theory of Elasticity," 2nd ed., McGraw-Hill, New York, 1956.
3. G. G. Love, "Structural Analysis of Orthotropic Shells--Minimizing Discontinuity Stresses in Filament-Wound Pressure Vessels," Paper No. 2919-63, Presented at the AIAA Launch and Space Vehicle Shell Structures Conference, April 1963.
4. R. Hartung, "Membrane Analysis of Filament-Wound Structures," Paper No. 2915-63, Presented at the AIAA Launch and Space Vehicle Shell Structures Conference, April 1963.
5. S. W. Tsai, "Structural Behavior of Composite Materials," NASA CR-71, July 1964.

6. S. W. Tsai, "Strength Characteristics of Composite Materials," NASA CR-224, April 1965.

7. S. Timoshenko and S. Woinowsky-Krieger, Theory of Plates and Shells, 2nd ed., McGraw-Hill, New York, 1959.

8. J. C. Halpin and N. J. Pagano, "The Laminate Approximation for Randomly Oriented Fibrous Composites," J. Composite Materials, 3, 720-724 (1969).

9. J. C. Halpin, K. Jerine, and J. M. Whitney, "The Laminate Analogy for 2- and 3-Dimensional Composite Materials," J. Composite Materials, 5, 36-49 (1971).

Chapter 6

NONSTATISTICAL MODELS OF COMPOSITE STRENGTH

I. INTRODUCTION

The strength of a composite is a complex function of the elastic and strength properties of the fiber and the matrix as well as of the fabrication process. Among the variables influenced by the fabrication process are void size and distribution, nonuniformity of fiber spacing, interface bond strength, and residual stresses. In general, the effect of the fabrication process must be considered on an empirical basis, through the use of theory-experiment correlation factors [1]. In this book, we will not concern ourselves with the effects that are primarily a function of the fabrication process but will limit our discussion to theories which attempt to relate the composite's strength characteristics to those of its components. Most of the research effort has been concentrated on gaining an understanding of the behavior of a unidirectionally fiber-reinforced composite (UFC). In this chapter, we will first outline the primary failure modes of a UFC and the simple theoretical models for determining the corresponding failure stresses. This will be followed by a discussion of the UFC under biaxial loading and the strength characteristics of laminated structures.

II. PRIMARY STRENGTHS OF A UFC

A unidirectionally fiber-reinforced composite, when subjected to an in-plane loading, exhibits five primary failure stresses, namely

(a) Longitudinal tensile (X).

(b) Longitudinal compressive (X′).

(c) Transverse tensile (Y).

(d) Transverse Compressive (Y′).

(e) In-plane shear (S).

These stresses are shown in Fig. 1. A UFC has maximum strength in the longitudinal direction and it is therefore not surprising that a large effort has been directed toward characterizing longitudinal tensile and compressive strengths. We will consider the various advanced theories for longitudinal tensile (Chapters 7-9) and compressive (Chapter 10) strengths in later chapters. Frequently, a composite is subjected to biaxial loading and in this case,

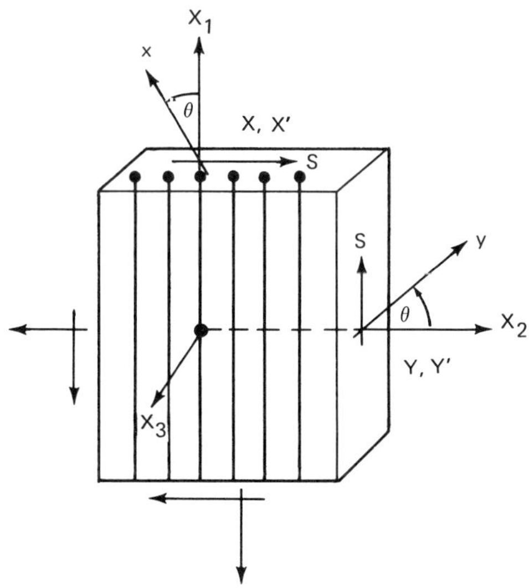

FIG. 1. Unidirectionally fiber-reinforced composite.

a knowledge of the transverse and in-plane shear strengths is essential. Unfortunately, at the present time, no completely satisfactory theories exist for these failure modes.

A. Longitudinal Tensile Strength

Consider a UFC loaded in tension in the direction of the fibers. Jech et al. [2] considered a "rule-of-mixtures" type of model in which during loading, a uniform state of strain exists up to the moment of fracture. When this uniform strain reaches the failure strain of the fibers (ε_{fT}), failure of the composite occurs. Mathematically, we have

$$\varepsilon_c = \varepsilon_f = \varepsilon_m \quad \text{and} \quad \varepsilon_{fT} \leq \varepsilon_{mT} \tag{1}$$

where ε_{mT} denotes the tensile failure strain of the matrix.

It is now assumed that both the fibers and the matrix remain linearly elastic up to the failure point. Then, from Eq. (1) and Hooke's law, there follows

$$\sigma_f/E_f = \sigma_m/E_m \tag{2a}$$

$$\sigma_{fT} \leq (E_f/E_m)\sigma_{mT} \tag{2b}$$

where σ_{fT} and σ_{mT} denote the tensile strengths of the fibers and the matrix, respectively. Equilibrium considerations lead to the following relationship for the composite stress σ_c:

$$\sigma_c = V_f\sigma_f + (1 - V_f)\sigma_m \tag{3}$$

where V_f denotes the volume fraction of the fibers. Combining Eqs. (2) and (3), we have for composite strength X

$$\frac{X}{V_f} = \sigma_{fT}\left(1 + \frac{1 - V_f}{V_f}\frac{E_m}{E_f}\right) \tag{4}$$

The quantity X/V_f $(= \bar{\sigma}_{fT})$ is frequently used as the composite tensile strength. Note that $\bar{\sigma}_{fT}$ is the equivalent fiber stress obtained by dividing the composite strength X by the total cross-sectional area (= volume fraction V_f) of the fibers in the composite.

Several objections may be raised at this point. First, in case Eq. (2b) is not satisfied, matrix yielding may occur (see Chapter 2). Second, most fibers of current interest, for example glass, display a large dispersion in their tensile strength. In this case, σ_{fT} in Eq. (4) must be taken to imply mean fiber strength. Finally, Eq. (4) predicts too large a composite strength [1, 3] as compared to experimental data. In recent years, several statistical models have been proposed for determining the tensile strength. We will discuss these in detail in Chapters 7-9.

B. Longitudinal Compressive Strength

By considerations similar to those for the longitudinal tensile strength X, the following expression may be derived for the longitudinal compressive strength X′:

$$\frac{X'}{V_f} = \sigma_{fC}\left(1 + \frac{1 - V_f}{V_f}\frac{E_m}{E_f}\right) \tag{5}$$

where σ_{fC} denotes the compressive strength of the fibers. Since it is not possible to test individual fibers in compression, σ_{fC} is usually taken as the compressive strength of the bulk material. The compressive strength given by Eq. (5) displays a very poor agreement with experimental data. There have been some attempts to continue to use equations of this type by inserting so-called "theory-experiment" correlation factors [1]. These attempts may be misdirected insofar as no consideration is given to the micromechanics of composite failure. Experimental observations reveal that compressive failure occurs as a result of the buckling of the fibers in the matrix or the debonding of the fibers from the matrix. We postpone a discussion of the more sophisticated models for compressive strength to Chapter 10.

C. Transverse Tensile, Transverse Compressive, and In-Plane Shear Strengths

The transverse tensile (Y), transverse compressive (Y′), and in-plane shear (S) strengths are assumed to be limited by the allowable strains or stresses in the matrix [1]. Mathematically, we have

$$\left\{\begin{matrix} Y \\ Y' \\ S \end{matrix}\right\} = \left\{\begin{matrix} E_2 & f(\varepsilon_{mT}) \\ E_2 & f(\varepsilon_{mC}) \\ G_{12} & f(\varepsilon_{mS}) \end{matrix}\right\} \tag{6}$$

or

$$\left\{\begin{matrix} Y \\ Y' \\ S \end{matrix}\right\} = \left\{\begin{matrix} \sigma_{mT} \\ \sigma_{mC} \\ \sigma_{mS} \end{matrix}\right\} \tag{7}$$

where E_2 and G_{12} denote the transverse Young's modulus and the shear modulus of the composite, respectively. In composites, application of relatively low strains (or stresses) can lead to locally large strains (or stresses) in the matrix. It is therefore necessary to evaluate the strain (and stress) concentration factors.

A simple discussion of strain magnification factors* based upon a strength-of-materials type of approach is given by Schulz [4]. Let us consider a UFC with all the fibers oriented in direction x_1. The composite is assumed to be thin in the x_3-direction (Fig. 2) and to

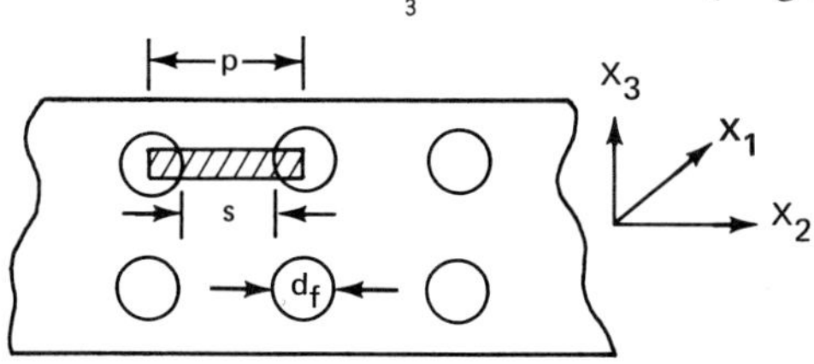

FIG. 2. UFC with fibers oriented in direction x_1.

*We will employ the term strain (stress) magnification factor to distinguish the results of this simple analysis from the more exact elasticity analyses.

be a state of plane stress (σ_{c_1}, σ_{c_2}, $\tau_{c_{12}}$). The stresses and strains in the matrix may be expected to be most severe in the strip shown in Fig. 2. The strip split into its component pieces is shown in Fig. 3. The stress-strain relationships for the fibers and the matrix are

$$E_f\varepsilon_{f_1} = \sigma_{f_1} - \nu_f\sigma_{f_2}, \quad E_f\varepsilon_{f_2} = \sigma_{f_2} - \nu_f\sigma_{f_1}$$

$$\tau_{f_{12}} = \tau_{f_{21}} = G_f\gamma_{f_{12}}$$

$$E_m\varepsilon_{m_1} = \sigma_{m_1} - \nu_m\sigma_{m_2}, \quad E_m\varepsilon_{m_2} = \sigma_{m_2} - \nu_m\sigma_{m_1}$$

$$\tau_{m_{12}} = \tau_{m_{21}} = G_m\gamma_{m_{12}} \tag{8}$$

The following relationships hold between the stresses and strains of the matrix, the fibers, and the composite material:

$$\varepsilon_{c_1} = \varepsilon_{f_1} = \varepsilon_{m_1}, \quad \varepsilon_{c_2} = (d_f\varepsilon_{f_2} + s\varepsilon_{m_2})/p$$

$$\gamma_{c_{12}} = (d_f\gamma_{f_{12}} + s\gamma_{m_{12}})/p$$

$$\sigma_{c_1} = (d_f\sigma_{f_1} + s\sigma_{m_1})/p, \quad \sigma_{c_2} = \sigma_{f_2} = \sigma_{m_2}$$

$$\tau_{c_{12}} = \tau_{f_{12}} = \tau_{m_{12}} \tag{9}$$

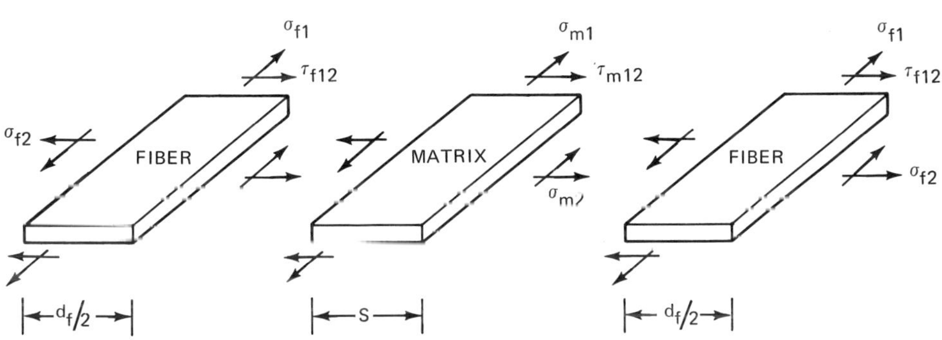

FIG. 3. Enlarged view of the strip length "p."

Equations (8) and (9) may be solved for the extensional and shearing strains in the matrix in terms of the overall strains of the composite material:

$$\varepsilon_{m1} = \varepsilon_{c1}$$

$$\varepsilon_{m2} = \left\{\frac{1}{(d_f/p)(E_m/E_f)\left[\left(1 - \nu_f^2\right)/\left(1 - \nu_m^2\right)\right] + (s/p)}\right\}\left\{\varepsilon_{c2}\right.$$

$$\left. + \varepsilon_{c1}\frac{d_f}{p}\left(\nu_f - \nu_m \frac{1 - \nu_f^2}{1 - \nu_m^2}\frac{E_m}{E_f}\right)\right\} \tag{10}$$

$$\gamma_{m12} = \frac{p\gamma_{c12}}{s + d_f(G_m/G_f)}$$

The strain magnification factors are defined as

$\varepsilon_{m1}/\varepsilon_{c1}$, $\varepsilon_{m2}/\varepsilon_{c2}$, and $\gamma_{m12}/\gamma_{c12}$

Schulz [4] gives numerical results for the following glass-resin composite:

$E_m/E_f = 1/20$, $G_m/G_f = 1/17$, $\nu_f = 0.23$, $\nu_m = 0.40$

$\varepsilon_{c1} = \varepsilon_{c2}$

The results are plotted in Fig. 4.

Equations (8) and (9) may also be manipulated to yield the matrix stresses in terms of the composite strains. It is now possible to define the stress magnification factor as the ratio of the stress produced in the matrix (σ_{mi}) under a given set of composite strains to the stress produced in the pure matrix (σ_{i0}) subjected to the same strains. These stress magnification factors are also shown in Fig. 4. It is seen from Fig. 4 that a relatively low

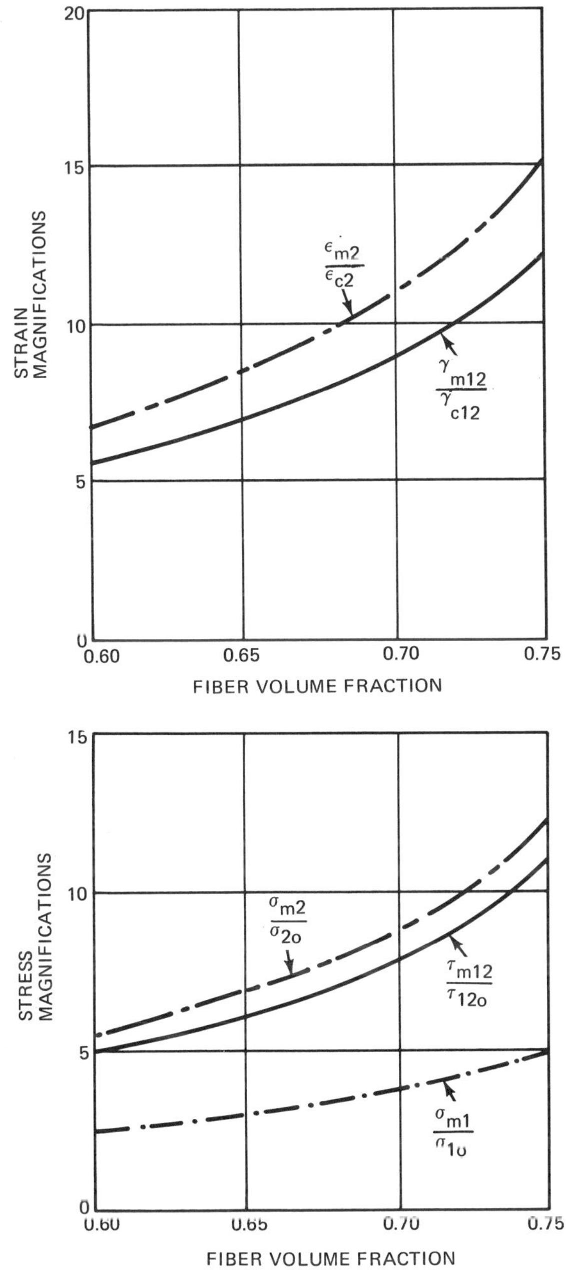

FIG. 4. Strain and stress magnification factors (from Ref. [4]).

composite strain (or stress) can lead to rather large strains (or stresses) in the matrix.

More exact numerical solutions [5], however, indicate that the maximum stress concentration in the matrix is far less than the one that may be inferred from the approximate approach outlined above. Foye [5] solved the stress concentration problem by the finite-element method for the following glass-epoxy composite:

$$E_f = 10.6 \times 10^6 \text{ psi}, \quad G_f = 4.17 \times 10^6 \text{ psi}, \quad \nu_f = 0.20$$

$$E_m = 0.5 \times 10^6 \text{ psi}, \quad G_m = 0.185 \times 10^6 \text{ psi}, \quad \nu_m = 0.34$$

The maximum stress concentrations for a square packed array are shown in Fig. 5. These results clearly indicate the occasional danger inherent in substituting untested physical intuition for more precise mathematics. The approximate analysis (Fig. 4) predicts stress magnification factors several times larger than the finite-element solution.

It is observed from Fig. 5 that for a typical composite (e.g., glass-epoxy), the transverse and shear stress concen-

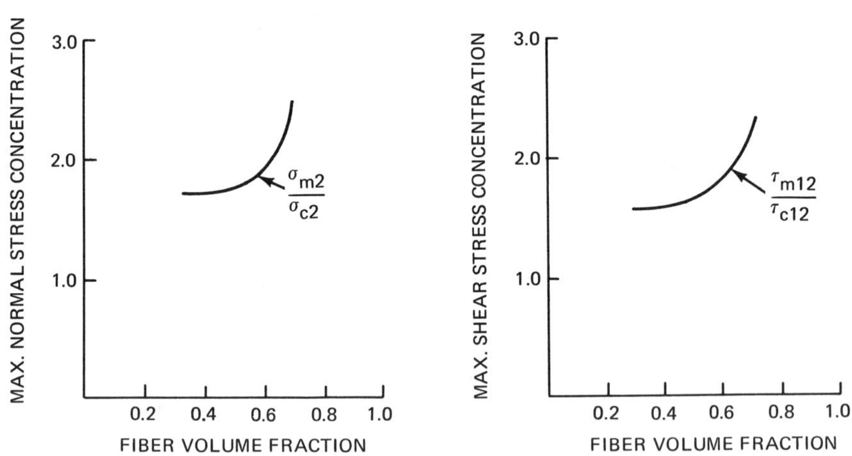

FIG. 5. Maximum stress concentration factors for a square packed array (from Ref. [5]).

tration factors lie in the ranges 1.75-2.5 and 1.5-2.5, respectively. The use of these stress concentration factors, however, does not lead to a correct prediction of transverse and in-plane shear strengths [6]. Thus, for example, for a glass-epoxy composite, matrix shear strength is of the order of 8 ksi. A stress concentration factor of 2.5 gives a composite shear strength ($S = \sigma_{mS}/2.5$) of 3 ksi. Measured values have been at least twice this value. In the preceding analysis, the fiber-matrix bond was assumed to remain intact during loading. A consideration of inelastic deformations together with imperfect interfacial bonds should lead to a better correlation between theoretically predicted and experimentally measured values for composite strength.

III. UNIDIRECTIONALLY FIBER-REINFORCED COMPOSITES SUBJECTED TO BIAXIAL LOADING

In the preceding section, we considered the primary strengths of a unidirectionally fiber-reinforced composite. In practice, a composite is frequently subjected to biaxial loading. Several theories have been proposed for describing the strength characteristics of a UFC under biaxial loading. In the following, we will discuss theories based on maximum stress, maximum strain, and maximum work. We will use the x_1-x_2-x_3 system of axes to denote the axes of symmetry of the composite (Fig. 1). The composite is assumed to be thin in the x_3-direction and subjected only to in-plane (x_1-x_2 plane) stresses.

A. Maximum Stress Theory

According to the maximum stress theory [7-9] failure of the composite will occur if one of the three ultimate strengths (X,Y,S) is reached. Mathematically, we have

$$\sigma_1 \leq X, \quad \sigma_2 \leq Y, \quad \tau_{12} \leq S \tag{11}$$

Let us now consider the x-y system of axes oriented at an angle θ, to the x_1-x_2 system of axes. The stresses in the x-y system may be expressed in terms of the stresses in the x_1-x_2 system through the following transformation relations:

$$\begin{Bmatrix} \sigma_1 \\ \sigma_2 \\ \tau_{12} \end{Bmatrix} = \begin{bmatrix} \cos^2\theta & \sin^2\theta & 2\sin\theta\cos\theta \\ \sin^2\theta & \cos^2\theta & -2\sin\theta\cos\theta \\ -\cos\theta\sin\theta & \cos\theta\sin\theta & \cos^2\theta - \sin^2\theta \end{bmatrix} \begin{Bmatrix} \sigma_x \\ \sigma_y \\ \tau_{xy} \end{Bmatrix} \tag{12}$$

As the fiber orientation varies from 0° to 90°, the stress components σ_1, σ_2, and τ_{12} may be calculated from Eq. (12). For simplicity, we consider the case of uniaxial stress, σ_x. In this case, we have

$$\sigma_1 = \sigma_x \cos^2\theta, \quad \sigma_2 = \sigma_x \sin^2\theta, \quad \tau_{12} = -\sigma_x \cos\theta \sin\theta \tag{13}$$

Combining Eqs. (11) and (13), the uniaxial strength is given by

$$\sigma_x \leq \text{Min}[X/\cos^2\theta,\ Y/\sin^2\theta,\ S/(\sin\theta\cos\theta)] \tag{14}$$

In the above discussion, we assumed the stress components σ_1 and σ_2 to be positive (tension). If σ_1 and σ_2 are negative, then X and Y must be replaced by the compressive strengths X' and Y'. Since the shear strength S has no directional property, it remains the same for both positive and negative shear components.

B. Maximum Strain Theory

In the maximum strain theory [10], it is assumed that associated with the three strain components $(\varepsilon_1, \varepsilon_2, \gamma_{12})$ there exist ultimate strains (X_e, Y_e, S_e). The maximum strain theory may be mathematically expressed as

$$\varepsilon_1 \leq X_e, \quad \varepsilon_2 \leq Y_e, \quad \gamma_{12} \leq S_e \tag{15}$$

In case ε_1 and ε_2 are negative, X_e and Y_e must be replaced by X_e' and Y_e'. If we introduce the additional assumption that the composite remains elastic up to failure, then the ultimate strains in Eq. (15) can be directly related to the strengths:

$$X_e = X/E_1, \quad Y_e = Y/E_2, \quad S_e = S/G_{12} \tag{16}$$

where E_1, E_2, and G_{12} denote the elastic constants of the composite material. The stress-strain relations for an orthotropic material are

$$\begin{aligned} \sigma_1 &= [E_1/(1 - \nu_{12}\nu_{21})][\varepsilon_1 + \nu_{21}\varepsilon_2] \\ \sigma_2 &= [E_2/(1 - \nu_{21}\nu_{12})][\varepsilon_2 + \nu_{12}\varepsilon_1] \\ \tau_{12} &= G_{12}\gamma_{12} \end{aligned} \tag{17}$$

where ν_{12} and ν_{21} denote the Poisson's ratios for the composite. Combining Eqs. (13) and (15)-(17), we obtain for the uniaxial strength

$$\sigma_x \leq \text{Min}[X/(\cos^2\theta - \nu_{12}\sin^2\theta), Y/(\sin^2\theta - \nu_{21}\cos^2\theta), S/(\sin\theta\cos\theta)] \tag{18}$$

C. Maximum Work Theory

The maximum work theory, based upon a yield criterion for anisotropic materials first proposed by Hill [11], has been used by several authors [10, 12-15] to analyze the strength characteristics of composite materials. The basic assumption involved in the use of a yield criterion is that the yield stress and ultimate strength are identical. Hill assumed the yield condition to be a quadratic in the stress components, i.e.,*

$$2f(\sigma_{ij}) = F(\sigma_2 - \sigma_3)^2 + G(\sigma_3 - \sigma_1)^2 + H(\sigma_1 - \sigma_2)^2$$

$$+ 2L\tau_{23}^2 + 2M\tau_{31}^2 + 2N\tau_{12}^2 = 1 \qquad (19)$$

where F, G, H, L, M, and N are parameters characterizing the anisotropy of the material and x_1, x_2, x_3 are the axes of anisotropic symmetry. In the general form (19), the yield condition is valid for a material with three axes of anisotropic symmetry. The yield condition may be represented as a six-dimensional surface in stress space $(\sigma_1, \sigma_2, \sigma_3, \tau_{23}, \tau_{31}, \tau_{12})$. Any point lying within the surface represents an admissible stress state, i.e., one which does not lead to yielding. A stress state corresponding to a point lying on the yield surface leads to failure.

We will denote by X, Y, and Z the normal strengths in the directions of anisotropic symmetry, and by R, T, and S the shear strengths with respect to the symmetry axes of anisotropy. It is straightforward to show that

$$2F = (1/Y^2) + (1/Z^2) - (1/X^2)$$

$$2G = (1/Z^2) + (1/X^2) - (1/Y^2)$$

*Note that the yield condition (19) reduces to the von Mises' yield condition for the case of an isotropic material.

$$2H = (1/X^2) + (1/Y^2) - (1/Z^2)$$

$$2L = 1/R^2, \quad 2M = 1/T^2, \quad 2N = 1/S^2 \tag{20}$$

The yield condition (19) can be specialized for the in-plane loading of a unidirectionally fiber-reinforced composite by setting $\sigma_3 = \tau_{23} = \tau_{31} = 0$ and utilizing the property of transverse isotropy, i.e., by taking $Z = Y$. The yield condition for a UFC becomes

$$\left(\frac{\sigma_1}{X}\right)^2 - \left(\frac{\sigma_1}{X}\right)\left(\frac{\sigma_2}{X}\right) + \left(\frac{\sigma_2}{Y}\right)^2 + \left(\frac{\tau_{12}}{S}\right)^2 = 1 \tag{21}$$

Also, combining Eqs. (13) and (21), we obtain for uniaxial strength σ_x

$$\frac{1}{\sigma_x{}^2} = \frac{\cos^4\theta}{X^2} + \left(\frac{1}{S^2} - \frac{1}{X^2}\right)\cos^2\theta \sin^2\theta + \frac{\sin^4\theta}{Y^2} \tag{22}$$

Tsai [10] compares the uniaxial strength predicted by maximum stress, maximum strain, and maximum work theories, with the test data obtained from uniaxial tensile and compressive tests on a unidirectional E-glass-epoxy composite. The glass-epoxy system has the following material constants:

$E_1 = 7.8 \times 10^6$ psi, $E_2 = 2.6 \times 10^6$ psi

$\nu_{12} = 0.25$, $G_{12} = 1.25 \times 10^6$ psi

$X - 150$ ksi, $X' = 150$ ksi, $Y = 4$ ksi

$Y' = 20$ ksi, $S = 8$ ksi

The predictions of various theories along with the experimental data are shown in Fig. 6. Solid circles and squares

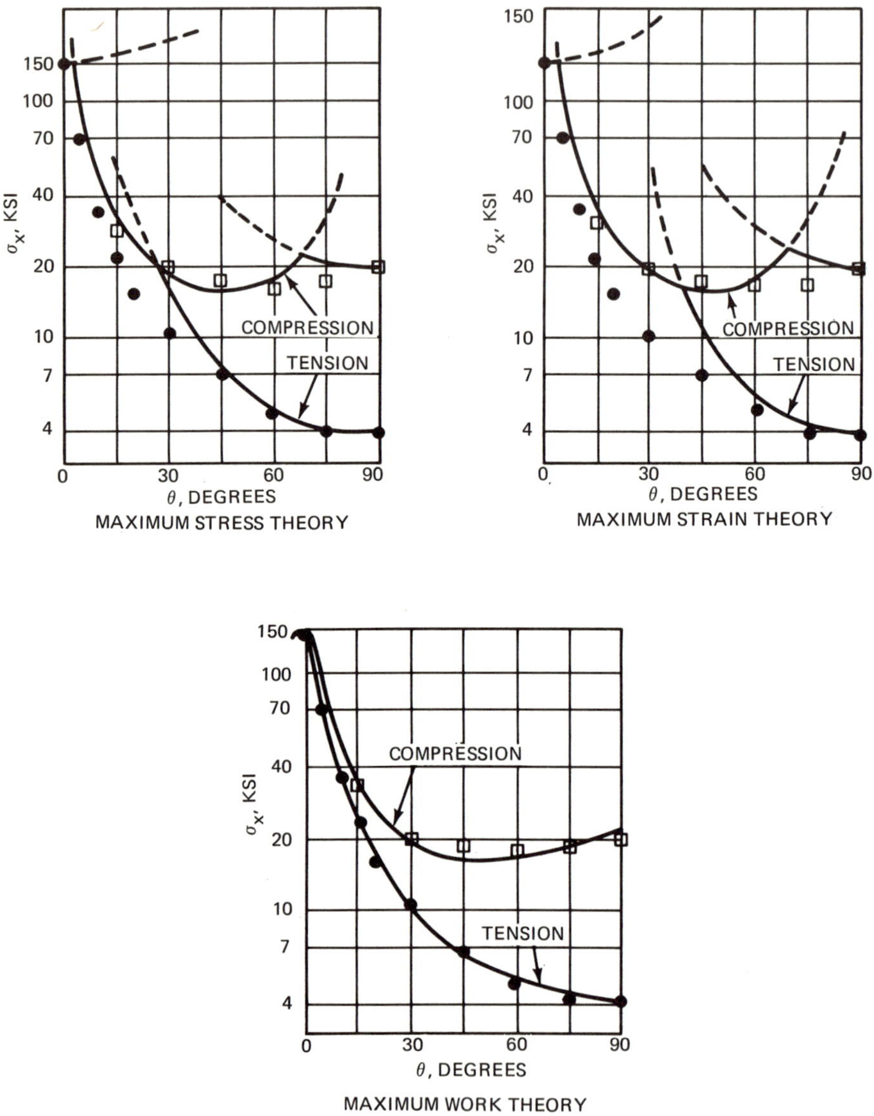

FIG. 6. Comparison between theoretical predictions for uniaxial composite strength and experimental data (from Ref. [10]).

represent the tensile and the compressive data, respectively. It can be seen from Fig. 6 that for the maximum work theory, there is a continuous variation of strength as a function of orientation angle θ and there is no abrupt rise in strength with θ as indicated by the maximum stress and maximum strain theories. Furthermore, it is evident from Fig. 6 that the maximum work theory offers better agreement with experimental data than do the other theories.

The yield condition (19) predicts the same yield stress in tension and compression. In order to realistically characterize tensile and compressive strength, it is necessary to introduce two separate yield conditions, i.e., one with tensile strengths (X,Y,Z) and the other with compressive strengths (X', Y', Z'). However, Hoffman [14] points out that it is possible to use a single yield condition by including terms in Eqs. (19) which are odd functions of $\sigma_1, \sigma_2, \sigma_3$. A choice of linear terms leads to the following yield condition:

$$C_1(\sigma_2 - \sigma_3)^2 + C_2(\sigma_3 - \sigma_1)^2 + C_3(\sigma_1 - \sigma_2)^2 + C_4\sigma_1 + C_5\sigma_2 + C_6\sigma_3 + C_7\tau_{23}^2 + C_8\tau_{31}^2 + C_9\tau_{12}^2 = 1 \tag{23}$$

where

$$C_1 = \frac{1}{2}\left(\frac{1}{YY'} + \frac{1}{ZZ'} - \frac{1}{XX'}\right), \quad C_2 = \frac{1}{2}\left(\frac{1}{ZZ'} + \frac{1}{XX'} - \frac{1}{YY'}\right)$$

$$C_3 = \frac{1}{2}\left(\frac{1}{XX'} + \frac{1}{YY'} - \frac{1}{ZZ'}\right), \quad C_4 = \frac{1}{X} - \frac{1}{X'}, \quad C_5 = \frac{1}{Y} - \frac{1}{Y'}$$

$$C_6 = \frac{1}{Z} - \frac{1}{Z'}, \quad C_7 = \frac{1}{R^2}, \quad C_8 = \frac{1}{T^2}, \quad C_9 = \frac{1}{S^2}$$

As before, the yield condition may be specialized to in-plane loading:

$$\frac{\sigma_1^2 - \sigma_1\sigma_2}{XX'} + \frac{\sigma_2^2}{YY'} + \frac{X' - X}{XX'}\sigma_1 + \frac{Y' - Y}{YY'}\sigma_2 + \frac{\tau_{12}^2}{S^2} = 1 \tag{24}$$

Hoffman also presents experimental data obtained from uniaxial tensile and compressive tests on variously oriented specimens. The experimental results agree reasonably well with the predictions of yield condition (24).

Despite the good agreement exhibited by the predictions of the maximum work theory with the experimental data, it would be wrong to conclude that there is any physical significance to this theory apart from a convenient empirical relation between basic strength data. In the preceding, we observed that the major assumption underlying the use of the maximum work theory is that failure by yielding and fracture are equivalent. In other words, the composite behaves in a linear elastic manner up to failure. This assumption is probably correct for glass-epoxy systems, but its use in the case of metal-matrix composites cannot be justified except on empirical grounds. Thus, the maximum work theory should be used with caution to analyze the strength behavior of composites which may exhibit inelastic behavior prior to failure.

IV. STRENGTH OF LAMINATED STRUCTURES*

We pointed out in Chapter 5 that two sources of interaction are induced in the process of lamination. First, a mechanical interaction is caused by the variation in the material properties across the thickness of the composite. This results in the distribution of the stress according to the relative stiffness of the constituent layers. Second, there exists a thermal interaction caused by the differential thermal expansions (or contractions) between different layers. Also, since most laminated composites are constructed at elevated temperatures,

*We will follow the treatment outlined by Tsai [13].

initial stresses are induced if the service temperature is different from the laminating temperature.

The stress components for the nth layer are given by

$$\sigma_i^{(n)} = C_{ij}^{(n)}\left(\varepsilon_j^{(n)} - \alpha_j^{(n)}T\right)$$

$$= C_{ij}^{(n)}\left[\left(A'_{jk} + x_3 B'_{jk}\right)\bar{N}_k + \left(B'_{jk} + x_3 D'_{jk}\right)\bar{M}_k - \alpha_j^{(n)}T\right] \qquad (25)$$

where $i, j, k = 1, 2\ 6$ and

$$\{\sigma\} = \begin{Bmatrix} \sigma_x \\ \sigma_y \\ \sigma_{xy} \end{Bmatrix}; \quad \{\bar{N}\} = \begin{Bmatrix} \bar{N}_x \\ \bar{N}_y \\ \bar{N}_{xy} \end{Bmatrix}; \quad \{\bar{M}\} = \begin{Bmatrix} \bar{M}_x \\ \bar{M}_y \\ \bar{M}_{xy} \end{Bmatrix}$$

and $C_{ij}^{(n)}$, $\alpha_j^{(n)}$, A'_{jk}, B'_{jk}, and D'_{jk} have been defined in Chapter 5. The strength analysis of the laminated composite is conducted by substituting the stress components of the nth constituent layer obtained from Eq. (25) into the yield condition (21).*

The yield condition (21) determines the maximum values of the stress components that the layer can sustain. When this maximum is reached, the particular layer is assumed to have failed. This failure results in a partial (or complete) uncoupling of the mechanical and thermal interactions and in a redistribution of loads. The physical and mathematical nature of this uncoupling is not easy to describe. Cracks transverse to the fibers may develop, causing a partial degradation of the effective stiffness.

*Note that the yield condition (21) is expressed in the x_1-x_2 coordinate system. In order to use it with stress components obtained from Eq. (25), it must first be transformed into the x-y system of coordinates.

It is also possible that a complete delamination may occur. In any case, failure of one or more layers results in a new effective stiffness for the composite. This is reflected in a change in A', B', and D' matrices and a "knee" in the stress-strain relationship for the composite. New stresses are then computed from the revised A', B', and D' matrices and substituted again into the yield condition (21). Then the layer or layers that would fail next are determined and the process is repeated until all layers have failed. The composite stress-strain curve thus appears as a piecewise linear relationship. We will illustrate this process by considering a cross-ply composite (see Ref. [13]).

We shall consider the particular case of three layers, with odd layers oriented at 0° and even layers at 90°. The cross-ply ratio m (ratio of total thickness of the odd layers to that of even layers) is taken as 0.2. The odd layers have the same thickness. Only uniaxial loading ($N_x \neq 0$; N_y, N_{xy}, M_x, M_y, and M_{xy} all zero) will be considered. The symmetry of the composite with respect to the centroidal axis, and the symmetrical loading (no bending moments) imply that the stress distribution in the odd layers will be identical. Thus we need to consider only the inner layer (layer 2) and one of the outer layers. The yield condition (21) in the x-y coordinate system becomes for the odd layers

$$\left(\frac{\sigma_x}{X}\right)^2 - \left(\frac{\sigma_x}{X}\right)\left(\frac{\sigma_y}{X}\right) + \left(\frac{\sigma_y}{Y}\right)^2 + \left(\frac{\tau_{xy}}{S}\right)^2 = 1 \tag{26}$$

For the even layers, we have

$$\left(\frac{\sigma_y}{X}\right)^2 - \left(\frac{\sigma_x}{X}\right)\left(\frac{\sigma_y}{X}\right) + \left(\frac{\sigma_x}{Y}\right)^2 + \left(\frac{\tau_{xy}}{S}\right)^2 = 1 \tag{27}$$

From Eq. (25), the stresses in the outer layer are

$$\sigma_x^{(1)} = C_{11}^{(1)}\left(A'_{11}\overline{N}_x + A'_{12}\overline{N}_y - \alpha_x^{(1)}T\right) + C_{12}^{(1)}\left(A'_{21}\overline{N}_x + A'_{22}\overline{N}_Y - \alpha_y^{(1)}T\right) = \left(C_{11}^{(1)}A'_{11} + C_{12}^{(1)}A'_{21}\right)N_x + \left(C_{11}^{(1)}A'_{11} + C_{12}^{(1)}A'_{21}\right)N_x^T + \left(C_{11}^{(1)}A'_{12} + C_{12}^{(1)}A'_{22}\right)N_y^T - \left(C_{11}^{(1)}\alpha_x^{(1)} + C_{12}^{(1)}\alpha_y^{(1)}\right)T \tag{28}$$

$$\sigma_y^{(1)} = \left(C_{21}^{(1)}A'_{11} + C_{22}^{(1)}A'_{21}\right)N_x + \left(C_{21}^{(1)}A'_{11} + C_{22}^{(1)}A'_{21}\right)N_x^T + \left(C_{21}^{(1)}A'_{12} + C_{22}^{(1)}A'_{22}\right)N_y^T - \left(C_{21}^{(1)}\alpha_x^{(1)} + C_{22}^{(1)}\alpha_y^{(1)}\right)T$$

$$\tau_{xy}^{(1)} = 0$$

where*

$$\overline{N}_x = N_x + N_x^T, \quad \overline{N}_y = N_y^T, \quad \overline{N}_{xy} = 0$$

To obtain the stresses in the inner layer, it is only necessary to replace superscript (1) by superscript (2) in Eqs. (28).

We will employ the following experimentally determined material properties for a typical unidirectional glass-epoxy composite:

$$C_{11}^{(1)} = C_{22}^{(2)} = 7.97 \times 10^6 \text{ psi}$$

$$C_{12}^{(1)} = C_{12}^{(2)} = 0.66 \times 10^6 \text{ psi}$$

$$C_{22}^{(1)} - C_{11}^{(2)} = 2.66 \times 10^6 \text{ psi}$$

*See Chapter 5 for definitions of A', N_x^T, N_y^T, etc.

$$C_{66}^{(1)} = C_{66}^{(2)} = 1.25 \times 10^6 \text{ psi}$$

$$C_{16}^{(1)} = C_{26}^{(1)} = C_{16}^{(2)} = C_{26}^{(2)} = 0$$

$$\alpha_x^{(1)} = \alpha_y^{(2)} = 3.5 \times 10^{-6}/°F$$

$$\alpha_y^{(1)} = \alpha_x^{(2)} = 11.4 \times 10^{-6}/°F$$

$$\alpha_{xy}^{(1)} = \alpha_{xy}^{(2)} = 0$$

These material properties lead to the following values for the elements of the A' matrix:

$$A'_{11} = 0.29 \times 10^{-6} \text{ in./lb}$$

$$A'_{12} = -0.03 \times 10^{-6} \text{ in./lb}$$

$$A'_{22} = 0.14 \times 10^{-6} \text{ in./lb}$$

Assuming a constant lamination temperature T, we obtain the equivalent thermal forces and moments:

$$N_x^T = 33.1T \text{ lb/in.,} \quad N_y^T = 35.0T \text{ lb/in.}$$

$$N_{xy}^T = M_i^T = 0$$

Substituting the computed values for A' and N_i^T into Eqs. (28), we obtain for the outer layer

$$\sigma_x^{(1)} = 2.27N_x + 35.5T, \quad \sigma_y^{(1)} = 0.12N_x - 16.0T$$

$$\tau_{xy}^{(1)} = 0 \qquad (29)$$

and for the inner layer

$$\sigma_x^{(2)} = 0.75N_x - 7.1T, \quad \sigma_y^{(2)} = 0.02N_x + 3.2T$$

$$\tau_{xy}^{(2)} = 0 \tag{30}$$

Taking the following experimentally determined strength values which represent a typical unidirectional glass-epoxy composite

axial strength = X = 150 ksi

transverse strength = Y = 4 ksi

shear strength = S = 6 ksi

and using Eqs. (26), (27), (29), and (30), we obtain the stress resultants that cause failure. For the inner layer,

$$N_x \approx 9.6T + 1.33Y$$

For the outer layer,

$$N_x \approx 110T + (57.5Y^2 - 3000T^2)^{1/2} \tag{31}$$

For a composite laminated at 270°F, we have

T = room temp. - lamination temp.

= 70°F - 270°F = -200°F

Substituting for T into Eqs. (31), we obtain for the inner layer,

$$N_x = 3400 \text{ lb/in.}$$

and for the outer layer,

$$N_x = 6300 \text{ lb/in.} \tag{32}$$

For a composite laminated at room temperature (T = 0), we have no thermal residual stress, and thus for the inner layer,

$$N_x = 5320 \text{ lb/in.}$$

and for the outer layer,

$$N_x = 30{,}400 \text{ lb/in.} \tag{33}$$

It is evident from Eqs. (32) and (33) that the inner layer fails first. Also, the first failure will occur at a higher stress if the lamination temperature is ambient. The effective stiffness of the laminated composite (unity thickness) up to first failure (knee) is the reciprocal of A'_{11}, i.e., 3.4×10^6 psi. Therefore, the in-plane strain at the "knee" is

$$\varepsilon_x^0 = N_x/A'_{11} = (3400)/[(3.4)10^6] = 0.1\%$$

The behavior of the composite after the first failure is quite complex and depends upon the uncoupling of thermal and mechanical interactions. Possibly, cracks transverse to the fibers are developed in the inner layer. This can be represented by letting $C_{22}^{(2)}$ remain constant and by putting the remaining components as zero or a very small fraction of their original values. The components of the A' matrix for the partially degraded composite (inner layer degraded) are thus

$$A'_{11} = 0.75 \times 10^{-6} \text{ in./lb}$$

$$A'_{12} = 0.01 \times 10^{-6} \text{ in./lb}$$

$$A'_{22} = 0.14 \times 10^{-6} \text{ in./lb}$$

Use of the new A' matrix leads to the following values for stress components:

$$\sigma_x^{(1)} = 6.00N_x, \quad \sigma_y^{(1)} = 0.47N_x - 19.3T$$

$$\tau_{xy}^{(1)} = 0, \quad \sigma_x^{(2)} = \tau_{xy}^{(2)} = 0, \quad \sigma_y^{(2)} = -0.09N_x + 3.9T$$

Substitution of the new stress components into the yield equation reveals maximum N_x for the outer layers to be considerably lower than the existing stress resultant of 3400 lb/in. What this implies is that the outer layers degrade immediately after the inner layer. This causes a complete uncoupling between thermal and mechanical interactions. In order to characterize this, we will assume that only the stiffnesses parallel to fibers remain intact, i.e., $C_{11}^{(1)}$ and $C_{22}^{(2)}$ are the only nonzero components. All other components of the C matrices are either put equal to zero or a very small fraction of their original values. The resulting A' matrix for the completely degraded composite becomes

$$A'_{11} = 0.77 \times 10^{-6} \text{ in./lb}, \quad A'_{12} = 0$$

$$A'_{22} = 0.15 \times 10^{-6} \text{ in./lb}$$

The only nonzero stress component is

$$\sigma_x^{(1)} = 6.00N_x$$

and the effective stiffness of the composite for a unit thickness is $1/A'_{11} = 1.3 \times 10^6$ psi. The stress in the outer layers immediately after degradation is computed by using $N_x = 3400$ lb/in. and $T = -200°F$.

$$\sigma_x^{(1)} \approx 600 \text{ psi}$$

The maximum value that $\sigma_x^{(1)}$ can reach is equal to 150 ksi (axial strength). Using Eq. (34), this yields a value of

28,300 lb/in. for N_x. The experimental measurement of the effective stress-strain relation of a three-layer composite, together with the theoretical predictions, is shown in Fig. 7. The agreement between theoretical and experimental results is observed to be quite good.

The above example is typical of the lengthy arithmetical operations involved in the strength analysis of even the simplest of composite structures. In general, it is necessary to resort to modern high-speed computers [13]. But apart from lengthy arithmetical computations, the theory has two other weak points. The first concerns our inability to accurately characterize the coupling between mechanical and thermal interactions. In the preceding, we saw that the problem is quite complex even for a simple three-layer cross-ply laminate. For more complicated laminates, it appears to be even more formidable. Second, the theory is unable to predict the separation of constituent layers due to the failure of the interlaminar bond. The failure of a layer may set up stress concentrations in

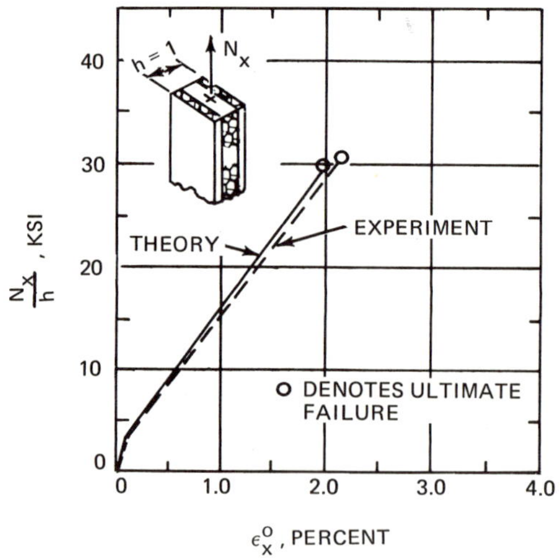

FIG. 7. Strength of a three-layer cross-ply laminate (from Ref. [13]).

the interlaminar bond and lead to debonding. Clearly, the situation after the failure of one or more constituent layers is very complex and inadequately modeled in the theory.

REFERENCES

1. C. C. Chamis, "Failure Criteria for Filamentary Composites," NASA TN D-5367, 1969.
2. R. W. Jech, D. L. McDaniels, and J. W. Weeton, "Fiber-Reinforced Metallic Composites," Proceedings of the 6th Sagamore Ordnance Materials Research Conference, August 1959.
3. B. W. Rosen, "Mechanics of Composite Strengthening," in Fiber Composite Materials, American Society for Metals, Metals Park, Ohio, 1965, pp. 37 75.
4. J. C. Schulz, "Maximum Stresses and Strains in the Resin of a Filament-Wound Structure," Paper presented at the Society of the Plastics Industry 18th Annual Conference, Section 7-D, 1963.
5. R. L. Foye, "An Evaluation of Various Engineering Estimates of the Transverse Properties of Uni-Directional Composites," Science of Advanced Materials and Process Engineering Series, Vol. 10, Western Periodicals Co., pp. G-31 to G-42, 1966.
6. S. W. Tsai, "Mechanics of Composite Materials, Part I--Introduction," Air Force Materials Laboratory Report AFML-TR-66-149, June 1966.
7. A. Kelly and G. Davies, "The Principles of the Fiber Reinforcement of Metals," Metallurgical Rev., 10, 1-77 (1965).
8. P. W. Jackson and D. Cratchley, "The Effect of Fiber Orientation on the Tensile Strength of Fiber-Reinforced Metals," J. Mech. Phys. Solids, 14, 49-64 (1966).

9. G. A. Cooper, "Orientation Effects in Fiber-Reinforced Metals," J. Mech. Phys. Solids, 14, 103-111 (1966).

10. S. W. Tsai, "Strength of Theories of Filamentary Structures," in Fundamental Aspects of Fiber-Reinforced Composites, edited by R. T. Schwartz and H. S. Schwartz, Interscience, New York, 1968, pp. 3-11.

11. R. Hill, The Mathematical Theory of Plasticity, Oxford Univ. Press, London, 1950, p. 317.

12. J. E. Griffith and W. M. Baldwin, "Failure Theories of Generally Orthotropic Materials," in Developments in Theoretical and Applied Mechanics, Vol. 1, Plenum Press, New York, 1963, pp. 410-420.

13. S. W. Tsai, "Strength Characteristics of Composite Materials," NASA CR-224, April 1965.

14. O. Hoffman, "The Brittle Strength of Orthotropic Materials," J. Composite Materials, 1, 200-206 (1967).

15. S. W. Tsai and E. M. Wu, "A General Theory of Strength for Anisotropic Materials," J. Composite Materials, 5, 58-80 (1971).

Chapter 7

STATISTICAL TENSILE STRENGTH OF FIBERS AND FIBER BUNDLES

I. INTRODUCTION

Most of the reinforcing fibers presently in use (glass, carbon, etc.) are known to be brittle with a wide scatter in their tensile strength values. Clearly, a statistical approach is called for in the description of the strength characteristics of these individual fibers, fiber bundles, and composite materials. The application of statistical strength theories to composite materials involves three steps. The first is to obtain a probability distribution function which adequately describes the strength characteristics of the individual fibers. The second step is to obtain statistical strength relationships for a group of these fibers (bundle theory). The last step is the application of the previous theory to an actual composite, and involves accounting for the contribution of the matrix to the composite strength. This can best be explained by saying that when a bundle of fibers is under load, a single break in a fiber renders it ineffective for carrying any load. In a composite, when a break occurs in a fiber, only a certain length of that fiber will be ineffective in carrying further load. The rest of the fiber can pick up load transferred to it (across the break) by shear in the matrix material. This defines a critical (ineffective) length which is essential to an understanding of the composite strength characteristics. In this chapter, we will consider the strength behavior of single fibers and fiber bundles. The tensile strength of the composite is considered in Chapters 8 and 9.

II. STRENGTH CHARACTERISTICS OF A SINGLE FIBER

Statistical theories of brittle fracture are mostly based on the Griffith fracture theory [1]. It is assumed that the strength of brittle materials is limited by the presence of microcracks, or flaws, distributed throughout the specimen. Fracture occurs when the local stress at the tip of a flaw (crack) exceeds a certain critical stress necessary to break the interatomic bonds. If it is further assumed that the flaws are distributed in a random manner and are noninteracting [2], then the fiber may be divided into a chain of ℓ links, each of unit length. When the stress at the root of a microcrack in any link reaches the theoretical cohesive strength, fracture ensues. It is thus evident that the most severe crack, or flaw, determines the failure of the entire fiber. In other words, statistical theories are based on the concept of the "weakest link," according to which the strength of a chain of ℓ links equals the strength of the weakest link. The concept of the weakest link was aptly stated by Peirce [3]: "It is a truism of which the mathematical implications are of no little interest, that the strength of a chain is that of its weakest link." It is to be noted that the weakest link theorem is a static hypothesis [4]. When the stress distribution along the specimen is changing with time, as in high strain-rate loading, the weakest link hypothesis cannot be applied without modification. However, the weakest link hypothesis may be extended to the case of quasistatic loading, where the stress distribution is approximately uniform along the length of the specimen (see Chapter 11).

Mathematically, if $G_1(\sigma)$ and $G_\ell(\sigma)$ denote the probability that the strength of the fibers of length unity and ℓ, respectively, does not exceed σ, the weakest link hypothesis yields

$$1 - G_\ell(\sigma) = [1 - G_1(\sigma)]^\ell \tag{1}$$

Equation (1) implies that the probability of the survival of a fiber of length ℓ subjected to a stress σ equals the probability that all of the ℓ links of unit length survive under the applied stress.

Coleman [5], using the weakest link and certain other very general hypotheses, shows that the cumulative distribution function $G_\ell(\sigma)$ of classical fibers is of the Weibull [6] type. A classical fiber is here defined as one whose strength is independent of the rate of loading and whose statistical properties are not a function of position along the length of a spool. In other words, if the fibers are taken from the same spool, then the statistical properties are free from any long-range effects. Coleman's derivation is discussed in the following paragraphs.

Apart from the weakest link hypothesis, the following two "self-evident" statements may be made regarding $G_\ell(\sigma)$:

(a) The fiber strength is greater than zero regardless of the fiber length ℓ.

(b) $G_\ell(\sigma)$ is a monotonically increasing function of σ.

Whenever ℓ in Eq. (1) is large, and $G_1(\sigma)$ differs appreciably from zero, the right-hand side of the equation will vanish. According to the second "self-evident" statement above, small $G_1(\sigma)$ corresponds to small σ. Thus, the form of $G_\ell(\sigma)$ for large ℓ, is governed by the behavior of $G_1(\sigma)$ as σ approaches zero. One must further assume that $G_1(\sigma)$ is well behaved near the origin, such that

$$\lim_{\sigma \to 0} G_1(\sigma) = \alpha\sigma^\beta \tag{2}$$

where α and β are positive constants. The Weibull distribution is derived from a slightly more general hypothesis than Eq. (2). One assumes that for all $X > 0$,

$$\lim_{\eta \to 0} \left[\frac{G_1(\eta X)}{G_1(\eta)}\right] = X^\beta \tag{3}$$

where η is any positive variable.

The quantity η_ℓ is determined such that

$$G_1(\eta_\ell) = 1/\ell \tag{4}$$

and a new random variable z_ℓ is defined such that

$$z_\ell = \eta_\ell \min(x_1, \ldots, x_i, \ldots, x_\ell) \tag{5}$$

where $x_1, x_2, \ldots, x_\ell$ denote the strengths of the ℓ links of unit length comprising the fiber. Also note that as $\ell \to \infty$, $G_1(\eta_\ell) \to 0$, and hence $\eta_\ell \to 0$. The probability that z_ℓ is greater than some fixed quantity X is

$$P\left\{z_\ell \geq X\right\} = [1 - G_1(\eta_\ell X)]^\ell \tag{6}$$

Substituting $\eta = \eta_\ell$, one obtains from Eqs. (3) and (6), for large ℓ,

$$\lim_{\ell\to\infty} P\{z_\ell > X\} = \lim_{\ell\to\infty}\left\{1 - G_1(\eta_\ell)X^\beta\right\}^\ell \tag{7}$$

Substitution of Eq. (4) in Eq. (7) results in

$$\lim_{\ell\to\infty} P\{z_\ell > X\} = \lim_{\ell\to\infty}\left[1 - (1/\ell)X^\beta\right]^\ell = \exp\{-X^\beta\} \tag{8}$$

Reversing the scale, $\sigma = \eta_\ell^{-1}X$, and noting that

$$P\{z_\ell > X\} = P\left\{\min(x_1, \ldots, x_\ell) > \sigma\right\}$$

and

$$P\left\{\min(x_1, \ldots, x_\ell) > \sigma\right\} = 1 - G_\ell(\sigma)$$

we see that Eq. (8) yields

$$G_\ell(\sigma) = 1 - \exp\left\{-(\sigma/\eta_\ell)^\beta\right\} \tag{9}$$

When Eq. (2) is applicable, $\eta_\ell \sim (\alpha\ell)^{-1/\beta}$, and Eq. (9) becomes

$$G_\ell(\sigma) = 1 - \exp\{-\ell\alpha\sigma^\beta\} \tag{10}$$

The above relation is the Weibull cumulative distribution function. Here, α and β are independent of the length ℓ, and are referred to as the distribution parameters. Substituting for $G_\ell(\sigma)$ from Eq. (10) into Eq. (1), it is readily seen that $G_1(\sigma)$ is also of the Weibull form.

Coleman considers another form of the behavior of $G_1(\sigma)$ near the origin. For small σ,

$$G_1(\sigma) \sim \exp\{-\phi(\sigma)\}$$

where

$$\lim_{\sigma\to\infty} \phi(\sigma) \to \infty \tag{11}$$

However, this assumption leads to the conclusion that there is no dispersion in tensile strength for long fibers, which is contrary to the observed behavior.

Five other possible assumptions about the form of the *a priori* strength distribution function for fibers, $G_1(\sigma)$, have been considered [4]. When $G_1(\sigma)$ is of the rectangular type, the strength is seen to be independent of length, which is contrary to physical observations. A Cauchy distribution for $G_1(\sigma)$ yields infinite mean value and variance for the strength. The other three probability density functions discussed by Epstein [4], namely the Gaussian, the Weibull, and the Laplacian, all yield the satisfying result that strength does increase with increasing length (increasing number of links). However, one is led to a contradiction when $G_1(\sigma)$ is of the Gaussian or the Laplacian type. There is found to exist a finite value of ℓ, say ℓ_c, such that for $\ell > \ell_c$, the mode of the strength distribution is negative. This cannot happen in real fibers. The foregoing should not be interpreted to imply that the Gaussian or the Laplacian distribution functions cannot be applied at all to describe the fiber strength distribution. They may indeed be of use in a limited range of the values of ℓ. As may be expected, the Weibull distribution function for $G_1(\sigma)$ leads to Eq. (10).

Thus, on theoretical grounds alone, the Weibull distribution function appears to be a logical choice to represent the strength characteristics of individual fibers.

The suitability of the Weibull distribution function for characterizing glass fiber strength distribution was verified experimentally by, among others, Wallhaus [7], and Armenàkas et al. [8]. The Weibull distribution function may also be employed to characterize the fiber strength behavior under quasistatic loading, in an approximate manner, by postulating the distribution parameters α and β as functions of the strain rate $\dot{\varepsilon}$ [8].

From the cumulative distribution function, Eq. (10), the following quantities, defined in the Appendix, may be calculated in a straightforward manner:

$$\langle\sigma\rangle_f = (\alpha\ell)^{-1/\beta}\Gamma(1 + 1/\beta) \tag{12a}$$

$$\Xi_f/\langle\sigma\rangle_f = \left(\frac{\Gamma(1 + 2/\beta)}{\Gamma^2(1 + 1/\beta)} - 1\right)^{1/2} \tag{12b}$$

$$\mu_f^3/\Xi_f^3 = \Big\{\Gamma(1 + 3/\beta) - 3\Gamma(1 + 1/\beta)\Gamma(1 + 2/\beta) + 2\Gamma^3(1 + 1/\beta)\Big\} \Big/ \Big\{\left[\Gamma(1 + 2/\beta) - \Gamma^2(1 + 1/\beta)\right]^{3/2}\Big\} \tag{12c}$$

where (12a) gives the average strength and (12b) and (12c) the coefficients of variation and of skewness, respectively. Here, $\Gamma(x)$ denotes the gamma function of x. Note that the coefficients of variation and skewness are independent of α and depend only upon β.

Equation (12a) predicts a linear relationship between $\log\{\langle\sigma\rangle_f\}$ and $\log(\ell)$ (see Fig. 1). This property leads to a simple graphical method for determining the Weibull parameters. The slope of the straight line $\log\{\langle\sigma\rangle_f\}$ - $\log(\ell)$ yields the value of $-1/\beta$. Knowing the parameter β, the parameter α may be obtained through use of Eq. (12a). This method is, however, subject to error.

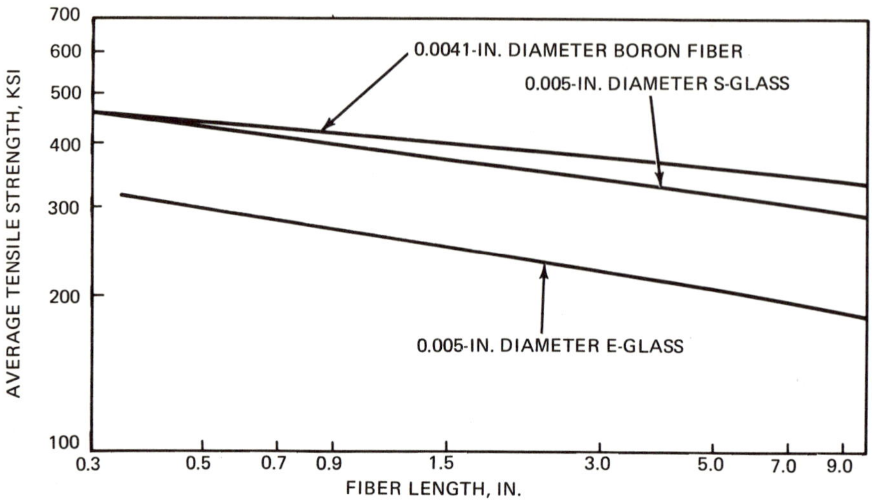

FIG. 1. Average tensile strength versus fiber length (from Ref. [9]).

In experimental work, one usually takes a sample from the entire population. The sample mean approaches the population mean as the sample size is increased indefinitely. Thus, in any reasonable experimental program, the mean strength $\langle\sigma\rangle_f$ for any given length can only be determined within certain statistical limits. This uncertainty as to $\langle\sigma\rangle_f$ will be magnified in the graphical evaluation of the parameter β.

A more reliable method for determining the parameters α and β is due to Freudenthal and Gumbel [10]. In this method, it is sufficient to run a number of tests for a single gauge length. Let us introduce a new variable $y(\sigma)$ such that

$$G_\ell(\sigma) = 1 - \exp\{-\ell\alpha\sigma^\beta\} = 1 - \exp\{-\sigma^y\} \tag{13}$$

If the failure stresses for N fibers are arranged in an ascending order, $\sigma_1, \sigma_2, \ldots, \sigma_i, \ldots, \sigma_N$, then it can be shown that for any σ_i, the corresponding y_i is given by

$$y_i = \log\left\{-\log\left[1 - G_i(\sigma)\right]\right\}$$

$$= \log\left\{-\log\left[(N + 1 - i)/(N + 1)\right]\right\} \tag{14}$$

Equation (13) may be alternatively written as

$$y = \log \ell + \log \alpha + \beta \log \sigma \tag{15}$$

Knowing σ_i $(i = 1, 2, \ldots, N)$ and hence y_i [from Eq. (14)], $\log \alpha$ and the parameter β are evaluated from Eq. (15) by the method of least squares:

$$\log \alpha = -\log \ell + \frac{\sum y_i \sum (\log \sigma_i)^2 - \sum \log \sigma_i \sum y_i \log \sigma_i}{N \sum (\log \sigma_i)^2 - \left(\sum \log \sigma_i\right)^2}$$

$$\beta = \frac{\sum y_i - \log \alpha - \log \ell}{\sum \log \sigma_i} \tag{16}$$

To make sure that the calculated theoretical distribution does indeed represent the experimental distribution, it is necessary to perform a goodness of fit (such as chi-squared) test [6].

III. STRENGTH OF A BUNDLE OF FIBERS

We consider a bundle made up of a very large number N of parallel fibers, all of equal length ℓ. The fibers are assumed to be clamped at the ends, such that all the unbroken elements have the same strain ε, and such that there is no frictional interaction between the individual fibers. The fibers remain elastic up to the point of rupture. It is evident that if all the fibers were of the same strength, i.e., if there was no dispersion in the strength of the fibers, the strength of the bundle would be equal to that of all its individual components. However, since there is a distribution of fiber strengths, the problem is more complex. It was first considered by Daniels [11]. In the following a brief outline of this work is given.

A large bundle is defined such that within it, the actual number of fibers with given properties can be treated as equal to their expectation in the total population. We also define the following symbols: L is the applied load on the bundle (function of ε and $\bar{\alpha}$); ε is the strain of the bundle fibers; $\bar{\alpha}$ is some parameter such as $E_f A_f$ (E_f is Young's modulus, A_f is the fiber cross-sectional area) which can be employed to define a particular load-strain relationship for a single fiber; L_B is the breaking load (function of ε and $\bar{\alpha}$); and ε_B is the breaking strain.

In Daniels' analysis, it is assumed that when a fiber breaks, the load it was carrying is instantaneously distributed equally among the surviving fibers. Thus neither stress concentrations nor dynamic wave propagation effects are considered.

The probability that for a fiber, ε_B and $\bar{\alpha}$ lie in specified small intervals ($d\varepsilon_B$, $d\bar{\alpha}$) is denoted by

$$\phi(\varepsilon_B, \bar{\alpha})\, d\varepsilon_B\, d\bar{\alpha} \tag{17}$$

where $\phi(\varepsilon_B, \bar{\alpha})$ is the joint probability density function for ε_B and $\bar{\alpha}$. Therefore, the probability density for ε_B for the fiber is

$$\psi(\varepsilon_B) = \int_0^\infty \phi(\varepsilon_B, \bar{\alpha})\, d\bar{\alpha} \tag{18}$$

The probability of a fiber having a breaking strain between ε and ∞, is given by

$$\int_{\varepsilon}^{\infty} \psi(\varepsilon_B)\, d\varepsilon_B = \int_{\varepsilon}^{\infty} d\varepsilon_B \int_{0}^{\infty} \phi(\varepsilon_B,\overline{\alpha})\, d\overline{\alpha} \tag{19}$$

Thus, if a load L is applied to a bundle of N fibers, causing strain ε, and if the bundle does not collapse, then the number of surviving fibers n is given by

$$n = N\int_{\varepsilon}^{\infty} d\varepsilon_B \int_{0}^{\infty} \phi(\varepsilon_B,\overline{\alpha})\, d\overline{\alpha} \tag{20}$$

Since the surviving fibers are each at a load $F(\varepsilon, \overline{\alpha})$ corresponding to strain ε, the total load L on the bundle is given by

$$L = N\int_{\varepsilon}^{\infty} d\varepsilon_B \int_{0}^{\infty} F(\varepsilon,\overline{\alpha})\phi(\varepsilon_B,\overline{\alpha})\, d\overline{\alpha} \tag{21}$$

The bundle strain at failure, ε_τ, corresponds to the maximum load L_τ and is therefore the root of the equation

$$\frac{d}{d\varepsilon_\tau}\int_{\varepsilon}^{\infty} d\varepsilon_B \int_{0}^{\infty} F(\varepsilon_\tau,\overline{\alpha})\phi(\varepsilon_B,\overline{\alpha})\, d\overline{\alpha} = 0 \tag{22}$$

where N, being a constant, has been dropped. Once ε_τ is known, the general expression for the breaking load L_τ of the bundle may be established. However, some practical simplifications may be obtained if certain assumptions regarding the behavior of the constituent fibers are allowed. For instance, if it is assumed that the dispersion in α is negligible compared to that in ε_τ, and further, that Hooke's law

$$F(\varepsilon, \bar{\alpha}) = \varepsilon\bar{\alpha} \tag{23}$$

applies, then Eqs. (21) and (22) can be reduced to the following simple forms:

$$L = N\bar{\alpha}\varepsilon \int_{\varepsilon}^{\infty} \psi(\varepsilon_B)\, d\varepsilon_B \tag{24}$$

and

$$\frac{d}{d\varepsilon_\tau}\left\{\varepsilon_\tau \int_{\varepsilon_\tau}^{\infty} \psi(\varepsilon_B) d\varepsilon_B\right\} = 0 \tag{25}$$

In the above development, the probability density function for the breaking strain ε_B was required. However, test data on fibers generally provide the probability density functions for the breaking load L_B. Thus, it seems desirable to recast the formulation of the bundle strength problem in terms of the probability density function $\theta(L_B)$ of the breaking load. Using the relationship

$$\theta(L_B) = (1/\bar{\alpha})\psi(\varepsilon_B) = (1/\bar{\alpha})\psi(L_B/\bar{\alpha}) \tag{26}$$

the equation for the total load on the bundle is obtained as

$$L = N\zeta \int_{\zeta}^{\infty} \theta(L_B)\, dL_B \tag{27}$$

where $\zeta = F(\varepsilon, \bar{\alpha})$ is the load on each surviving fiber. The breaking load L_τ is obtained by finding ζ_τ from the equation

$$\frac{d}{d\zeta_\tau}\left\{\zeta_\tau \int_{\zeta_\tau}^{\infty} \theta(L_B) dL_B\right\} = 0 \tag{28}$$

IV. A GRAPHICAL REPRESENTATION OF BUNDLE STRENGTH

We now digress to outline a graphical presentation of the results of bundle strength analysis. This is done in terms of the so-called "weakness formulation." The probability of a fiber's strength not exceeding ζ is defined by

$$a(w) = \int_0^{\zeta} \theta(L_B)\ dL_B \tag{29}$$

where w is the weakness of the fiber ($w = 1/\zeta$). Thus, the formulas for tensile strength may be rewritten as

$$L = N\zeta \int_{\zeta}^{\infty} \theta(L_B)\ dL_B = N\zeta \left\{ \int_0^{\infty} \theta(L_B)\ dL_B - \int_0^{\zeta} \theta(L_B)\ dL_B \right\}$$

$$= (N/w)[1 - a(w)] \tag{30}$$

and

$$\frac{d}{dw_\tau} \left\{ \left[1 - a(w_\tau)\right] / w_\tau \right\} = 0 \tag{31}$$

For an arbitrary function $a(w)$, one has the end points (i) $w = 0$ ($\zeta = \infty$), $a(w) = 1$; and (ii) $w = \infty$($\zeta = 0$), $a(w) = 0$. Equation (30) represents a straight line in the $[a(w), w]$ plane. Referring to Fig. 2, R_1 and R_2, the points of intersection of the straight line [Eq. (30)] with an arbitrary curve $a(w)$, are the end points of a range of possible equilibrium states for the bundle under a given load L. The sequence of events is as follows:

(a) Under an applied load L, the fibers break until $a(w_2)$ of them have given way. At this point, equilibrium is established.

(b) With an increase in L ($L \to L_\tau$), additional fibers are broken until $a(w_\tau)$ of them are removed.

(c) Any further increase in load leads to a collapse of the bundle.

(d) Under the original load L, additional fibers could have been cut until $a(w_1)$ of them were removed before collapse occurred.

Therefore, the maximum load the bundle can support is reached when both the upper and lower end points of the equilibrium range coincide at the tangent point, as denoted by T in Fig. 2. For a small bundle (N small), a(w) is a step function rather than a smooth curve. However, the rest of the analysis remains valid.

V. TENSILE STRENGTH DISTRIBUTION OF LARGE BUNDLES

In the case where the load-strain curves of all the fibers are similar, it can be shown that the distribution

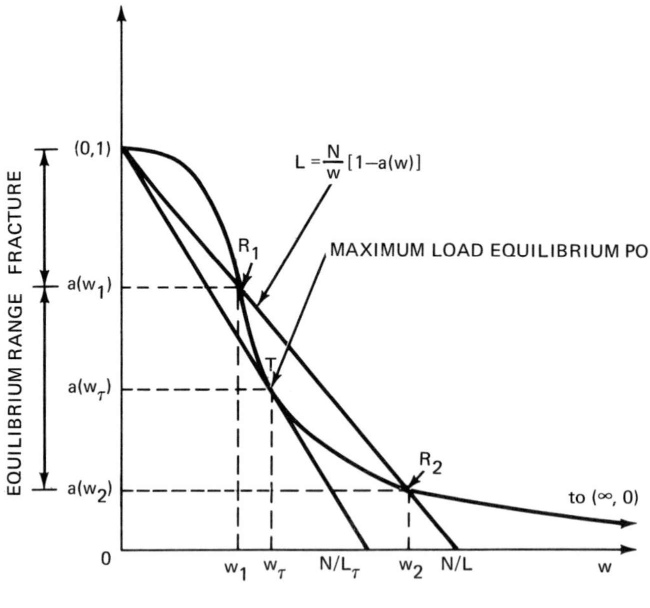

FIG. 2. Strength of large bundles.

of the tensile strength $\sigma = L/(NA_f)$ (A_f is the cross-sectional area of a fiber in the bundle) tends toward a normal distribution, with an expectation of

$$\langle\sigma\rangle_B = \sigma_\tau\left[1 - G_\ell(\sigma_\tau)\right] \tag{32}$$

and a standard deviation of

$$\Xi_B = \sigma_\tau\left\{G_\ell(\sigma_\tau)\left[1 - G_\ell(\sigma_\tau)\right]\right\}^{1/2} N^{-1/2} \tag{33}$$

where σ_τ is given by the equation

$$\frac{d}{d\sigma}\left\{\sigma\left[1 - G_\ell(\sigma)\right]\right\}_{\sigma=\sigma_\tau} = 0 \tag{34}$$

Note that $\langle\sigma\rangle_B$, σ_τ, and Ξ_B are all in units of force per unit area. Equations (32)-(34) can be derived directly from Eqs. (27) and (28). The proof is, however, quite complicated [11] and shall be omitted here.

If the cumulative distribution function $G_\ell(\sigma)$ may be represented by Eq. (10), Eqs. (32)-(34) reduce to

$$\sigma_\tau = (\ell\alpha\beta)^{-1/\beta} \tag{35}$$

$$\langle\sigma\rangle_B = (\ell\alpha\beta)^{-1/\beta} e^{-1/\beta} \tag{36}$$

$$\Xi_B = (\ell\alpha\beta)^{-1/\beta}\left\{\left[1 - e^{-1/\beta}\right]e^{-1/\beta}\right\}^{1/2} N^{-1/2} \tag{37}$$

Combining Eq. (36) with Eq. (12), it can be shown that the ratio of the tensile strength of the bundle to that of its fibers is [5]

$$\omega = \langle\sigma\rangle_B/\langle\sigma\rangle_f = \left[\beta^{1/\beta}e^{1/\beta}\Gamma(1 + 1/\beta)\right]^{-1} \tag{38}$$

Note that ω is independent of both α and ℓ. The ratio ω is 1 when there is no dispersion ($\beta = \infty$) and

TABLE 1

Strength of 0.00485 in. Diameter S-glass Fiber Bundles (N = 31)[a]

Strain rate, in./in./min	Mean bundle strength, $\langle\sigma\rangle_B \times 10^{-5}$ psi	Standard deviation, $\Xi_B \times 10^{-5}$ psi	Theoretical values, 10^5 psi		Experimental mean fiber strength, $\langle\sigma\rangle_f \times 10^{-5}$ psi
			$\langle\sigma\rangle_B$	Ξ_B	
0.0265	1.22 ± 0.26	$0.25 < \Xi_B < 0.66$	1.27	0.156	2.37
0.66	1.57 ± 0.30	$0.28 < \Xi_B < 0.75$	1.17	0.151	2.26
26.5	1.21 ± 0.26	$0.25 < \Xi_B < 0.66$	1.05	0.145	2.15

[a]Based on [12].

is zero for infinite dispersion ($\beta = 0$). Thus, for real fibers, the tensile strength of large bundles has the same order of magnitude, but is less than, the average tensile strength of the constituent fibers.

Experimental work [12] on 0.00485-in.-diameter S-glass fiber bundles confirms the preceding theory in some respects. From Table 1, it is seen that the average bundle strength agrees, within statistical error, with that predicted by Eq. (36). However, the experimental standard deviation is much larger than that given by Eq. (37). This difference may be attributed to the small size ($N \approx 31$) of the experimental bundles. Daniels' bundle theory is an asymptotic one, in the sense that it is valid only for large bundles. It is possible that, for small bundles, the standard deviation decreases as $1/N^k$ ($k > 0.5$) and not as $1/N^{1/2}$. This conjecture can be verified only by further experimental work on bundles containing different numbers of fibers. Finally, it is to be observed from Table 1 that the bundle strength is of the same order, but is less than, the mean fiber strength.

REFERENCES

1. A. A. Griffith, "The Phenomena of Rupture and Flow in Solids," Trans. Roy. Soc. (London), Series A, 221, 163-198 (1920).

2. A. M. Freudenthal, The Inelastic Behavior of Engineering Materials and Structures, John Wiley, New York, 1950.

3. F. Peirce, "Tensile Tests for Cotton Yarns. V.—'The Weakest Link'. Theorems on the Strength of Long and of Composite Specimens," J. Textile Institute, 17, T355-T368 (1926).

4. B. Epstein, "Statistical Aspects of Fracture Problems, J. Appl. Phys., 19, 140-147 (1948).

5. B. D. Coleman, "On the Strength of Classical Fibers and Fiber-Bundles," J. Mech. Phys. Solids, 7, 60-70 (1958).

6. W. Weibull, "A Statistical Distribution Function of Wide Applicability," J. Appl. Mech., 18, 293-297 (1951).

7. R. A. Wallhaus, "A Statistical Study of Factors Influencing the Strength of Glass Fibers," Theoretical and Applied Mechanics Report No. 217, University of Illinois (Urbana), May 1962.

8. A. E. Armenàkas, S. K. Garg, and C. A. Sciammarella, "Strength Characteristics of Glass Fibers under Dynamic Loading," J. Appl. Phys., 41, 1657-1664 (1970).

9. V. G. Grinius, "Analytical and Experimental Micro-Mechanics Studies," Science of Advanced Materials and Process Engineering Series, Vol. 10, Western Periodicals Co., pp. G.1 to G.12, 1966.

10. A. M. Freudenthal and E. J. Gumbel, "On the Statistical Interpretation of Fatigue Tests," Proc. Roy. Soc. (London), Series A, 216, 309-331 (1953).

11. H. E. Daniels, "The Statistical Theory of the Strength of Bundles of Threads. I," Proc. Roy. Soc. (London), Series A, 183, 405-435 (1945).

12. A. E. Armenàkas, S. K. Garg, C. A. Sciammarella, and V. Svalbonas, "Experimental Study of the Failure Mechanism and Strength Characteristics of Fiber Bundles and Composites," Expt. Mech., 12, 1-10 (1972).

CHAPTER 8

COMPOSITE TENSILE-STRENGTH MODELS

I. INTRODUCTION

In the preceding chapter, it was established that a good probability density function to represent the strength distribution of a single fiber is of the Weibull form; and that, for large bundles, the strength distribution is of the Gaussian (normal) type. In this chapter, the above information is utilized to form theoretical models of composite strength.

Composite specimens exhibit various modes of failure under tensile loads. In this chapter, we shall consider two modes of failure. The first (the noncumulative mode) is associated with the occurrence of the first fiber break, and fracture is probably the result of a propagating stress wave or a matrix crack parallel to the fiber direction leading to a loss of structural integrity. The other mode (cumulative mode) is associated with the accumulation of scattered fiber breaks throughout the composite, and fracture results due to the inability of the weakest cross section to sustain the applied load.

II. NONCUMULATIVE FRACTURE MODEL

In some composites, failure results after the first fiber break, or at most a few isolated fiber breaks [1]. In such cases, it is instructive to study the correlation between the theoretical strength of the weakest fiber and the observed composite failure loads. Mathematically, the problem of finding the probability distribution function for the weakest fiber out of a sample of N fibers is equivalent to finding the strength distribution of a fiber made up of N links.*

Consider a population N of fibers of length ℓ whose strength is characterized by a Weibull distribution function (see Chapter 7). Then, the probability density function of the strength of the weakest fiber is

$$W_{\ell}(\sigma) = Ng(\sigma)[1 - G(\sigma)]^{N-1} \tag{1}$$

The mode of this distribution is

$$\sigma_W = [(\beta - 1)/N\ell\alpha\beta]^{1/\beta} \tag{2}$$

The structural details of the failure mechanism for this noncumulative model have not been studied on any significant scale. Several different possibilities may occur.

*See Chapter 7 and the Appendix for a discussion of these questions.

The fracture of a single fiber may initiate fracture propagation in the matrix, resulting in the failure of additional fibers, and hence collapse of the structure. Alternately, it may cause the debonding of the fiber from the matrix and lead to the loss of structural integrity. For efficient utilization of high-strength fibers, this matrix-dependent mode of fracture must be prevented, and the weaker fibers must be permitted to break without causing catastrophic matrix or bond failure.

An interesting situation arises when the coefficient of strength variation for the fibers is small, i.e., all the fibers have approximately the same strength. In this case, when the weakest fiber reaches its breaking load, the combination of load redistribution and the static and dynamic stress concentration factors on the adjacent fibers (to be discussed in the following chapter) is sufficient to continue breaking additional fibers, and finally the composite, at the same overall load. For this to be the case, the variation of fiber strength along its length also has to be small. Theoretically, the cumulative weakening model, discussed below, should predict the same collapse load as the noncumulative model for this type of fibers. Zweben [1] discusses some examples for which the noncumulative model gives reasonable agreement between theory and experiment. These comparisons are presented in Table 1.

TABLE 1

Specimens	Noncumulative model fracture predictions	Test values
2-in. boron-epoxy unilayer plates	172 ksi	193 ksi, 215 ksi [2]
Three-dimensional boron-epoxy	160 ksi	197 ksi (average value) [3]

III. CUMULATIVE WEAKENING MODEL

It was noted in Chapter 7 that most brittle materials display a large variation in strength. In composites reinforced with these materials, a large number of isolated fiber breaks are observed prior to composite failure. This mode of failure cannot be treated by the noncumulative fracture model discussed above. Gücer and Gurland [4] first proposed the so-called "cumulative weakening model" to account for this failure mode. In their model, the composite is divided into a series of n layers of unit thickness (say δ). Each layer consists of N elements uniformly loaded in parallel. Thus, the composite becomes a chain of bundles (Fig. 1). Each fiber of length δ

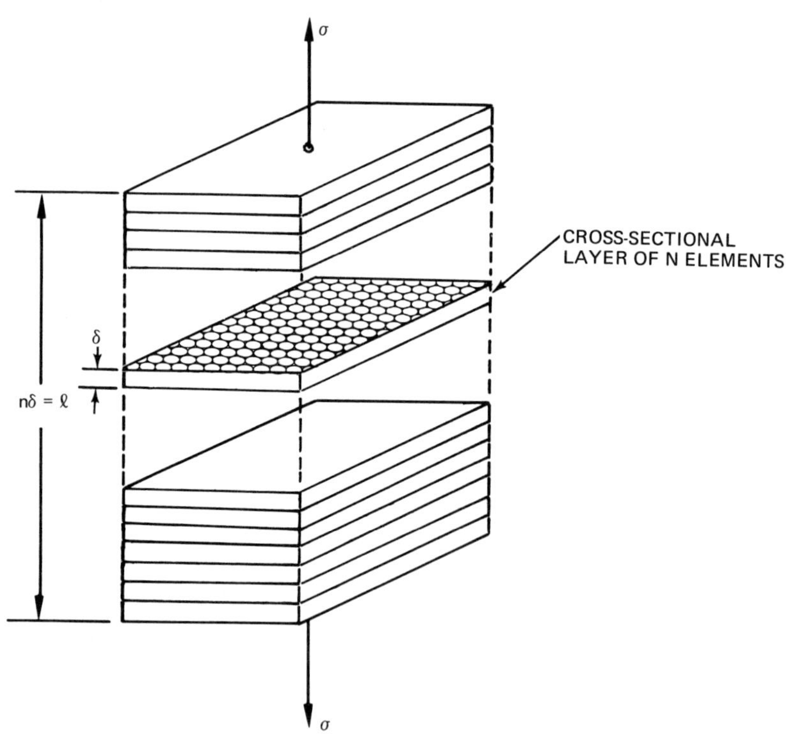

FIG. 1. Model of aggregate body.

(basic element) will fracture independently, and the composite will fail at a critical load, i.e., when a sufficient number of these fibers in a layer have broken so that the surviving fibers of the layer can no longer support the applied load. A crack propagation barrier is assumed to exist, so that the aggregate body (composite) fails by independent, simultaneous or consecutive, fracture of the individual fibers. The mathematical analysis of this model involves the following steps.

(a) Establishing a strength distribution for the fibers of length δ.

(b) Finding a strength distribution of a bundle containing N fibers of length δ.

(c) Application of the weakest link theorem to obtain the composite strength.

Each of these steps is considered in detail in the following.

The cumulative distribution function for the strength of the fibers of length δ is obtained by replacing ℓ by δ in Eq.(10) of Chapter 7:

$$G_\delta(\sigma) = 1 - \exp\{-\delta\alpha\sigma^\beta\} \qquad (3)$$

Corresponding to the cumulative distribution function, Eq. (3), one has the probability density function

$$g_\delta(\sigma) = (\alpha\beta\delta\sigma^{\beta-1}) \exp\{-\delta\alpha\sigma^\beta\} \qquad (4)$$

The average fiber strength is given by

$$\langle\sigma\rangle_{f\delta} = (\alpha\delta)^{-1/\beta}\Gamma(1 + 1/\beta) \qquad (5)$$

It was shown by Daniels [5] that for large bundles (large N), the distribution of bundle strength approaches a Gaussian distribution:

$$h(\sigma) = \left[1/\left\{(2\pi)^{1/2}\Xi_B\right\}\right]\exp\left[-\left(\sigma - \langle\sigma\rangle_B\right)^2/2\Xi_B{}^2\right] \tag{6}$$

with the cumulative distribution function being

$$H(\sigma) = \left[1/\left\{(2\pi)^{1/2}\Xi_B\right\}\right]\int_0^{\sigma} \exp\left\{-\left[\left(\sigma - \langle\sigma\rangle_B\right)^2/2\Xi_B{}^2\right]\right\}d\sigma \tag{7}$$

and where $\langle\sigma\rangle_B$ is the strength of a bundle of length δ:

$$\langle\sigma\rangle_B = \sigma_\tau\left[1 - G_\delta(\sigma_\tau)\right] \tag{8}$$

and Ξ_B is the standard deviation:

$$\Xi_B = \sigma_\tau\left\{G_\delta(\sigma_\tau)\left[1 - G_\delta(\sigma_\tau)\right]\right\}^{1/2} N^{-1/2} \tag{9}$$

The quantity σ_τ is obtained by maximizing $\sigma[1 - G_\delta(\sigma)]$, i.e., by solving the equation

$$\frac{d}{d\sigma}\left\{\sigma\left[1 - G_\delta(\sigma)\right]\right\}_{\sigma=\sigma_\tau} = 0 \tag{10}$$

Substituting for $G_\delta(\sigma)$ from Eq. (3) into Eqs. (8)-(10), we obtain

$$\langle\sigma\rangle_B = (\alpha\delta\beta)^{-1/\beta} \exp(-1/\beta) \tag{11}$$

$$\Xi_B = (\alpha\delta\beta)^{-1/\beta}\left\{\left[\exp(-1/\beta)\right]\left[1 - \exp(-1/\beta)\right]\right\}^{1/2} N^{-1/2} \tag{12}$$

Thus the strength characteristics of the individual layers have been established.

In the Gücer-Gurland model, each of these layers forms one of n links in a chain $(n = \ell/\delta)$. The composite fails when any one of the n links gives way. In other words, the composite failure is governed by the strength of its weakest link. For large n, the problem becomes one of finding the distribution of extreme values in the sample. The probability that one of n links fails between σ and $\sigma + d\sigma$ in strength, while all the others survive, is

$$S_c(\sigma)d\sigma = nh(\sigma)\ d\sigma[1 - H(\sigma)]^{n-1} \tag{13}$$

Therefore, the distribution of strength of the composite is given by

$$S_c(\sigma) = nh(\sigma)[1 - H(\sigma)]^{n-1} \tag{14}$$

The mean and the variance for this distribution, are given by [6]

$$\begin{aligned} \langle\sigma\rangle_c = {} & \langle\sigma\rangle_B - \Xi_B(2 \log n)^{1/2} \\ & + \Xi_B \frac{\log(\log n) + \log 4\pi}{2(2 \log n)^{1/2}} \\ & + 0.5772 \frac{\Xi_B}{(2 \log n)^{1/2}} \end{aligned} \tag{15}$$

and

$$\Xi_c{}^2 = \pi^2 \Xi_B{}^2/(12 \log n) \tag{16}$$

The above description of the cumulative weakening model is complete, except that no method has been given for calculating the layer thickness δ. As mentioned earlier, the size of the layer is dependent upon the interaction between the fibers and the matrix material. In Daniels' bundle theory, a single break in a fiber renders it ineffective for carrying any further load. Thus, the size of a layer in a composite must be such that a break in any fiber within that layer renders the fiber ineffective only within that layer. The thickness of the layer is obtained by the calculation of this "ineffective length."

At the fiber break, the stress in the fiber is zero. It then builds up with increasing distance to the break to its original value, due to the shear transfer from the matrix material, provided there is no bond failure between the fiber and the matrix. This buildup length, on each side of the fiber break, is the ineffective length of the fiber. It is assumed for this model that the load carried by the broken fibers is instantaneously and uniformly transferred to the surviving fibers in the layer. The stress in any fiber in this layer is then given by

$$\sigma_{layer} = [N/(N - r)]\sigma_{average} \tag{17}$$

where N is the total number of fibers and r is the number of broken fibers in the layer.

Several things are implied about the matrix material in this model. First, the matrix carries no tensile load. All the tensile load is carried only by the fibers. The matrix must also be strong enough in shear to transmit the load from a broken fiber to the surviving fibers in a layer. If this is not the case, there is the possibility of shear failure in the matrix resulting in the collapse of the composite.

IV. CALCULATION OF THE INEFFECTIVE LENGTH

To calculate the ineffective length δ, it is necessary to determine the shear stress distribution along the fiber-matrix interface. This problem has been solved by an approximate shear-lag type analysis by Rosen [7]. The model employed is shown in Fig. 2. It consists of a fiber surrounded by a matrix. This, in turn, is imbedded within a composite material having the overall average properties

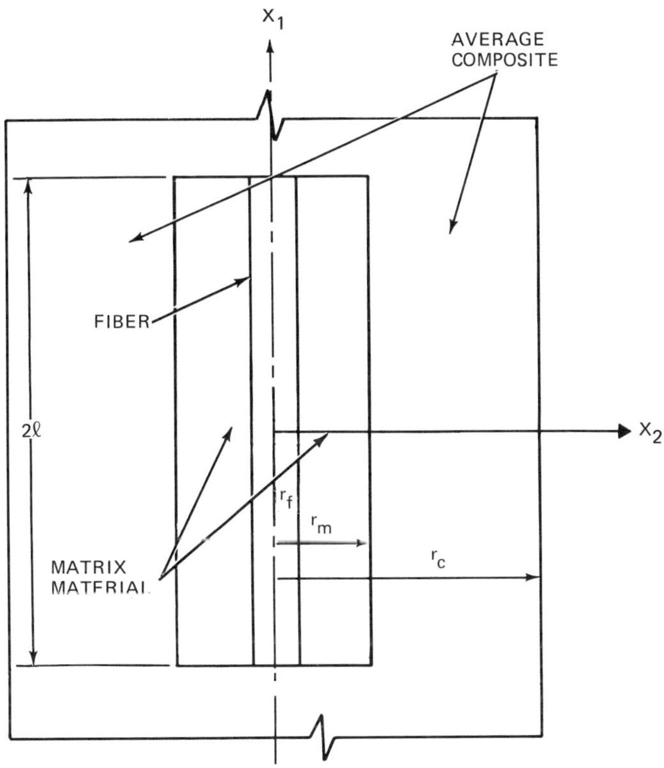

FIG. 2. Model for fiber ineffective length calculations.

of the material under consideration. The fibers may carry only tensile loads and the matrix transmits only shear. The fiber ends can transmit no shear to the average material. The shear stresses are assumed to decay at a negligible distance from the matrix-average material interface.

For equilibrium of the fiber in the axial direction, we have

$$2\pi r_f \tau_m + \pi r_f^2 (d\sigma_{f_1}/dx_1) = 0 \tag{18}$$

where τ_m is the shear stress in the matrix material and σ_{f_1} is the axial stress in the fiber.

For equilibrium of the composite in the axial direction, we have

$$\pi r_f^2 \sigma_{f_1} + \pi\left(r_c^2 - r_m^2\right)\sigma_{c_1} = \pi r_c^2 \bar{\sigma}_1 \tag{19}$$

where σ_{c_1} is the axial stress in the average material and $\bar{\sigma}_1$ is the applied stress. The displacements of the fiber and of the average material are denoted by u_f and u_c respectively. The matrix shear strain γ_m is defined as

$$\gamma_m = \Delta u/\Delta r = (u_c - u_f)/(r_m - r_f) \tag{20}$$

Differentiating the above equation twice with respect to x_1, and using Hooke's law, we obtain

$$\frac{1}{E_c}\frac{d\sigma_{c_1}}{dx_1} - \frac{1}{E_f}\frac{d\sigma_{f_1}}{dx_1} = \frac{r_m - r_f}{G_m}\frac{d^2\tau_m}{dx_1^2} \tag{21}$$

where E_c is the effective Young's modulus of the composite, E_f is the Young's modulus of the fiber, and G_m is the shear modulus of the matrix. Differentiating Eq. (19), and substituting the result into Eqs. (18) and (21), we obtain

$$\left(d^2\tau_m/dx_1^2\right) - \eta^2\tau_m = 0 \tag{22}$$

with the boundary conditions

$$\tau_m(0) = 0, \quad \sigma_{f_1}(\ell) = 0$$

where

$$\eta^2 = \frac{2G_m}{E_f(r_m - r_f)r_f}\left[1 + \frac{E_f}{E_c}\left(\frac{r_f^2}{r_c^2 - r_m^2}\right)\right]$$

The solution to Eq. (22) is of the form

$$\tau_m = A \sinh \eta x_1 + B \cosh \eta x_1 \tag{23}$$

Solution for the constants A and B, by using the boundary conditions, leads to

$$\tau_m = \frac{G_m \bar{\sigma}_1 r_c^2 \sinh \eta x_1}{\eta E_c (r_m - r_f)\left(r_c^2 - r_m^2\right)\cosh \eta \ell}$$

and

$$\sigma_{f_1} = - \frac{\bar{\sigma}_1 r_c^2 E_f}{E_c\left(r_c^2 - r_m^2\right) + E_f r_f^2}\left(\frac{\cosh \eta x_1}{\cosh \eta \ell} - 1\right) \tag{24}$$

If one assumes that $r_c >> r_m$, and thus $r_c >> r_f$, then

$$\eta^2 \approx 2G_m / [r_f E_f (r_m - r_f)] \tag{25a}$$

and

$$\tau_m = \frac{\bar{\sigma}_1 G_m}{\eta E_c (r_m - r_f)}\left(\frac{\sinh \eta x_1}{\cosh \eta \ell}\right) \tag{25b}$$

$$\sigma_{f_1} = -\bar{\sigma}_1 \frac{E_f}{E_c}\left(\frac{\cosh \eta x_1}{\cosh \eta \ell} - 1\right) \tag{25c}$$

The maximum $\sigma_{f_1}(0)$ occurs for $\ell \to \infty$, and is given by

$$\sigma_{f_1}(0)\Big|_{\ell \to \infty} = - \bar{\sigma}_1 \frac{E_f}{E_c} \tag{26}$$

The stress at a point a distance δ from the fiber end is

$$\sigma_{f_1}(\ell - \delta) = -\bar{\sigma}_1 \frac{E_f}{E_c}\left[\frac{\cosh \eta(\ell - \delta)}{\cosh \eta\ell} - 1\right] \qquad (27)$$

We now define the fiber efficiency ϕ as

$$\phi = [\sigma_{f_1}(\ell - \delta)/\sigma_{f_1}(0)]\Big|_{\ell\to\infty} = 1$$

$$- \cosh \eta\delta + \tanh \eta\ell \sinh \eta\delta \qquad (28)$$

Since, for large ℓ, $\tanh \eta\ell \approx 1$, we have

$$\phi = 1 - \cosh \eta\ell + (\cosh^2\eta\delta - 1)^{1/2}$$

and so,

$$\cosh \eta\delta = (1 + [1 - \phi]^2)/[2(1 - \phi)] \qquad (29)$$

Finally, substituting for η, from Eq. (25a), we obtain

$$\frac{\delta}{d_f} = \frac{1}{2}\left[(V_f^{-1/2} - 1)\frac{E_f}{2G_m}\right]^{1/2} \cosh^{-1}\left[\frac{1 + (1 - \phi)^2}{2(1 - \phi)}\right] \qquad (30)$$

where V_f is the fiber volume fraction and d_f is the fiber diameter. Therefore, if we set a value of $\phi = 0.9$, this means that we are considering that portion of the fiber as ineffective which carries less than 90% of the stress existing at infinity.

The δ calculated above is a measure of only one side of the fiber break, so that the thickness of a layer for the Gücer-Gurland model is 2δ. Figure 3, reproduced from Ref. [7], shows the variation of the ineffective length with constituent moduli for various fiber concentrations. In Fig. 4, theoretical predictions for composite strength

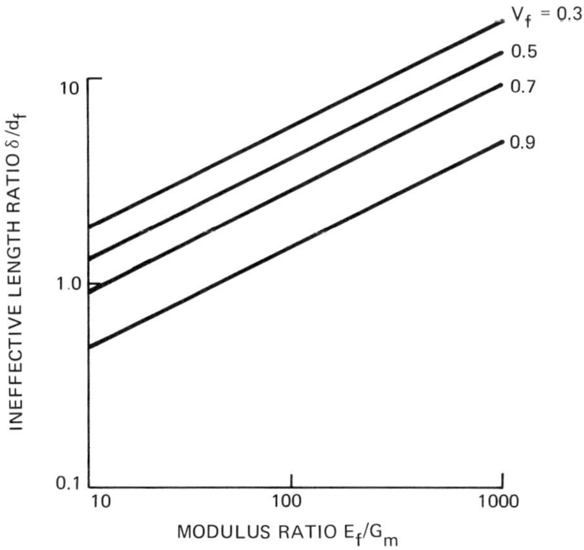

FIG. 3. Ineffective length of fibers in composites (from Ref. [7]).

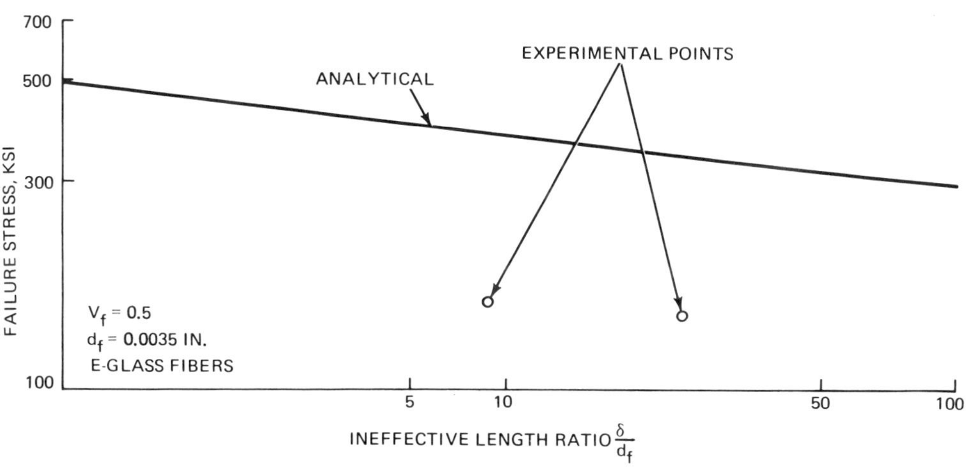

FIG. 4. Composite tensile strength versus fiber ineffective length (from Ref. [7]).

are compared with experimental results. As can be seen, the experimental results only substantiate the trend of the analytical results. There is no quantitative agreement.

V. EVALUATION OF THE CUMULATIVE WEAKENING MODEL

Despite the poor agreement between the theoretical and experimental results, Rosen [7] is able to reach the following conclusions.

(a) Randomly distributed fiber fractures occur well below the ultimate composite strength.

(b) The strength of the composite is determined by the frequency of the fiber breaks and by the efficiency with which the matrix limits the effect of the perturbation of the local stress field produced by a fiber break.

The second conclusion leads us directly to the examination of the basic model assumptions regarding crack propagation and stress concentrations adjacent to a broken fiber. In the model discussed above, the matrix is assumed to provide a crack propagation barrier, and the load carried by the broken fiber is assumed to be uniformly distributed among the surviving fibers. Both these assumptions are somewhat related, and are perhaps inherent in the use of Daniels' bundle theory to describe a layer. When a fiber breaks in a bundle, the load it was carrying is instantantaneously and uniformly distributed among the surviving members. However, in a composite, the case is quite different. The matrix, which transfers the load, gives rise to stress concentrations in the fibers adjacent to the broken fiber. Thus, after a single fiber breaks, it becomes possible for crack propagation to occur, since the adjacent fibers may break due to stress concentrations without any increment in the externally applied load. Such a propagating crack will be arrested only when all the possible paths are blocked by fibers strong enough to survive under the various stress concentrations that might occur.

Another criticism of the Gücer-Gurland-Rosen model

is that it predicts no size effect. In order to illustrate this point, note that as the number of fibers N approaches infinity, the bundle standard deviation Ξ_B approaches zero. The composite strength $\langle\sigma\rangle_c$ and standard deviation Ξ_c then become equal to bundle strength and standard deviation, respectively. In other words, the composite strength becomes independent of its length.

There is also some experimental evidence indicating that the ineffective length is variable, and in fact, increases with an increase in the applied load [8]. The constant value of δ calculated previously is a particular consequence of the assumed elastic behavior of the matrix. To illustrate the preceding statement, let us assume the matrix to be rigid plastic, with yield strength Y_m. Considering, once again, the equilibrium of the fiber in the axial direction (Fig. 2), we obtain [9]

$$\delta = (r_f/2)(\bar{\sigma}_1/Y_m) \tag{31}$$

Equation (31) implies that the ineffective length δ is directly proportional to the applied stress $\bar{\sigma}_1$. Therefore, the number of layers n is inversely proportional to the applied stress. We must, however, point out that Eq. (31) represents merely another idealization of the matrix behavior. In reality, the resin matrices presently in wide use behave viscoelastically (see, for example, Ref. [8]). Finally, we observe that one can possibly sidestep this uncertainty regarding the behavior of the matrix by determining the ineffective length experimentally (say, from photoelastic measurements).

REFERENCES

1. C. Zweben, "Tensile Failure of Fibrous Composites," AIAA J., 6, 2325-2331 (1968).

2. E. Friedman, D. G. Flom, and V. F. Mazzio, "High-Strength, High-Modulus, Whisker-Reinforced Plastic Composites," Air Force Materials Laboratory Report AFML-TR-66-362, February 1967.

3. V. G. Grinius, "Micromechanics-Experimental and Analytical Studies," Air Force Materials Laboratory Report AFML-TR-67-148, April 1967.

4. D. E. Gücer, and J. Gurland, "Comparison of the Statistics of Two Fracture Modes," J. Mech. Phys. Solids, 10, 365-373 (1962).

5. H. E. Daniels, "The Statistical Theory of the Strength of Bundles of Threads—I," Proc. Roy. Soc. (London), Series A, 183, 405-435 (1945).

6. B. Epstein, "Statistical Aspects of Fracture Problems," J. Appl. Phys., 19, 140-147 (1948).

7. B. W. Rosen, "Mechanics of Composite Strengthening," in Fiber Composite Materials, American Society for Metals, 1965, pp. 37-75.

8. A. E. Armenàkas, S. K. Garg, C. A. Sciammarella, and V. Svalbonas, "Experimental Study of the Failure Mechanism and Strength Characteristics of Fiber Bundles and Composites," Expt. Mech., 12, 1-10 (1972).

9. P. M. Scop and A. S. Argon, "Statistical Theory of Strength of Laminated Composites," J. Comp. Materials, 1, 92-99 (1967).

Chapter 9

CUMULATIVE WEAKENING MODEL INCLUDING STRESS CONCENTRATIONS

I. INTRODUCTION

In Chapter 8, we discussed the cumulative weakening or Gücer-Gurland-Rosen model of composite strength without stress concentrations. It was assumed there that when a fiber breaks in a layer, the load it was carrying is instantaneously and uniformly distributed among the surviving members of the layer. Experimental observations [1] reveal, however, that a complex stress situation is created around a broken fiber (see Fig. 1). If the induced shear stresses in the matrix around the broken fiber are sufficiently high, then the matrix may debond from the fiber and a crack may propagate through the matrix in a direction parallel to that of the fibers. This failure mode was considered in Chapter 8 (noncumulative mode). For the purposes of the present discussion, it will be assumed that the matrix and the fiber/matrix bond are sufficiently strong to prevent this mode of fracture. In Fig. 2, the strain distribution is shown along the line normal to the direction of the fibers at the point where a fiber failure occurred. Large strain concentrations are induced in the fibers adjoining the broken fiber. It can be seen that the strain concentration in both the direction of the fibers and the direction perpendicular to the fibers decays to zero and the strain becomes uniform at some distance (approximately 20 fiber diameters in this particular case) from the point of a fiber break. However, it is clear from Fig. 2 that the load previously carried by a broken fiber is distributed in a highly nonuniform manner among the surviving fibers of that cross section.

Presently, we shall consider a model including the effect of such stress and strain concentrations. The application of this model [2] to a particular material involves the following five steps:

(a) Definition of the representative volume element.

(b) Determination of strength distribution of the elements.

(c) Determination of stress (or load) concentration factors.

(d) Analysis of crack growth.

(e) Establishment of a fracture criterion.

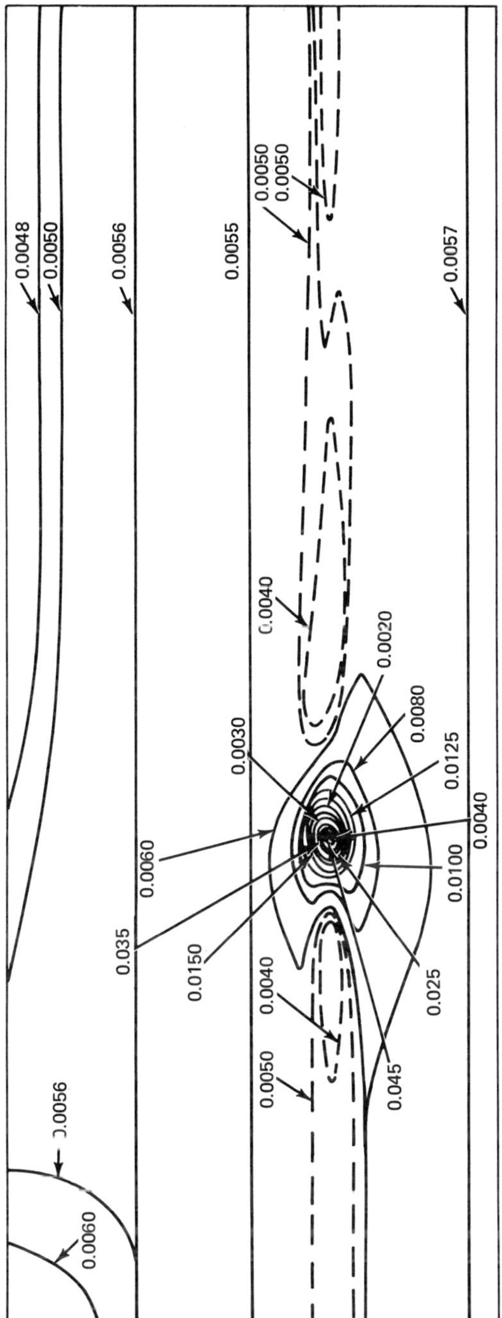

FIG. 1. Contours of the strain distribution at the surface of the specimen around a fiber failure (average strain 0.0055). Broken fiber end location is indicated by X. (Reproduced from Ref. [1].)

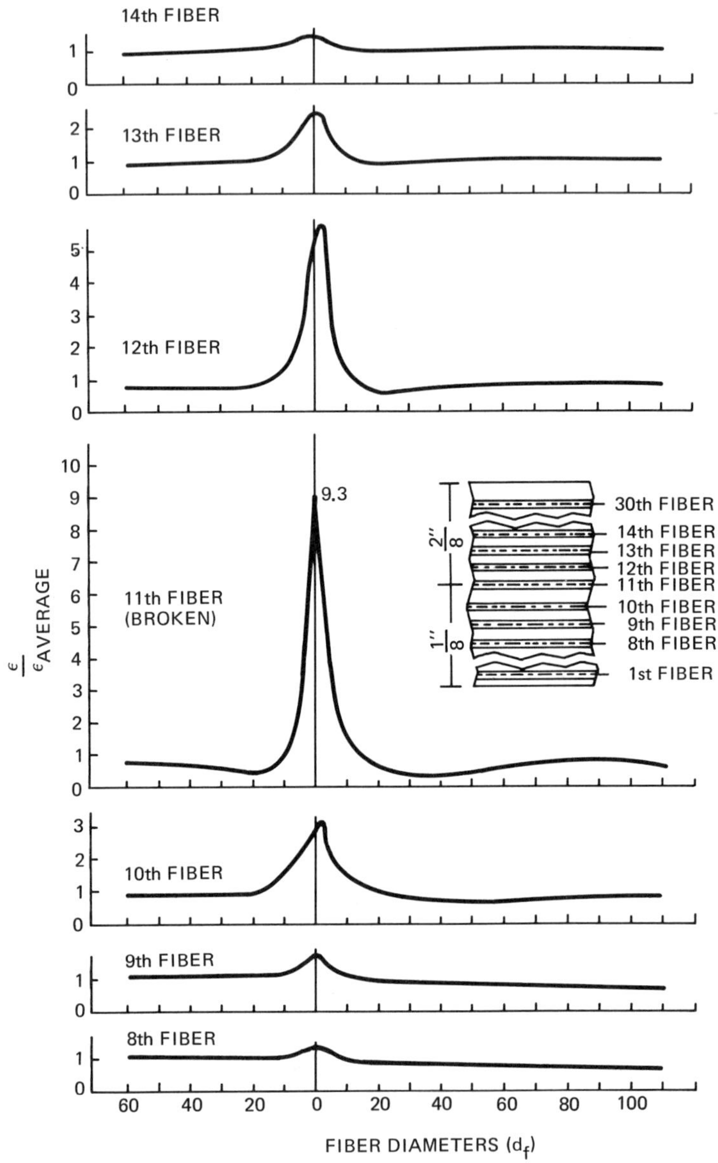

FIG. 2. Variation of the strain at the surface of the specimen along the broken fiber and its neighboring fibers (from Ref. [1]).

The geometrical model to be used here is identical with that of Chapter 8. The composite made of N fibers of length ℓ is considered to be composed of layers of thickness δ (ineffective length). The basic element is thus a fiber of length δ. The total number of such basic elements is nN, where $n = \ell/\delta$ denotes the number of layers. Once again, the Weibull distribution is employed to characterize the strength distribution of the fibers of length δ. Before proceeding further, it is necessary to calculate the stress (or load) concentrations. This is accomplished in the following section.

II. CALCULATION OF LOAD CONCENTRATIONS*

The load concentrations in two- and three-dimensional structures have been investigated by Hedgepeth [3], Fichter [4], and Hedgepeth and Van Dyke [5] through the use of approximate shear-lag type analyses. In the following, we will begin with a discussion of the two-dimensional case and then outline the theoretical approach for the three-dimensional composite structure. The implications of this analysis for the definition of ineffective length δ are of special interest. It will be shown that the ineffective length is a function of the number of broken fibers. This result suggests that the mathematical composite model for calculating the ineffective length, as discussed in Chapter 8, is somewhat inadequate.

*The term "load concentration factor" is employed to denote the "stress concentration factor" obtained on the basis of approximate shear-lag type analyses in contradistinction to the exact elasticity analyses.

A. Two-Dimensional Elastic Model

The two-dimensional composite configuration [3,4] considered is shown in Fig. 3. The fibers are separated by a constant distance and are numbered from $-\infty$ to ∞. The loading is assumed to be at infinity, with the fibers taking only tensile loads and the matrix only transmitting shear. The force on the nth fiber is given by

$$p_n(x_1,t) = E_f A_f \, \partial u_n(x_1,t)/\partial x_1 \tag{1}$$

where E_f is the modulus of elasticity of the fiber, A_f is the cross-sectional area of the fiber ($=\pi d_f^2/4$), d_f is the fiber diameter, $u_n(x_1,t)$ is the displacement of the nth fiber, and $p_n(x_1,t)$ is the force on the nth fiber.

The shear force per unit length in the bay between the nth and (n + 1)th fibers is given by

$$s_{n,n+1} = G_m h(u_{n+1} - u_n)/d \tag{2}$$

where G_m is the shear modulus of the matrix, $h = \pi d_f$, and d is the distance between the fibers.

Equilibrium of an element of the nth fiber yields

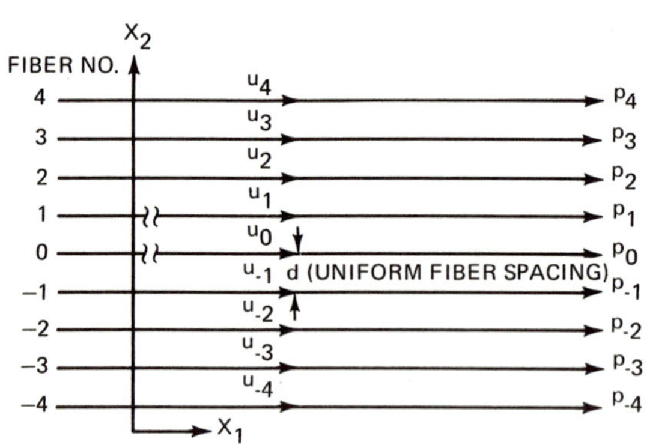

FIG. 3. Two-dimensional model for load concentration factors.

$$E_f A_f \frac{\partial^2 u_n}{\partial x_1^2} + \frac{G_m h}{d} (u_{n+1} - 2u_n + u_{n-1}) = m \frac{\partial^2 u_n}{\partial t^2} \tag{3}$$

where the assumption has been made that the mass per unit length m associated with the nth fiber is concentrated at that fiber [3].

For r broken fibers ($0 \leq n \leq r - 1$), the appropriate boundary conditions are

$$\begin{aligned} p_n(0,t) &= 0 && 0 \leq n \leq r - 1 \\ u_n(0,t) &= 0 && n < 0 \text{ and } n \geq r \\ p_n(\pm\infty, t) &= p && \text{all } n \end{aligned} \tag{4}$$

In addition to the boundary conditions (4), we have the following initial conditions:

$$p_n(x_1,0) = p, \quad (\partial u_n/\partial t)(x_1,0) = 0 \tag{5}$$

The problem can be recast into a more convenient form by introducing the following dimensionless variables:

$$\begin{aligned} &P_n = p_n/p, \quad U_n = (u_n/p)(E_f A_f G_m h/d)^{1/2} \\ &S_{n,n+1} = (s_{n,n+1}/p)(E_f A_f d/G_m h)^{1/2} \\ &\xi = (G_m h/E_f A_f d)^{1/2} x_1, \quad \bar{t} = (G_m h/md)^{1/2} t \end{aligned} \tag{6}$$

Substitution of Eqs. (6) into Eqs. (1)-(5) yields the partial differential difference equations

$$(\partial^2 U_n/\partial\xi^2) + U_{n+1} - 2U_n + U_{n-1} = \partial^2 U_n/\partial\bar{t}^2 \tag{7}$$

$$P_n(\xi,\bar{t}) = \frac{\partial U_n}{\partial \xi} \tag{8a}$$

$$S_{n,n+1} = U_{n+1} - U_n \tag{8b}$$

with boundary conditions

$$\begin{aligned} U_n(0,\ \bar{t}) &= 0 \qquad n < 0 \quad \text{and} \quad n \geq r \\ P_n(0,\ \bar{t}) &= 0 \qquad 0 \leq n \leq r - 1 \\ P_n(\pm\infty,\ \bar{t}) &= 1 \qquad \text{all} \quad n \end{aligned} \tag{9}$$

and initial conditions

$$P_n(\xi,\ 0) = 1, \quad (\partial U_n/\partial\bar{t})(\xi,\ 0) = 0 \tag{10}$$

The boundary problem for static loading is obtained by setting the right-hand side of Eq. (7) equal to zero and by ignoring the initial conditions (10). The resulting problem has mixed boundary conditions at $\xi = 0$ and may be solved by the technique of influence coefficients. Consider a composite sheet in which all the fibers except the zeroth are intact, and displace the zeroth fiber by a unit amount while keeping zero displacement at $\xi = 0$ for all of the others. Denote the resulting forces and displacements by $L_n(\xi)$ and $V_n(\xi)$. The forces $P_n(\xi)$ and the displacements $U_n(\xi)$ for r broken fibers may then be expressed as follows by superposition of influence coefficients:

$$P_n(\xi) = 1 + \sum_{m=0}^{r-1} L_{n-m}(\xi)U_m(0) \tag{11}$$

$$U_n(\xi) = \xi + \sum_{m=0}^{r-1} V_{n-m}(\xi) U_m(0) \tag{12}$$

In Eqs. (11) and (12), it is necessary to sum over only the broken fibers ($0 \leq m \leq r - 1$), since $U_m(0) = 0$ for all other values of m. The second of the boundary conditions (9) yields

$$0 = 1 + \sum_{m=0}^{r-1} L_{n-m}(0) U_m(0), \quad 0 \leq n \leq r - 1 \tag{13}$$

Eq. (13) may be solved for the r unknowns $U_m(0)$ which can then be substituted back into Eqs. (11) and (12) to yield the entire solution. First, however, it is necessary to determine the influence functions $L_n(\xi)$ and $V_n(\xi)$. These functions are obtained by solving the differential equation ($\xi > 0$)

$$(d^2 V_n / d\xi^2) + V_{n+1} - 2V_n + V_{n-1} = 0 \tag{14}$$

with the boundary conditions

$$\begin{aligned} V_n(0) &= 1, & n &= 0 \\ V_n(0) &= 0, & n &\neq 0 \\ (dV_n/d\xi)(\infty) &= 0, & &\text{all } n \end{aligned} \tag{15}$$

The solution to Eq. (14) is [3]:

$$V_n(\zeta) = (1/\pi) \int_0^{\pi} (\cos n\theta) \exp[-2\xi \sin(\theta/2)] \, d\theta \tag{16}$$

The functions $L_n(\xi)$ are given by

$$L_n(\xi) = dV_n(\xi)/d\xi \tag{17}$$

An examination of the problem shows that the maximum force occurs at $\xi = 0$ in the first intact fiber adjacent to the r broken ones. The load in the remaining intact fibers ($n < -1$ and $n > r$) rapidly decays to a uniform value. Thus the load concentration factor K_r for r broken fibers ($0 \leq n \leq r - 1$) is simply $P_r(0)$. Note that, by virtue of symmetry, $P_r(0) = P_{-1}(0)$. The load concentration factor K_r is given by

$$K_r = [4 \cdot 6 \cdot 8 \cdots (2r + 2)]/[3 \cdot 5 \cdot 7 \cdots (2r + 1)] \qquad (18)$$

In Fig. 4, the load concentration factor K_r is plotted as a function of the number of broken fibers r. Also given in Fig. 4 is the absolute value of the maximum shear force S_{max}:

$$S_{max} = \left| S_r(0) \right| = U_0(0) \qquad (19)$$

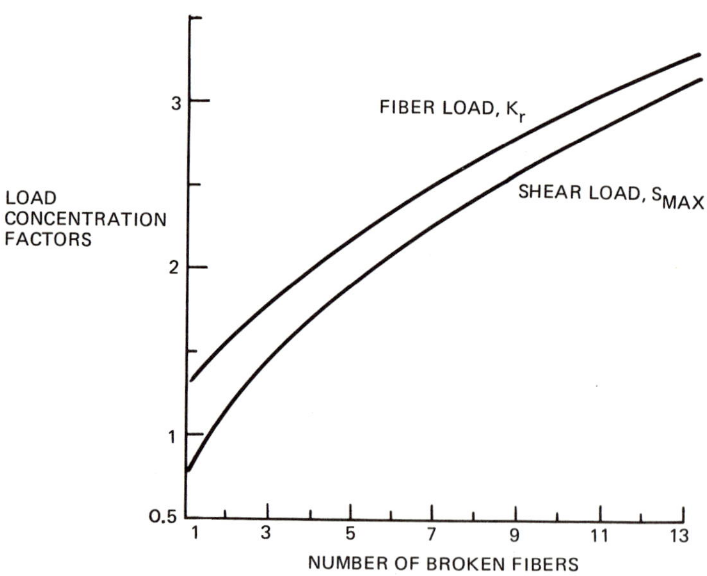

FIG. 4. Load concentration factors for fiber load and matrix shear force (from Ref. [4]).

Note that the maximum shear force and the load concentration factor increase with an increase in the number of broken fibers.

For a cut across r consecutive fibers, the load in the nth fiber is given by Eq. (11). In Fig. 5 we plot the load growth in the broken fibers for a cut across one, two, and three fibers. It is seen from this plot that the length over which a broken fiber recovers a particular fraction of its load (say 90%) varies considerably with the number of fibers that are broken. In Chapter 8, the ineffective length was defined as that portion of a broken fiber which carries less than 90% of the load applied at infinity. The ineffective length calculated for the composite mathematical model employed in Chapter 8 is independent of the number of broken fibers. This assumption is somewhat in contradiction with the results shown in Fig. 5, and suggests that the mathematical composite model discussed in Chapter 8 is inadequate. Although this difference between the two composite models is of a basic nature, it is perhaps of little practical significance since whenever the proba-

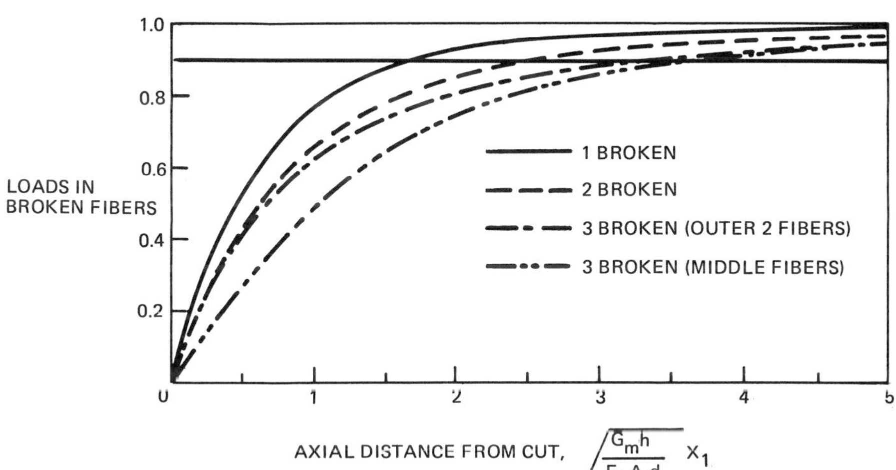

FIG. 5. Loads in broken fibers near a cut (from Ref. [4]).

bility of secondary failures becomes significant, failure of the composite results.*

The comparable dynamic problem is solved by taking a Laplace transform of the time-dependent differential equation (7) and boundary conditions (9). The resulting equations are similar in form to the static equations and can also be solved by the technique of influence functions [3]. The dynamic response factor η_r is defined as the ratio between the maximum load and the static load. The values of η_r for one, two, and three broken fibers are given in Table 1.

The dynamic response factor increases with an increasing number of broken fibers, and may therefore be reasonably expected to be maximum in the limit for an infinite number of broken fibers. Its value for this limiting case was shown to be 1.27 [3].

B. Two-Dimensional Plasticity Problem

In the preceding analysis, both the fibers and the matrix were assumed to behave elastically. The shear stress

TABLE 1

Dynamic Response Factors η_r

r	η_r
1	1.15
2	1.19
3	1.20

*See Section 3 for a fuller discussion of failure criteria.

in the low modulus matrix may, however, exceed its yield stress, causing it to behave plastically (or perhaps viscoplastically). Hedgepeth and Van Dyke [5] have extended the solution for the static loading case to take into account the ideally elastic-plastic behavior of the matrix. The shear stress is allowed to increase elastically to a limit plastic stress Y_m after which the stress remains constant. The configuration for this problem is shown in Fig. 6. The limit shear force region a extends in both positive and negative x_1 directions and is shown as the cross-hatched region in the figure. We are interested in the load concentrations in the elements $n = \pm 1$.

Introducing a dimensionless limit shear force per unit width and a plastic region designated by $\overline{Y}_m$ and $\overline{a}$, respectively,

$$\overline{Y}_m = (E_f A_f d/G_m h)^{1/2} Y_m h/p, \quad \overline{a} = (G_m h/E_f A_f d)^{1/2} a \tag{20}$$

the equilibrium equation for the nth fiber may be written as

$$(\partial^2 U_n/\partial \xi^2) + (U_{n+1} + U_{n-1} - 2U_n) = kf(\xi) \tag{21}$$

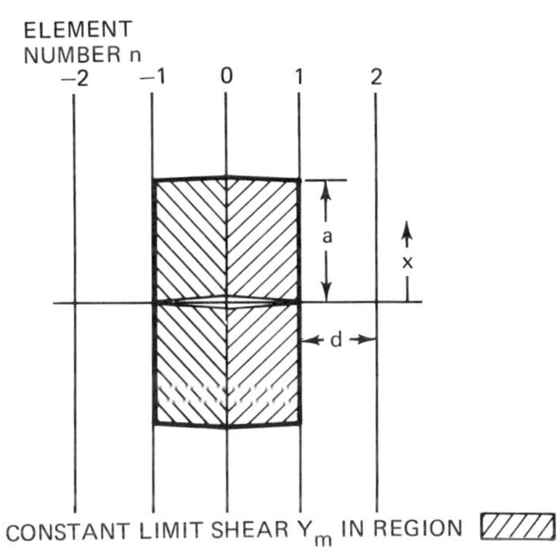

FIG. 6. Two-dimensional plasticity configuration (from Ref. [5]).

where $k = 2$ for $n = 0$, $k = 1$ for $n = 1$, and $k = 0$ for all other n; and

$$f(\xi) = U_0 - U_1 - \Upsilon_m \tag{22}$$

The appropriate boundary conditions are

$$U_n(0) = 0 \qquad n \neq 0$$

$$P_n(0) = 0 \qquad n = 0$$

$$P_n(\pm\infty) = 1 \qquad \text{all } n$$

and

$$f(\xi) = 0, \qquad \xi > \bar{a} \tag{23}$$

The solution to Eq. (21) with the associated boundary conditions (23) can be expressed in the form of Eqs. (11)-(12). In this case, the influence coefficients V_n and L_n are given by

$$V_n = B_n + \int_0^{\bar{a}} \left\{ f(t) \left[C_n(|\xi - t|) - C_n(\xi + t) \right] / 2 \right\} dt$$

$$L_n = \partial V_n(\xi) / \partial \xi \tag{24}$$

where

$$B_n(\xi) = \int_0^{\pi} \exp\left(-[2\sin(\theta/2)]\xi \right) \cos(n\theta/\pi) d\theta$$

$$C_n(\xi) = \int_0^{\pi} 2\sin(\theta/2) \exp$$

$$\left(-[2 \sin(\theta/2)]\xi \right) \cos(n\theta/\pi) d\theta \tag{25}$$

For the single broken fiber (n = 0), the single unknown displacement U_0 and the load in the first intact fiber are [from Eqs. (13) and (11)]

$$U_0(0) = -1/L_0(0), \quad P_1(\xi) = 1 - [L_1(\xi)/L_0(0)] \tag{26}$$

In order to determine the influence functions, the integral equation (24) is solved numerically [5]. In Figs. 7 and 8 we give the numerical results for the extent of the plastic region and the load concentration factor as functions of a load factor (P/P_{limit}) which is defined as the ratio of the actual load at infinity to the load at infinity required to raise the matrix stress to the limit value only. It should be noted that the plastic region increases almost linearly with the load factor. The load concentration factor at first decreases sharply with an increase in P/P_{limit}, but later it levels off at higher load factor values.

C. Three-Dimensional Elastic Load Concentrations

The three-dimensional elastic load concentration problem can be formulated in an analogous manner to the corresponding two-dimensional case [5]. In the following, we will consider the fibers arranged in square and hexagonal arrays and spaced an equal distance d apart. Denoting by $U_{n,m}(\xi)$ the dimensionless displacement in the (n,m)th element, the equilibrium equations may be written as

$$(\partial^2 U_{n,m}/\partial\xi^2) + \Big[U_{n+1,m} + U_{n,m+1} + U_{n-1,m} + U_{n,m-1} + k_1(U_{n+1,m-1} + U_{n-1,m+1}) - k_2 U_{n,m}\Big] = 0 \tag{27}$$

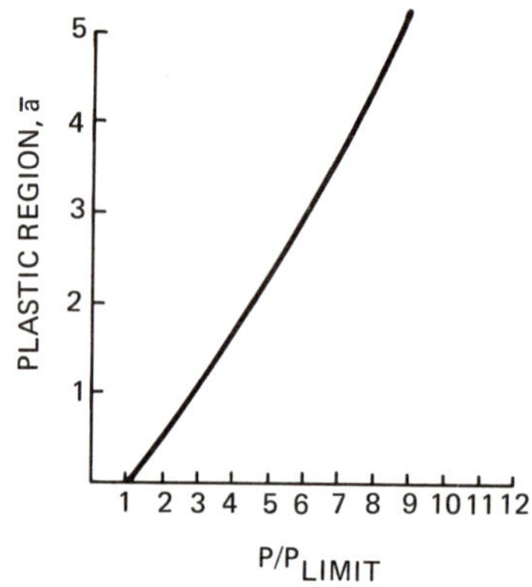

FIG. 7. Extent of plastic region $\bar{a}$ (from Ref. [5]).

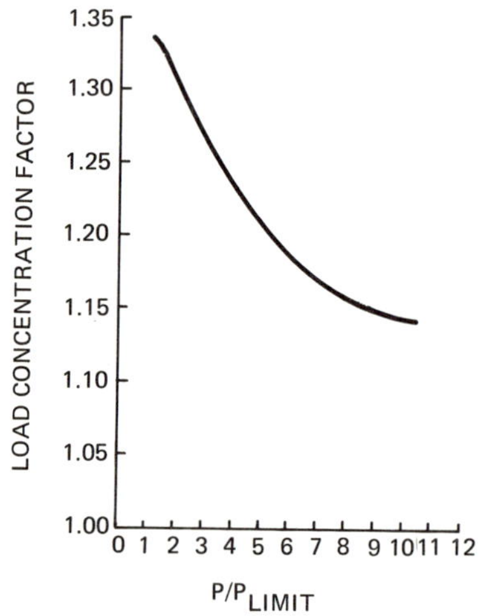

FIG. 8. Load concentration factor (from Ref. [5]).

where $k_1 = 0$ and $k_2 = 4$ for a square array and $k_1 = 1$ and $k_2 = 6$ for a hexagonal array. The boundary conditions are

$$P_{n,m}(0) = 0, \quad n,m \text{ broken}$$

$$U_{n,m}(0) = 0, \quad \text{all other } n,m \tag{28a}$$

and

$$P_{n,m}(\infty) = 1 \quad \text{for all } n,m \tag{28b}$$

As for the two-dimensional problem, the solution to Eq. (27) is obtained by introducing influence coefficients $V_{n,m}$ and $L_{n,m}$:

$$P_{n,m}(\xi) = 1 + \sum_i \sum_j L_{n-i,m-j}(\xi)\, U_{i,j}(0)$$

$$U_{n,m}(\xi) = \xi + \sum_i \sum_j V_{n-i,m-j}(\xi)\, U_{i,j}(0) \tag{29}$$

where

$$L_{n,m} = \partial V_{n,m}/\partial \xi \tag{30}$$

The influence coefficients $V_{n,m}$ and $L_{n,m}$ are determined by solving Eq. (27) with $U_{n,m}$ replaced by $V_{n,m}$ under the following boundary conditions:

$$V_{n,m}(0) = 1 \quad \text{for } n = m = 0$$

$$V_{n,m}(0) = 0 \quad \text{for all other } n,m \tag{31a}$$

$$(\partial V_{n,m}/\partial \xi)(\infty) = 0 \quad \text{for all } n,m \tag{31b}$$

Hedgepeth and Van Dyke [5] present results for load concentrations for square and hexagonal arrays of fibers. The solutions are dependent on the assumed configuration of broken fibers. For the square fiber array, results have been obtained for two cases: (i) the fibers are broken to form circular regions (Fig. 9), and (ii) the fibers are broken in a square configuration. In the first case, M_1

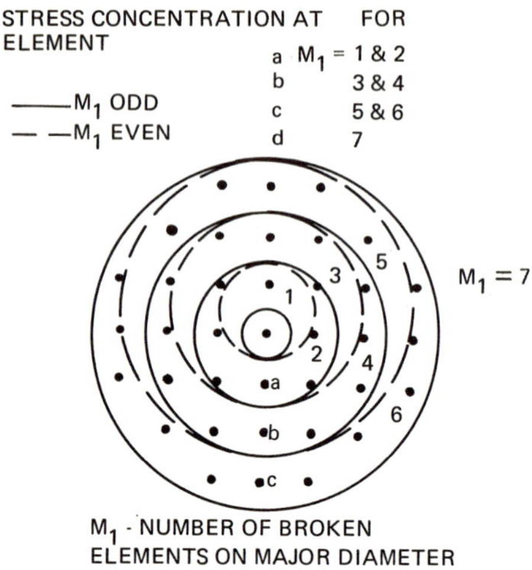

FIG. 9. Circular broken element configuration for square array.

represents the number of broken fibers on the major diameter and M_2 denotes the total number of fibers (broken) within a circle of diameter $M_1 d$. The load concentration factors for this case are given in Table 2. For the second case, M_1 denotes the number of broken fibers on one side of the square. The results for this configuration are given in Table 3. It should be noted that the three dimensional load concentration factors for the square array are considerably less than the corresponding two-dimensional factors.

The configuration of broken fibers for the hexagonal array is shown in Fig. 10. Here, M_1 denotes the number of broken fibers on a major diagonal of the array. In Table 4, load concentrations are presented for: (i) an element adjacent to the broken fibers on the major diagonal, and (ii) an element in the hexagonal ring of fibers surrounding the broken fibers, where the load concentration is the greatest. Again, the load concentrations are observed to be smaller than those of the corresponding two-dimensional cases.

TABLE 2

Square Array Load Concentration Factors for Circular Broken Fiber Configuration[a]

M_1	M_2	LCF
1	1	1.146
2	2	1.188
3	9	1.456
4	12	1.491
5	21	1.619
6	26	1.653
7	37	1.732

[a] M_1 is the number of broken fibers on major diameter, M_2 is the total number of broken fibers, and LCF is the load concentration at adjacent element on major diameter.

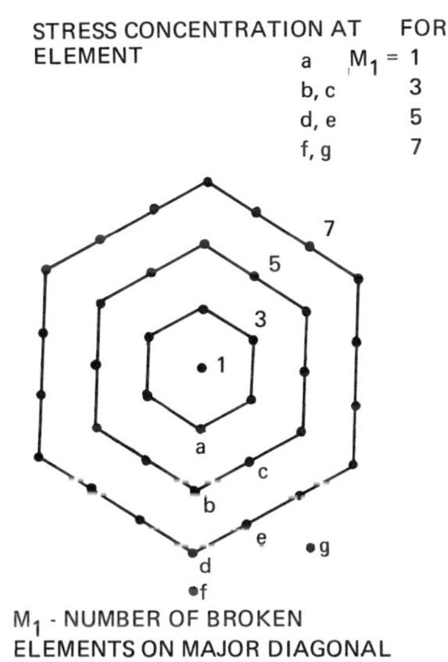

FIG. 10. Hexagonal array broken fibers.

TABLE 3

Square Array Load Concentration Factors for Square Broken Fiber Configuration[a]

M_1	M_2	LCF
1	1	1.146
2	4	1.281
3	9	1.456
4	16	1.582
5	25	1.728
6	36	1.841

[a]M_1 is the number of broken fibers on a side and LCF is the load concentration at adjacent element on the center of a side.

TABLE 4

Hexagonal Array Load Concentration Factors for Hexagonal Broken Fiber Configuration[b]

M_1	M_2	LCF_a	LCF_b
1	1	1.104	1.104
3	7	1.252	1.410
5	19	1.374	1.630
7	37	1.476	1.874

[b]M_1 is the number of broken fibers on major diagonal, $LSCF_a$ is the load concentration at adjacent element on major diagonal, and $LSCF_b$ is the maximum load concentration in an adjacent element.

III. ANALYSIS OF CRACK PROPAGATION

In recent years, models incorporating the effect of two- and three-dimensional load concentrations have been developed by several authors [2, 6-8]. We shall first discuss the analysis for a two-dimensional composite structure made up of continuous fibers and then outline the extensions to three-dimensional and whisker (discontinuous fibers) reinforced structures.

A. Analysis for Two-Dimensional Structures

In the following discussion of crack propagation, we will assume that the matrix behaves elastically and that the ineffective length δ is given by the expression derived in Chapter 8. The latter assumption implies that the ineffective length is independent of the number of fiber breaks. If $G_\delta(\sigma)$ denotes the probability that the strength of a fiber of length δ does not exceed σ, then the expected number of single failures in a group of nN elements under this stress is

$$I_1 = nNG_\delta(\sigma) \tag{32}$$

The breaks represented by the above equation are all single breaks. At this point, the static load concentration factors, K_r, are introduced.* The fracture of a single

*Scop and Argon [8] employ the so-called "stress enhancement factors" which are given by

$$\frac{1}{2}\,\delta\left[\int_0^{\delta/2} \sigma(x_1)^\beta \, dx_1\right]^{1/\beta}$$

where $\sigma(x_1)$ represents the stress in a fiber adjacent to a broken one, and β denotes the Weibull distribution parameter. The use of load concentration factors leads to conservative results since $\sigma(0) \geq \sigma(x_1)$ for $0 < x_1 < \delta/2$.

fiber at a stress σ in the given layer subjects the two adjoining fibers in that layer to an increased stress $K_1\sigma$. The probability that a fiber adjacent to a broken one will fail due to the load concentration is equal to

$$G_\delta(K_1\sigma) - G_\delta(\sigma)$$

where σ denotes the nominal fiber stress (= applied load/ total fiber cross-sectional area). Any increase in average fiber stress caused by the initial fractures is neglected. Additionally, the probability of two adjacent elements having strength less than σ is assumed to be negligibly small. These assumptions are justified since the number of broken fibers in a single layer prior to failure is likely to be rather small. Therefore, given that a single fiber is broken, the probability that one and only one of the two adjacent fibers will fail due to the load concentration is

$$p_{2/1} = 2\left[G_\delta(K_1\sigma) - G_\delta(\sigma)\right]\left\{1 - \left[G_\delta(K_1\sigma) - G_\delta(\sigma)\right]\right\} \tag{33}$$

The factor 2 in the expression for $p_{2/1}$ is to account for the fact that either of the two adjacent overstressed fibers may fail. The probability that both of the adjacent fibers break simultaneously is

$$p_{3/1} = \left[G_\delta(K_1\sigma) - G_\delta(\sigma)\right]^2 \tag{34}$$

Therefore, the probability that an element will fracture followed by the failure of at least one adjacent element is

$$p_2 = G_\delta(\sigma)(p_{2/1} + p_{3/1}) \tag{35}$$

In the total composite, then, the expected number of such groups of two or more broken fibers can be written as

$$I_2 = nNp_2 \tag{36}$$

and the associated probability of having at least one such group is

$$P_2 = 1 - (1 - p_2)^{nN} \tag{37}$$

The two fibers adjacent to two broken fibers are now subjected to a stress $K_2\sigma$. One of these unbroken fibers was previously exposed to a stress $K_1\sigma$, while the other was subjected to a stress σ. The probability that one of the two fibers adjacent to two broken ones will break is

$$p_{3/2} = \left[G_\delta(K_2\sigma) - G_\delta(K_1\sigma)\right] + \left[G_\delta(K_2\sigma) - G_\delta(\sigma)\right]$$
$$- 2\left[G_\delta(K_2\sigma) - G_\delta(K_1\sigma)\right]\left[G_\delta(K_2\sigma) - G_\delta(\sigma)\right] \tag{38}$$

where the first and second terms on the right-hand side denote the probability of fiber strength lying between $K_2\sigma$ and $K_1\sigma$, and $K_2\sigma$ and σ, respectively. The probability that both adjacent fibers will break simultaneously is

$$p_{4/2} = \left[G_\delta(K_2\sigma) - G_\delta(K_1\sigma)\right]\left[G_\delta(K_2\sigma) - G_\delta(\sigma)\right] \tag{39}$$

The probability that a fiber will break followed by the breaking of at least two other fibers is equal to the probability of only one fracture following the first subtracted from the probability of at least one fracture following the first. Thus

$$p_3 = p_2 - G_\delta(\sigma)p_{2/1}\left[1 - (p_{3/2} + p_{4/2})\right] \tag{40}$$

where $G_\delta(\sigma)p_{2/1}$ represents the probability of the fracture of one element followed by that of another element, and $G_\delta(\sigma)p_{2/1}(p_{3/2} + p_{4/2})$ represents the probability of the fracture of two elements followed by the fracture of at least one more element. Thus the expected number of groups containing at least three broken fibers in the composite is

$$I_3 = nNp_3 \tag{41}$$

The associated probability of having at least one group with three or more fractures is

$$P_3 = 1 - (1 - p_3)^{nN} \tag{42}$$

This procedure can be repeated for groups of four or more broken fibers. However, the complexity greatly increases due to various possible paths for crack propagation and the increase in the number of possible load concentration factors. Fortunately, it is unnecessary to carry this calculation further since composite fracture is usually observed to result whenever the probability of secondary breaks becomes significant.

B. Experimental Data for Two-Dimensional Composite Structures

On the basis of the test data of Rosen [9] on two-dimensional composite structures, Zweben [6] is able to state that for composite specimens exhibiting a significant number of scattered fiber breaks before failure, a conservative prediction of the failure load is that load for which the first multiple break is to be expected. In other words, the equation to be used for composite strength calculation is

$$I_2(\sigma) = 1$$

or,

$$nNG_\delta(\sigma)\left\{2\left[G_\delta(K_1\sigma) - G_\delta(\sigma)\right] - \left[G_\delta(K_1\sigma) - G_\delta(\sigma)\right]^2\right\} = 1 \tag{43}$$

Other experimental work on S-glass fiber-reinforced composites [1] shows that the fracture criterion of Eq. (43) occasionally provides too low a fracture load (see Fig. 11). A better estimate of the fracture load for that particular test series was provided by

$$I_4(\sigma) = 1$$

The fracture criterion is governed by specimen geometry and the fiber and matrix properties.

The cumulative weakening model [Eq. (43)] with stress concentrations usually predicts a lower failure stress than one without stress concentrations (Gücer-Gurland-Rosen

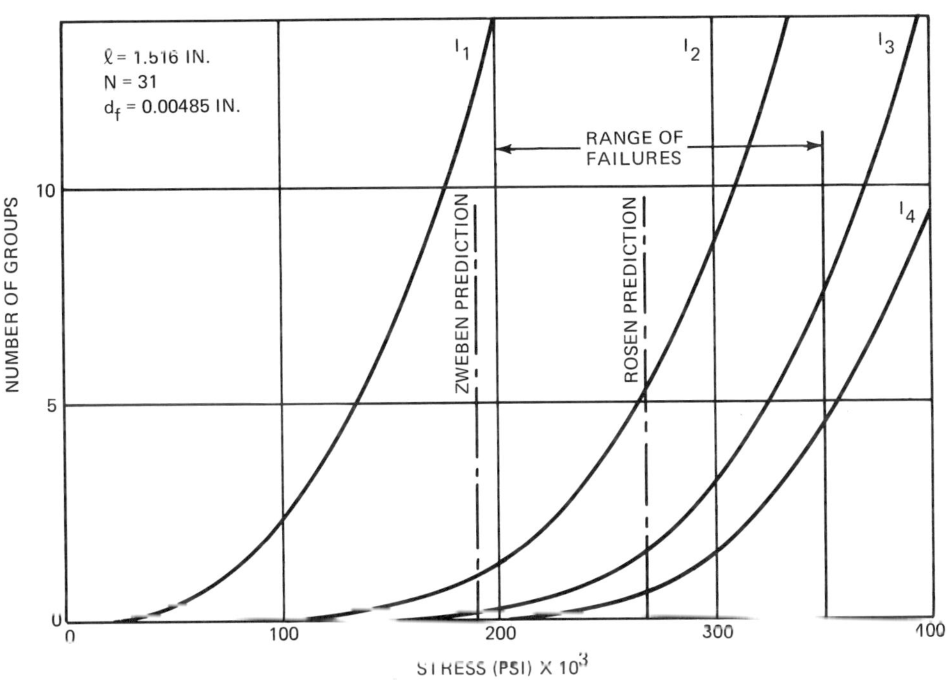

FIG. 11. Expected number of fracture groups as a function of fiber stress for two-dimensional S-glass/epoxy composite ($\delta = 30\ d_f$, $\dot{\varepsilon} = 0.0265$ in./in./min).

model). A comparison of the qualitative predictions of the various theories is shown in Fig. 12.

C. Crack Analysis for Three-Dimensional Composite Structures

The analysis of three-dimensional composite structures has been developed by Zweben and Rosen [2] and McKee and Sines [7]. The approach is analogous to that for two-dimensional structures. In both models, a square fiber array is assumed. One model [2] used for the analysis is presented in Fig. 13, where the solid line shows the influence zone for the first fiber break and the broken line represents the influence zone for the second fiber break. The total number of elements is $nN(n = \ell/\delta)$ and the number of single breaks in the composite is once again given by Eq. (32).

The break of a single fiber subjects four surrounding fibers to a stress level $K_1\sigma$ ($K_1 = 1.146$). The probabilities of failure of one, two, three and four elements surrounding a broken one are

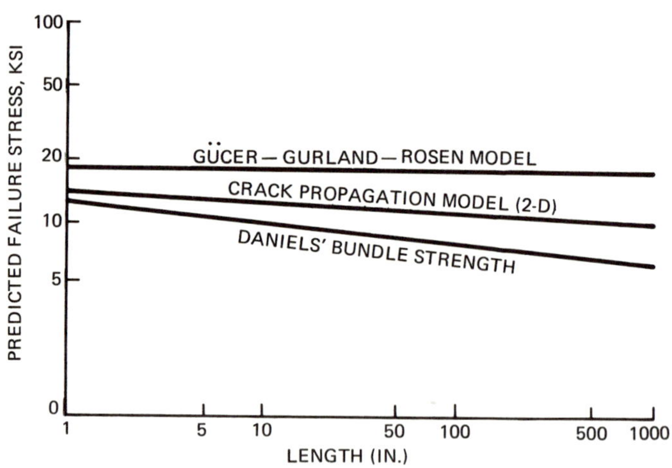

FIG. 12. Variation of predicted failure stresses with length (from Ref. [2]).

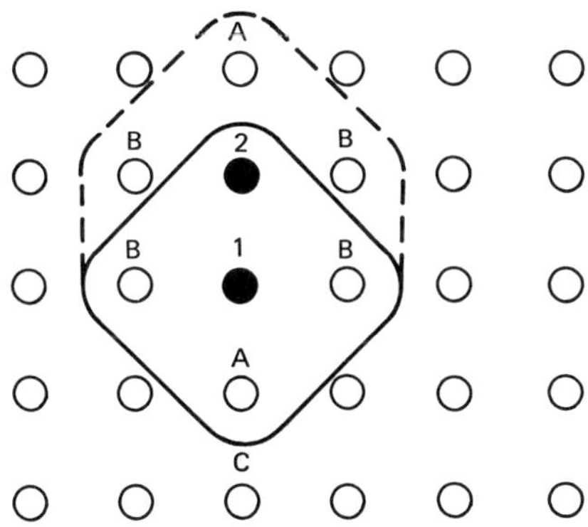

FIG. 13. Square fiber array.

$$
\begin{aligned}
e_{2/1} &= 4\left[G_\delta(K_1\sigma) - G_\delta(\sigma)\right]\left[1 - G_\delta(K_1\sigma) + G_\delta(\sigma)\right]^3 \\
e_{3/1} &= 6\left[G_\delta(K_1\sigma) - G_\delta(\sigma)\right]^2\left[1 - G_\delta(K_1\sigma) + G_\delta(\sigma)\right]^2 \\
e_{4/1} &= 4\left[G_\delta(K_1\sigma) - G_\delta(\sigma)\right]^3\left[1 - G_\delta(K_1\sigma) + G_\delta(\sigma)\right] \\
e_{5/1} &= \left[G_\delta(K_1\sigma) - G_\delta(\sigma)\right]^4
\end{aligned}
\tag{44}
$$

The probability that at least one of the four elements will fracture is

$$p_{2/1} = 1 - \left[1 - G_\delta(K_1\sigma) + G_\delta(\sigma)\right]^4 \tag{45}$$

Therefore the expected number of groups in the composite containing at least two fractures is

$$I_2(\sigma) = I_1(\sigma)p_{2/1} \tag{46}$$

The associated probability of having at least one group containing a minimum of two broken elements is

$$P_2 = 1 - \left[1 - G_\delta(\sigma)p_{2/1}\right]^{nN} \tag{47}$$

Let us now consider groups of fractures containing at least three broken elements. This can happen in two distinct ways: (1) by the simultaneous fracture of two overstressed elements following a single isolated break, or (2) by the successive breaking of one element (after a single isolated break) followed by the failure of at least one other element.

The probability that at least two of the overstressed elements adjoining a broken one will fracture simultaneously is given by the probability of at least one fiber break less the probability that exactly one will break:

$$p_{3/1} = p_{2/1} - e_{2/1} \tag{48}$$

The zone of influence for two broken elements is shown in Fig. 13. Note that the fibers marked A and B are subjected to different stress levels. Zweben and Rosen [2] however, employ $K_2 = 1.188$ for both elements as a first approximation. Three of the six elements adjoining the broken ones were previously at a stress level $K_1\sigma$ and the remaining ones at σ. Therefore, the probability that at least one of these elements will fracture is

$$p_{3/2} = 1 - \left[1 - G_\delta(K_2\sigma) + G_\delta(\sigma)\right]^3\left[1 - G_\delta(K_2\sigma) + G_\delta(K_1\sigma)\right]^3 \tag{49}$$

Hence the expected number of fractures containing at least three broken elements is

$$I_3 = I_1(p_{3/1} + e_{2/1}p_{3/2}) \tag{50}$$

Again the associated probability of having at least one such group is

$$P_3 = 1 - \left[1 - G_\delta(\sigma)(p_{3/1} + e_{2/1}p_{3/2})\right]^{nN} \tag{51}$$

The failure criterion set for three-dimensional structures is nominally the same as for the earlier two-dimensional composites, i.e.,

$$I_2(\sigma) = 1$$

Figure 14 reproduces some experimental data for a boron-aluminum composite. The failure criterion can be seen to give a conservative estimate of the measured ultimate load.

The preceding analysis may be generalized to whisker-reinforced composite structures by introducing the following

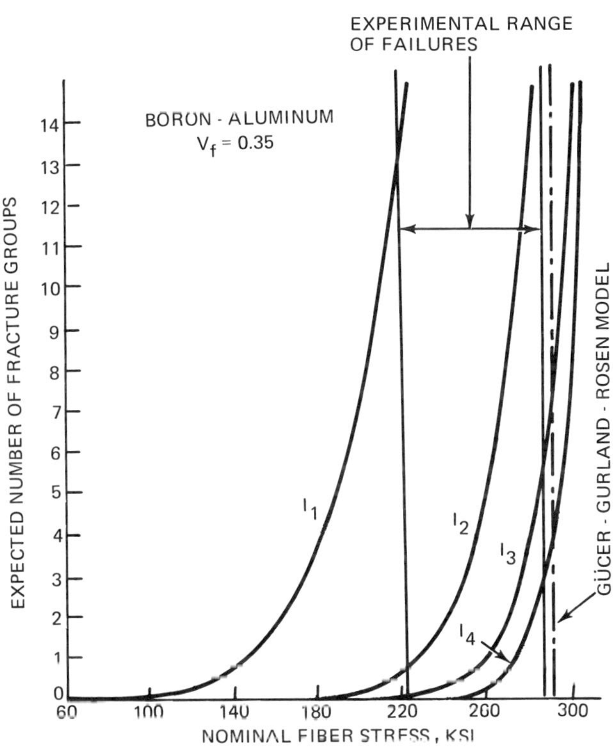

FIG. 14. Expected number of fracture groups as a function of fiber stress for a three-dimensional composite (from Ref. [2]).

probability density and cumulative density functions for the strength characteristics of the fibers [2]*:

$$q_\delta(\sigma) = \zeta\ \Delta(\sigma - 0) + (1 - \zeta)g_\delta(\sigma)$$

$$Q_\delta(\sigma) = \zeta + (1 - \zeta)G_\delta(\sigma) \qquad (52)$$

where Δ is the Dirac delta function, ζ is the ineffective fiber length δ divided by the whisker length w, and $g_\delta(f)$ and $G_\delta(f)$ are the usual Weibull density and cumulative distribution functions employed for fiber strength characteristics. The calculations for the expected number of groups containing one, two, or three broken elements are carried out in the previously demonstrated manner. Analysis of experimental data [2] for whisker-reinforced composites shows that a different failure criterion must be entertained, namely

$$I_3(\sigma) = 1$$

Fig. 15 presents some experimental data for whisker-reinforced composites. Note that I_1 and I_2 are rather close and increase rapidly at low stress. However, I_2 is quite distant from I_3 and I_4. It is of interest to compare Fig. 15 with Figs. 11 and 14. In Figs. 11 and 14, I_1 and I_2 are quite far apart. However, I_2, I_3, and I_4 are close together. Thus for whisker-reinforced structures, the role of I_3 is similar to that of I_2 for composites reinforced with continuous fibers.

*The total number of discontinuities in the composite is given by $N(\ell/w - 1) \approx N\ell/w$ (for $\ell >> w$). Therefore, the number of discontinuities in a single layer is $(N\ell/w)/(\ell/\delta) = N\delta/w = N\zeta$. Hence the assumed forms of probability density and cumulative density functions.

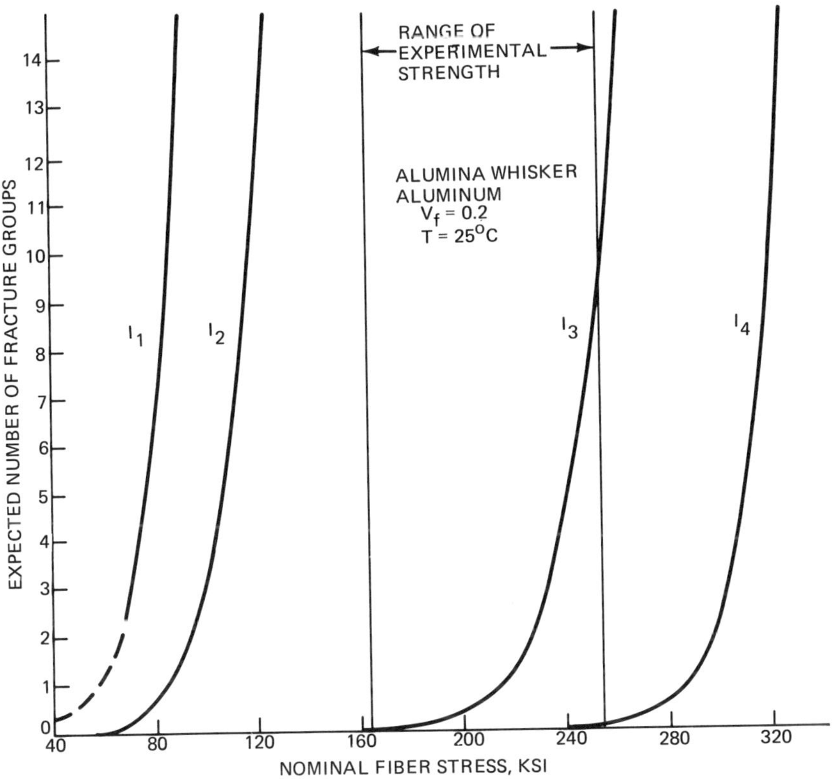

FIG. 15. Expected number of fracture groups as a function of fiber stress for a whisker reinforced composite (from Ref. [2]).

IV. NOTCHED COMPOSITES

The crack propagation analysis developed in the preceding section may also be applied to study the effect of notches on the tensile strength of composite materials. Let us consider an infinite two-dimensional plate with a central notch (Fig. 16) and a semiinfinite two-dimensional plate (Fig. 17) with an edge notch. The composite is once again divided into slabs of width δ. It is assumed that

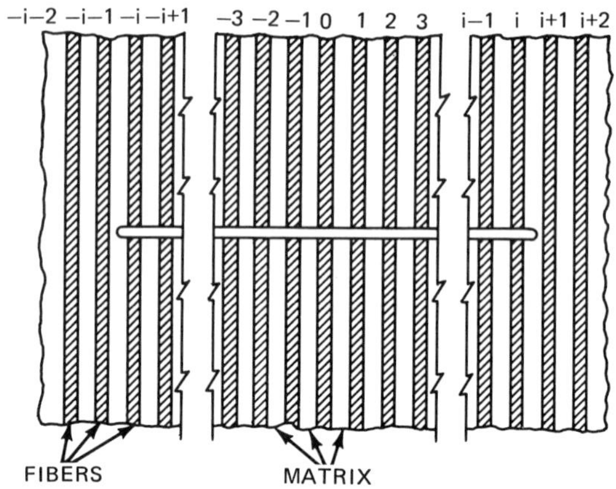

FIG. 16. Central notch.

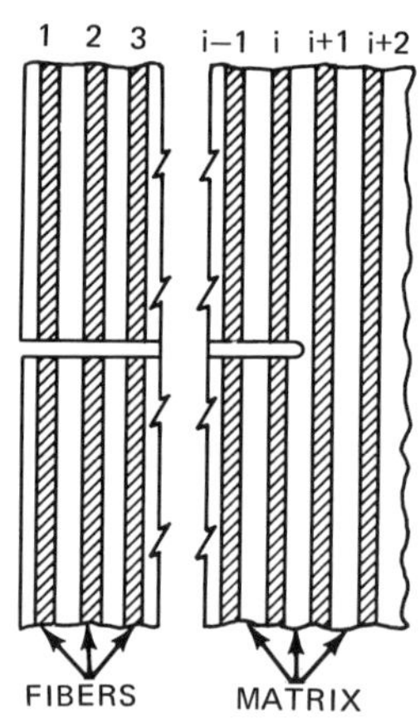

FIG. 17. Edge notch.

the notch is contained in one of these slabs. The fibers adjacent to the notch root ($n \leq -i - 1$ and $n \geq i + 1$ for the central notch and $n \geq i + 1$ for the edge notch) are subjected to local increases in stress. Due to the scatter in strength values, fibers labeled $-i - 2$ and $i + 2$ (Fig. 16) may fail prior to the notch root fibers ($-i - 1$, i + 1). For the sake of simplicity, we will ignore the stress concentrations in fibers other than those at the notch root. In other words, we will restrict our consideration to the situation in which at least one of the notch-root fibers (i + 1, -i - 1) fails prior to the failure of any other fibers in the vicinity of the notch. The stress in the (i + 1)th or (-i - 1)th fiber is equal to $K_i\sigma$, where K_i denote the previously defined load concentration factors. For the edge-notched specimen, the probability of failure of the (i + 1)th fiber is

$$P_i' = G_\delta(K_i\sigma) \tag{53}$$

For the central notched plate, the probability that at least one of the root fibers (i + 1 or -1 - 1) will fail is

$$P_i = 2G_\delta(K_i\sigma) - G_\delta^2(K_i\sigma) \tag{54}$$

If we assume that $G_\delta(\sigma)$ is given by the Weibull distribution

$$G_\delta(\sigma) = 1 - \exp(-\alpha\delta\sigma^\beta) \tag{55}$$

then Eqs. (53) and (54) become

$$P_i' = 1 - \exp\left(-\alpha\delta K_i^{\beta}\sigma^{\beta}\right) \tag{53'}$$

$$P_i = 1 - \exp\left(-2\alpha\delta K_i^{\beta}\sigma^{\beta}\right) \tag{54'}$$

The average stress associated with the break of at least one of the notch-root fibers is, for a central notch,

$$\bar{\sigma}_i = \int_0^\infty \sigma(dP_i/d\sigma)d\sigma = (1/2)^{1/\beta}(1/K_i)\langle\sigma\rangle_{f\delta} \tag{56}$$

where $\langle\sigma\rangle_{f\delta} = (\alpha\delta)^{-1/\beta}\Gamma(1 + 1/\beta)$ denotes the average strength of fibers of length δ, and for an edge notch,

$$\bar{\sigma}_i' = (1/K_i)\langle\sigma\rangle_{f\delta} \tag{57}$$

Since $K_{i+1} > K_i$, the average stress associated with the breakdown of any other fibers will be less than $\bar{\sigma}_i$ or $\bar{\sigma}_i'$. Therefore, Eqs. (56) and (57) may be regarded as the average stresses for which a notched composite specimen will fail by crack propagation from the notch.

To illustrate the effect of notch size, Fig. 18 shows Zweben's [10] experimental results for a glass-epoxy composite. Figure 18 also shows the range of failure of unnotched specimens [9]. Experimental observations reveal a significant number of scattered, single breaks in unnotched specimens prior to failure. It is clear from Fig. 18 that notches cutting a single fiber propagate in the range of failures associated with unnotched specimens. We also conclude that a notch cutting of the order of ten fibers has a significant influence on the composite strength. However, the scatter in fiber strength can result in modes of failure, in unnotched specimens, associated with stress levels required to propagate a notch cutting a small number of fibers. The analysis of crack propagation for three-dimensional notches is similar to that for two-dimensional notches and is outlined in Ref. [10].

V. EVALUATION OF THE CRACK PROPAGATION THEORY

On the basis of the preceding, it is possible to make the following general comments.

(a) Discontinuous fiber composites will always have

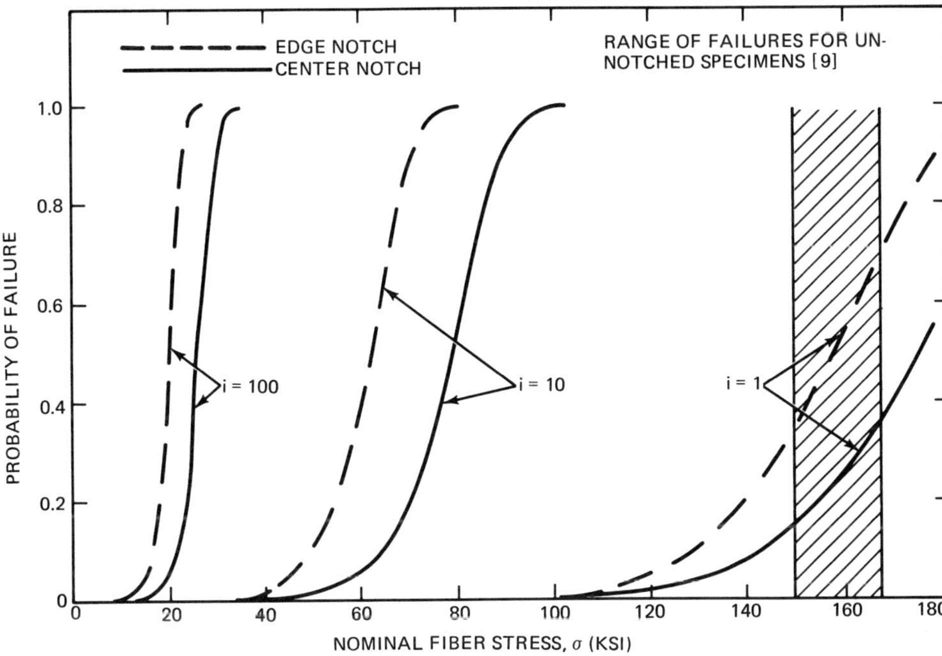

FIG. 18. Probability of failure of at least one notch-root fiber for notches of varying depth in a glass-epoxy composite (from Ref. [10]).

lower strengths than those of continuous fibers, due to increased number of stress concentrations.

(b) A three-dimensional composite will have higher strength than a two-dimensional structure with the same number of fibers since the stress concentration factors are smaller for three-dimensional structures.

(c) Composite strength decreases with the size of the body, i.e., with the number of elements (nN).

Two criticisms may be leveled against the preceding models. The first one concerns the use of static stress concentration factors, since the effect of the dynamic response (fiber debonding and crack propagation in the matrix) on stress concentration is considered to be of second order. Depending upon the type of loading, the actual stress concentration factors may be larger than the static stress concentration factors by 15-27% [3]. The second criticism involves the establishment of failure criteria such as $I_2(\sigma) = 1$. There is no inherent mechanism in this theory for predicting failure. Fracture criteria must be imposed on the basis of empirical evidence and as such cannot be universal. At best, a particular criterion may be employed to predict composite failure within a specified range of specimen size and fiber and matrix properties. A complete theory can only be built by considering the value of stress σ in a particular slab as a random variable. The specific functional form for such a random variable is, however, not easy to visualize.

One of the crucial assumptions made in the present analysis is that composite failure occurs by crack extension and is contained in a single slab of width δ. In reality, of course, composite materials may fail in several modes. As an example, fiber debonding and matrix cracking parallel to the fiber direction may occur, leading to the loss of structural integrity. Such failure need not affect the tensile load-carrying capacity of the fibers; the transverse properties are, however, degraded. This failure mode is observed in composites with brittle matrices and can be associated with the occurrence of the first fiber break (noncumulative weakening model, Chapter 8). Another possible mode of failure is a combination of fiber debonding/matrix cracking and random fiber fractures [1]. In this case, the composite failure may not be contained in a single slab. This mode has not been theoretically treated as yet. Therefore, whenever one wishes to apply a specific model to calculate composite strength, it is first necessary to examine the fracture morphology and to ascertain if the model in question is applicable.

REFERENCES

1. A. E. Armenàkas, S. K. Garg, C. A. Sciammarella, and V. Svalbonas, "Experimental Study of the Failure Mechanism and Strength Characteristic of Fiber Bundles and Composites," Expt. Mech., 12, 1-10 (1972).

2. C. Zweben, and B. W. Rosen, "A Statistical Theory of Material Strength with Application to Composite Materials," J. Mech. Phys. Solids, 18, 189-206 (1970).

3. J. M. Hedgepeth, "Stress Concentrations in Filamentary Structures," NASA TN D-882, May 1961.

4. W. B. Fichter, "Stress Concentrations around Broken Filaments in a Filament-Stiffened Sheet," NASA TN D-5453, October 1969.

5. J. M. Hedgepeth and P. Van Dyke, "Local Stress Concentrations in Imperfect Filamentary Composite Materials," J. Composite Materials, 1, 294-309 (1967).

6. C. Zweben, "Tensile Failure of Fibrous Composites," AIAA J., 6 (12),2325-2331 (1968).

7. R. B. McKee, Jr. and G. Sines, "A Statistical Model for the Tensile Fracture of Parallel Fiber Composites," presented at the ASME Winter Annual Meeting, Dec. 1968.

8. P. M. Scop and A. S. Argon, "Statistical Theory of Strength of Laminated Composites II," J. Composite Materials, 3, 30-47 (1969).

9. B. W. Rosen, "Mechanics of Composite Strengthening," in Fiber Composite Materials, American Society of Metals, 1965, pp. 37-75.

10. C. Zweben, "On the Strength of Notched Composites," AIAA Paper No. 71-369, presented at AIAA/ASME 12th Structures, Structural Dynamics and Materials Conference, April 1970.

CHAPTER 10

COMPRESSIVE STRENGTH OF COMPOSITES

I. INTRODUCTION

In Chapters 7-9, we considered the strength behavior of a unidirectionally fiber-reinforced composite under tensile loads. Presently, we shall investigate the strength characteristics of the composite under compressive loads. While a single fiber can support a significant amount of load in tension, it is obvious that it would immediately buckle under compression. Thus, while the primary function of the matrix in tension is to transfer loads from a broken fiber to the adjoining unbroken ones, in compression the matrix contribution is even more important. The matrix provides the necessary lateral support to the fibers, without which they would be useless under compressive loads. Dow et al. [1] have suggested that the mode of failure for a unidirectionally fiber-reinforced composite in compression is similar to the buckling of a column on an elastic foundation. The motivation for such a failure model is provided by photoelastic observations of the buckling of fibers in glass/epoxy composites caused by the shrinkage of the epoxy during cooling from its cure temperature down to room temperature. Similar behavior has also been observed in tungsten/silicon rubber and boron/silicon rubber composites [1]. Utilizing this information as a departure point, Rosen [2], Schuerch [3], and Herrmann, et al. [4] have developed analytical models for predicting this instability failure.

II. MATHEMATICAL MODEL

Both Rosen [2] and Schuerch [3] consider a two-dimensional model of a uniaxial composite (see Fig. 1). The model consists of alternating layers of fibers of thickness h and a matrix of thickness 2c. The composite material is subjected to a compressive load parallel to the direction of the fibers. For this idealized model, we will consider two forms of instability failure (see Fig. 2). First, the fibers may buckle in opposite directions in adjacent fibers, as shown in Fig. 2a. In this case, the matrix is extended

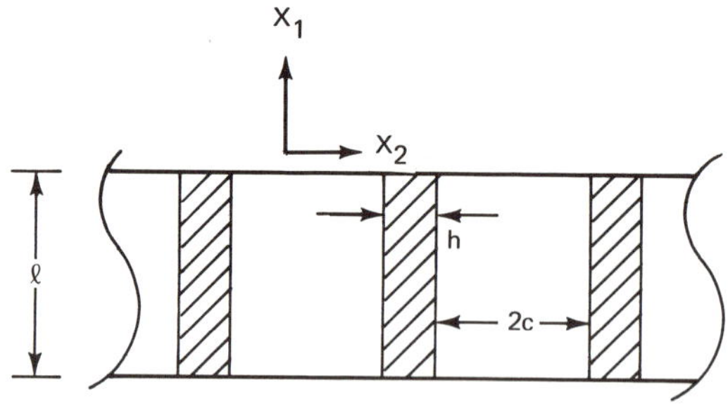

FIG. 1. Two-dimensional model.

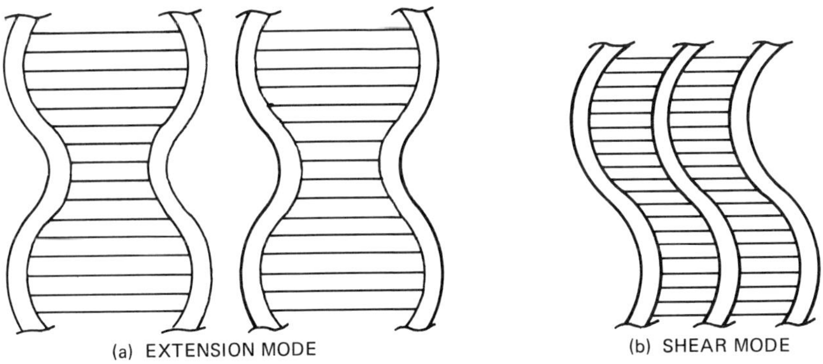

FIG. 2. Failure modes.

in a direction transverse to the fibers. The shear deformations in the fibers relative to those in the matrix are neglected. Second, the fibers may buckle in the same direction, as shown in Fig. 2b, subjecting the matrix material to a shear deformation.

In order to compute the buckling loads for these failure modes, both Rosen [2] and Schuerch [3] utilize an energy principle. The composite is considered to be subjected to the buckling load. The change in the strain energy of the composite consists of two parts: ΔU_f in the fibers and ΔU_m in the matrix. This strain energy change is equal to the change in the potential energy ΔT due to the shortening of the distance between the applied loads at the ends of the fibers. Thus

$$\Delta U_f + \Delta U_m = \Delta T \tag{1}$$

The procedure is to assume a particular buckling mode and to calculate the corresponding lowest buckling load satisfying Eq. (1). The "lowest buckling load" will be hereafter denoted as the buckling load.

III. EXTENSION MODE

We assume that each fiber is subjected to a compressive load L. The load L per running inch on each fiber can be expressed as

$$L = \sigma_f h \tag{2}$$

where σ_f denotes the fiber stress. Each fiber is now considered to buckle into a sinusoidal pattern expressed by the series

$$u_{f_2} = \sum_n a_n \sin(n\pi x_1/\ell) \tag{3}$$

where u_{f_2} denotes the displacement of the fiber in the x_2 direction. Assuming that the transverse strain in the binder material is independent of the x_2 direction, there follows

$$\varepsilon_{m_2} = u_{f_2}/c \tag{4}$$

and

$$\sigma_{m_2} = E_m \varepsilon_{m_2} = E_m u_{f_2}/c \tag{5}$$

Neglecting the changes in strain energy due to axial and shear stresses, one obtains

$$\Delta U_m = \frac{1}{2} \int \sigma_{m_2} \varepsilon_{m_2}\, dV = (E_m \ell/2c) \sum_n a_n^2 \tag{6}$$

The change in strain energy in the fibers is given by [5]

$$\Delta U_f = \frac{E_f I}{2} \int_0^{\ell} \left(\frac{d^2 u_{f_2}}{dx_1^2}\right)^2 dx_1 = \frac{\pi^4 E_f I}{4\ell^3} \sum_n n^4 a_n^2 \tag{7}$$

where I denotes the moment of inertia of the fiber. Taking the moment of inertia for a unit width to be $h^3/12$, there follows

$$\Delta U_f = \left(\pi^4 E_f h^3/48\ell^3\right) \sum_n n^4 a_n{}^2 \tag{8}$$

In order to evaluate the change in potential energy ΔT, an expression must be obtained for the shortening of the distance between the applied loads at the ends of the fibers. The difference between the length of an element ds of the curve and the corresponding element dx_1 (see Fig. 3) is

$$\begin{aligned} ds - dx_1 &= dx_1\left[1 + (du_{f_2}/dx_1)^2\right]^{1/2} - dx_1 \\ &\approx \frac{1}{2}\,(du_{f_2}/dx_1)^2\,dx_1 \end{aligned} \tag{9}$$

where a truncated Taylor expansion has been used. Therefore the shortening of the fiber λ is given by

$$\lambda = \int_0^\ell \tfrac{1}{2}(du_{f_2}/dx_1)^2\,dx_1 = (\pi^2/4\ell) \sum_n n^2 a_n{}^2 \tag{10}$$

Hence we have

$$\Delta T = L\lambda = (\pi^2 L/4\ell) \sum_n n^2 a_n{}^2 \tag{11}$$

Substituting from Eqs. (6), (8), and (11) into Eq. (1), and solving for the load L, we find

$$L = \frac{\pi^2 E_f h^3}{12\ell^2}\,\frac{\sum_n n^4 a_n{}^2 + \left[24\ell^4 E_m/\left(\pi^4 c h^3 E_f\right)\right]\sum_n a_n{}^2}{\sum_n n^2 a_n{}^2} \tag{12}$$

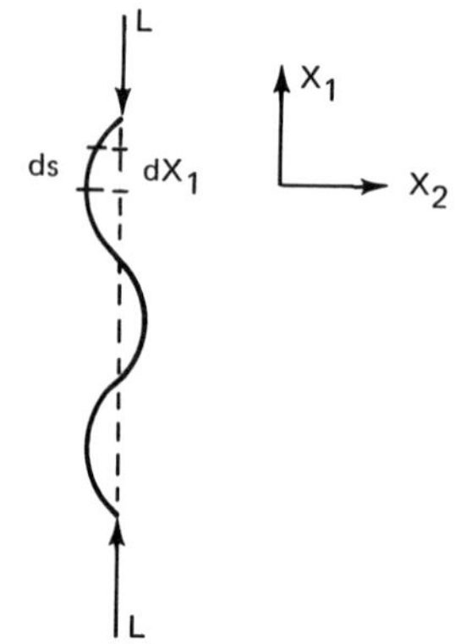

FIG. 3. Buckled fiber.

The applied load per running inch L is minimum for some value of n, say m. Equation (12) can be minimized by setting

$$\partial L/\partial (n^2)\Big|_{n=m} = 0 \tag{13}$$

This yields

$$\sigma_{f,cr} = L_{cr}/h = 2\left[V_f\ E_m E_f/\left\{3(1 - V_f)\right\}\right]^{1/2} \tag{14}$$

where $V_f = h/(h + 2c)$ denotes the fiber volume fraction. Also, the critical strain ε_{cr} is given by

$$\varepsilon_{cr} = \sigma_{f,cr}/E_f = 2\left[V_f/\left\{3(1 - V_f)\right\}\right]^{1/2}(E_m/E_f)^{1/2} \tag{15}$$

The critical composite stress at failure σ_c is

$$\sigma_c = \sigma_{f,cr}\left[V_f + (1 - V_f)E_m/E_f\right] \tag{16}$$

Assuming that $E_m << E_f$, it follows from Eqs. (14) and (16) that

$$\sigma_c = 2V_f\left[V_fE_mE_f/\left\{3(1 - V_f)\right\}\right]^{1/2} \tag{17}$$

IV. SHEAR MODE

In the shear mode, the shear deformation of the matrix constitutes the principal source of strain energy. All the fiber displacements are taken to be of the same amplitude and in phase with one another [Eq. (3)]. Furthermore, fiber shear strains are assumed to be negligible, while the matrix shear strain is considered to be a function of the x_1 coordinate alone.

The shear strain in the matrix is given by

$$\gamma_{m_{12}} = (\partial u_{m_2}/\partial x_1) + (\partial u_{m_1}/\partial x_2) \tag{18}$$

Since the displacement in the x_2 direction depends only on the x_1 coordinate, we have

$$\partial u_{m_2}/\partial x_1 = du_{f_2}/dx_1 \tag{19}$$

Further, since the matrix shear strain is independent of x_2, we have

$$\begin{aligned}\partial u_{m_1}/\partial x_2 &= (1/2c)\left[u_{m_1}(c) - u_{m_1}(-c)\right]\\ &= (1/2c)\left[u_{f_1}(c) - u_{f_1}(-c)\right]\\ &= u_{f_1}(c)/c\end{aligned} \tag{20}$$

If we now introduce the assumption that the fiber shear strain is negligible, there follows

$$\partial u_{f_1}/\partial x_2 = -du_{f_2}/dx_1 \tag{21}$$

Noting that

$$\partial u_{f_1}/\partial x_2 = (1/h)[u_{f_1}(c + h) - u_{f_1}(c)]$$
$$= -2u_{f_1}(c)/h$$

we obtain from Eq. (21)

$$u_{f_1}(c) = (h/2)\ du_{f_2}/dx_1 \tag{22}$$

Substitution of Eqs. (19)-(22) into Eq. (18) yields

$$\gamma_{m_{12}} = [1 + (h/2c)]\ du_{f_2}/dx_1$$

and

$$\tau_{m_{12}} = G_m \gamma_{m_{12}} \tag{23}$$

Neglecting extensional stresses in the matrix, ΔU_m becomes

$$\Delta U_m = \frac{1}{2}\int \tau_{m_{12}} \gamma_{m_{12}} dV = G_m c\left(1 + \frac{h}{2c}\right)^2 \left(\frac{\pi^2}{2\ell}\right) \sum_n a_n{}^2 n^2 \tag{24}$$

Substituting Eq. (24) instead of Eq. (6) into Eq. (1) and proceeding as for the extension mode, it follows that

$$\sigma_{f,cr} = \frac{G_m}{V_f(1 - V_f)} + \frac{\pi^2 E_f}{12} (mh/\ell)^2 \tag{25}$$

The second term in the above equation is negligible for buckling wave lengths ℓ/m which are large compared to the fiber width h. Therefore, the critical buckling stress is approximately

$$\sigma_{f,cr} = G_m/V_f(1 - V_f) \tag{26}$$

and

$$\sigma_c = G_m/(1 - V_f) \tag{27}$$

The critical strain ε_{cr} is given by

$$\varepsilon_{cr} = \sigma_{f,cr}/E_f = [1/\{V_f(1 - V_f)\}]\ G_m/E_f \tag{28}$$

V. INELASTIC ANALYSIS AND EXPERIMENTAL RESULTS

In the preceding analysis a simplification was made by assuming both constituents to be perfectly elastic. As long as the deformations remain small, the above analysis may also be applied when the matrix has exceeded its yield point due to compressive strains. This involves replacing the matrix Young's modulus E_m and shear modulus G_m with the appropriate tangent moduli corresponding to the axial compressive strain in the matrix.

In Fig. 4, the results for E-glass/epoxy composites as a function of the fiber volume fraction V_f are shown. These curves suggest that for low-fiber-volume fractions the extension mode gives lower failure stresses. For higher volume fractions, the shear mode prevails. "Shear mode" results are given for both elastic and inelastic

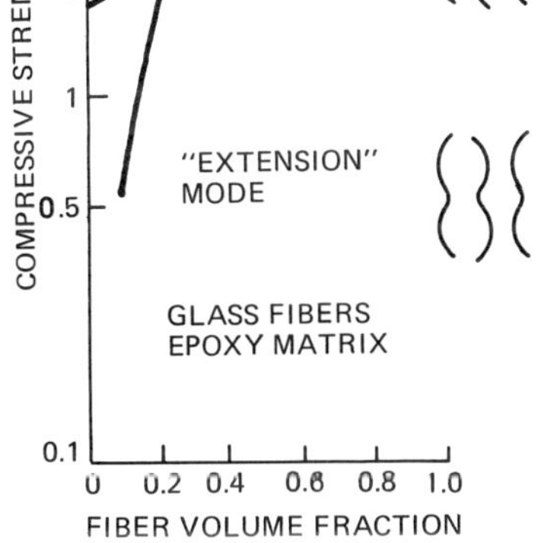

FIG. 4. Compressive strength of E-glass/epoxy composites (from Ref. [2]).

matrix behavior. The inelastic portion of the curve is obtained by assuming that the shear modulus G_m varies linearly from its elastic value at 1% strain to zero at 5% strain. For a composite with 60% fiber volume, these results give a compressive strength of the order of 450,000-500,000 psi. Such strengths have not been experimentally obtained in actual composites. We note here that Eqs. (17) and (27) become unbounded as V_f approaches unity. Thus, these equations are valid only for moderate fiber volume fractions. Equations (17) and (27) are, however, bounded when the matrix Young's and shear moduli are replaced by the appropriate tangent moduli. Furthermore, it should be noted that the case $V_f = 1$ breaks down because the matrix compliance vanishes. The case $V_f = 0$ implies that axially inextensible fibers of unlimited compressive strength are present.

Experimental results for a hollow glass fiber/epoxy composite are given in Fig. 5 in terms of the ratio of strength to density for the hollow fiber composite normalized with respect to a composite containing solid E-glass fibers with the same matrix volume fraction (30%). Each point represents the average of at least five tests of nominally identical specimens. The ratio of the inner

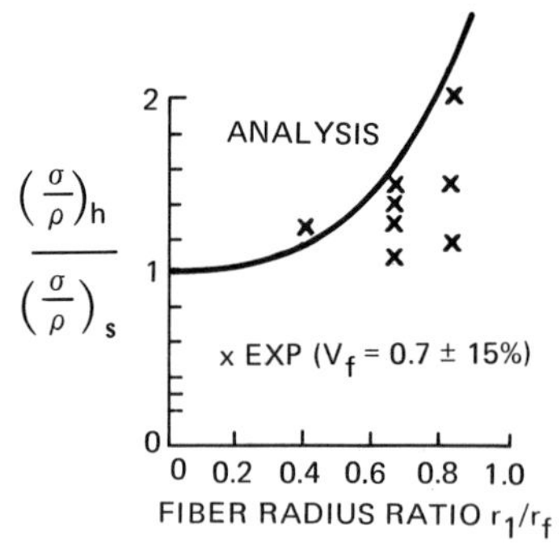

FIG. 5. Comparison of theory and experiment for hollow glass/epoxy composites (from Ref. [2]).

radius r_i to the outer radius r_f of the hollow glass fibers is the independent variable. The analytical curve is obtained from Eq. (27) with the matrix shear modulus replaced by the tangent modulus in the appropriate strain regime. The analytical results seem to provide an upper bound for the experimental results. In view of the various analytical simplifications, the agreement is rather remarkable. Similar agreement between experiment and theory for boron/magnesium composites is reported by Schuerch [3].

VI. EVALUATION OF THE THEORETICAL MODEL

Although both Rosen [2] and Schuerch [3] cite good agreement between theoretical results and experimental data, there exists considerable evidence that the model consistently predicts higher failure stresses than those obtained experimentally. Lager and June [6] report, for a boron/epoxy composite, a strength value which is only 63% of the theoretically calculated failure stress. Approximately the same discrepancy between measured and calculated strength values was observed for a cold-drawn steel wire/polyester composite [7]. Broutman [8] found only 80% of the predicted failure strengths for glass/resin composites. Alexander, et al. [9] report strength values for three types of composites, Ni/W wire, Al/B fiber, and Mg/B fiber, which are at considerable variance with theoretical failure stress predictions. This disagreement is not really surprising in view of the two-dimensional character of the model. Strictly speaking, the theoretical model is valid only for laminated plates. There is little reason to expect that the fiber buckling would be planar; in fact, the fibers may buckle in a nonplanar, or even in a helical form [7]. It has been suggested that the difference between the elastic energy stored in the planar buckled fiber and that in the nonplanar-buckled fiber could account for this disagreement between theoretical and experimental strength values [6,7]. The problem of nonplanar or helical buckling pattern has yet to be treated theoretically.

The buckling model, whether planar or not, implicitly assumes that the fibers have a finite bending or shear strength. This model is inappropriate for brittle fibers with an exceedingly low shear strength. De Ferran and Harris [7] tested a single carbon fiber embedded in a resin matrix. The failure occurred by regularly spaced breaks in the fiber and subsequent gross lateral splitting of the specimen. This indicates that the failure mode for composites reinforced with low-shear-strength fibers may very well be fiber failure resulting in a delamination of the matrix.

REFERENCES

1. N. F. Dow, B. W. Rosen, and Z. Hashin, "Studies of Mechanics of Filament Composites," NASA Report CR-492, 1966.

2. B. W. Rosen, "Mechanics of Composite Strengthening," in Fiber Composite Materials, American Society for Metals, 1965, pp. 37-75.

3. H. Schuerch, "Prediction of Compressive Strength in Uniaxial Boron Fiber-Metal Matrix Composite Materials," AIAA J., 4, 102-106 (1966).

4. L. R. Herrmann, W. E. Mason, and S. T. K. Chan, "Response of Reinforcing Wires to Compressive States of Stress," J. Composite Materials, 1, 212-226 (1967).

5. S. P. Timoshenko and J. M. Gere, Theory of Elastic Stability, 2nd Ed., McGraw-Hill, New York, 1961, pp. 24-31.

6. J. R. Lager and R. R. June, "Compressive Strength of Boron-Epoxy Composites," J. Composite Materials, 3, 48-56 (1969).

7. M. E. De Ferran and B. Harris, "Compression Strength of Polyester Resin Reinforced with Steel Wires," J. Composite Materials, 4, 63-72 (1970).

8. L. J. Broutman, "Failure Mechanisms for Filament-Reinforced Plastics Subjected to Static Compression," Paper 9-C, 19th Annual Meeting of Reinforced Plastics/Composites Division of the Society for Plastics Industry, 1964.

9. J. A. Alexander, J. C. Withers, and B. A. Macklin, "Investigation of Three Classes of Composite Materials for Space Vehicle Applications," NASA Report CR-785, 1967.

CHAPTER 11

THEORY OF BREAKING KINETICS

I. INTRODUCTION

In Chapters 7-9, statistical theories of tensile strength were presented for conditions of static loading. It was noted there that the treatment may be extended to conditions of slow dynamic or quasistatic loading by taking the Weibull distribution parameters as functions of the strain rate. This, however, is an approximate technique. A more elegant and exact approach is to consider the loading duration explicitly. A great deal of our present knowledge in this area is due to Coleman [1-7], whose primary interest was in cotton, rayon, and polyester fibers. These fibers exhibit a strong dependence of tensile strength on strain rate. In developing his "theory of breaking kinetics," Coleman employs certain *a priori* assumptions that account for the time dependence of breaking strengths. The dynamic behavior studied is of the slow dynamic, or quasistatic, type, which is applicable to the study of creep failure but excludes impulsive loading or wave propagation effects. Although the concepts used here are different from the static statistical theories, the two approaches yield compatible results. In the following, the theory of breaking kinetics for a first-order ensemble of fibers is presented in Sections II-IV. The "first-order ensemble" fibers have the characteristic that the probability of one of them breaking in a small time interval dt, at time t, depends only on the tensile stress at t and is independent of the past loading history. In other words, the "first-order ensemble" fibers exhibit no memory effects. The generalized version of the theory (fibers with memory) is discussed in Section V.

II. FIRST-ORDER ENSEMBLE OF FIBERS

In this section, we will discuss the strength characteristics of "first-order ensemble" fibers under both constant and time-dependent stresses. It will further be shown that "first-order ensemble" theory is consistent with the "weakest-link" hypothesis. Consider an ensemble of fibers carrying equal but separate stresses of magnitude σ. Let us assume σ to be sufficiently large so that some fibers will eventually break, resulting in a distribution of breaking times t_B. The term "equal stresses" means that the fiber stress σ is the same for all unbroken fibers, and the term "separate stresses" implies that the stress is independent of either the original or the surviving number of fibers. The ensemble initially contains N unbroken fibers. After the stress σ is applied at time $t = 0$ and as time t increases, the number of unbroken fibers decreases. Let $n(t)$ denote the number of unbroken fibers still remaining at time t. The time required for $n(t)$ to reach zero is the lifetime of the ensemble. We define by $\varepsilon_{N,n}$ the state of the ensemble of N fibers at time t if it contains n unbroken fibers at that time.

The ensemble is assumed to be subjected to a constant stress σ. The only allowable transitions for the ensemble are $\varepsilon_{N,N} \to \varepsilon_{N,N-1} \cdots \varepsilon_{N,0}$. This statement about transitions has the physical significance of excluding any healing processes. The probability $\lambda_{N,n}\,dt$ of the transition $\varepsilon_{N,n} \to \varepsilon_{N,n-1}$ occurring in a small time interval dt may be written as

$$\lambda_{N,n}\,dt = nk(\sigma)\,dt \tag{1}$$

where k is a function of both the stress σ and the fiber length ℓ. If we assume that for small dt the probability of an m-tuple break in a fiber of length $m\ell$ is the same as for m single breaks in separate fibers of length ℓ, we may write

$$k(\sigma) = \ell K(\sigma) \tag{2}$$

where K is independent of ℓ. Specific functional forms for $K(\sigma)$ are to be obtained on the basis of experimental data. In the following discussion, we shall consider two relationships that have been found satisfactory for a large class of polymers, i.e.,

$$K(\sigma) = \xi_1^{-1}\exp(\xi_2\sigma) \tag{3}$$

and

$$K(\sigma) = \xi_3 \sigma^{\xi_4} \tag{4}$$

where ξ_1, ξ_2, ξ_3, and ξ_4 are constants. Details of suitable experimental procedures for verifying the applicability of these forms will be discussed later.

Let us now consider the process $\varepsilon_{N,N} \rightarrow \varepsilon_{N,N-1}, \ldots, \varepsilon_{N,0}$. If the probability of the ensemble being in a state $\varepsilon_{N,n}$ at time t under a constant stress σ is given by $P_{N,n}(t/\sigma)$, the equations that this process must obey are

$$\begin{aligned} P_{N,N}^{(1)} &= -\lambda_{N,N} P_{N,N}, \ldots, \\ P_{N,n}^{(1)} &= -\lambda_{N,n} P_{N,n} + \lambda_{N,n+1} P_{N,n+1}, \ldots, \\ P_{N,0}^{(1)} &= \lambda_{N,1} P_{N,1} \end{aligned} \tag{5}$$

where $P^{(m)}$ denotes the mth partial derivative of P with respect to time. Equations (5) merely state that the time rate of change in the probability $P_{N,n}(t/\sigma)$ of the ensemble being in the state $\varepsilon_{N,n}$ at t equals the probability of state $\varepsilon_{N,n+1}$ passing into $\varepsilon_{N,n}$ minus the probability of state $\varepsilon_{N,n}$ becoming $\varepsilon_{N,n-1}$. The differential equations (5) are subject to the boundary conditions

$$0 \leq P_{N,n}(t/\sigma) \leq 1 \quad \text{(for all } n,t) \tag{6a}$$

$$P_{N,n}(0/\sigma) = 0 \qquad (n \neq N) \tag{6b}$$

$$P_{N,N}(0/\sigma) = 1 \tag{6c}$$

Equation (6a) is the statement of the physical fact that the probability of the ensemble being in any state $\varepsilon_{N,n}$

at any time t lies between zero and one. Equations (6b) and (6c) imply that at time $t = 0$ the ensemble contains N unbroken fibers. Equations (5) and (6) yield for all $m < N$

$$P_{N,0}^{(m)}(0/\sigma) = 0, \quad P_{N,0}^{(N)}(0/\sigma) = N!\ \Lambda_N$$

where*

$$\Lambda_N \equiv (N!)^{-1} \prod_{i=1}^{N} \lambda_{N,i} \tag{7}$$

For constant stress σ, $\lambda_{N,n}$ is not an explicit function of t, and the solution of Eqs. (5) and (6) may be expressed as

$$P_{N,n}(t/\sigma) = \lambda_{N,n}^{-1} \sum_{j=N}^{n} \left[\phi_{N,n,j}\lambda_{N,j} \exp(-\lambda_{N,j}t)\right] \tag{8}$$

where

$$\phi_{N,n,j} = \prod_{\substack{i=N \\ i\neq j}}^{n} \left[1 - (\lambda_{N,j}/\lambda_{N,i})\right]^{-1}$$

The probability $p_{N,n}(t/\sigma)\ dt$ that the transition $\varepsilon_{N,n} \to \varepsilon_{N,n-1}$ occurs in a small time interval dt at t is given by

---

*Note that $\prod_{i=1}^{N} \lambda_{N,i} = (\lambda_{N,1})(\lambda_{N,2}) \cdots (\lambda_{N,N})$.

$$p_{N,n}(t/\sigma)\ dt = \lambda_{N,n} P_{N,n}(t/\sigma)\ dt$$

$$= dt \sum_{j=N}^{n} \left[\phi_{N,n,j}\lambda_{N,j} \exp(-\lambda_{N,j} t)\right] \tag{9}$$

In order to obtain expressions for the average time $E_{N,n}\{t\}$ at which the transition $\varepsilon_{N,n} \to \varepsilon_{N,n-1}$ occurs, and the variance $\Xi^2_{N,n}$ in this time, we will first compute the characteristic function $E_{N,n}\{e^{i\psi t}\}$, $(i = \sqrt{-1},\ \psi$ real):

$$E_{N,n}\{e^{i\psi t}\} = \int_0^\infty e^{i\psi t} p_{N,n}(t/\sigma)\ dt$$

$$= \sum_{j=N}^{n} \phi_{N,n,j}\left[1 - i\psi/\lambda_{N,j}\right]^{-1}$$

$$= \prod_{j=N}^{n} \left[1 - i\psi/\lambda_{N,j}\right]^{-1} \tag{10}$$

The mth moment $E_{N,n}\{t^m\}$ and the mth semiinvariant $\chi_{N,n,m}$ are given by

$$i^m E_{N,n}\{t^m\} = \left.\frac{\partial^m E_{N,n}\{e^{i\psi t}\}}{\partial\psi^m}\right]_{\psi=0}$$

$$i^m \chi_{N,n,m} = \left.\frac{\partial^M \log E_{N,n}\{e^{i\psi t}\}}{\partial\psi^m}\right]_{\psi=0} \tag{11}$$

Noting that the Maclaurin series expansion of $\log E_{N,n}\{e^{i\psi t}\}$ is given by

$$\log E_{N,n}\{e^{i\psi t}\} = \sum_{r=1}^{\infty} \left\{\frac{(i\psi)^r}{r} \sum_{j=N}^{n} \lambda_{N,j}^{-r}\right\}$$

we obtain the following expressions for $E_{N,n}\{t\}$, $\Xi^2_{N,n}$, and the third moment about the mean, $\mu^3_{N,n}$

$$E_{N,n}\{t\} \equiv \chi_{N,n,1} = \sum_{j=N}^{n} \lambda^{-1}_{N,j}$$

$$\Xi^2_{N,n} \equiv \chi_{N,n,2} = \sum_{j=N}^{n} \lambda^{-2}_{N,j} \tag{12}$$

$$\mu^3_{N,n} \equiv \chi_{N,n,3} = 2\sum_{j=N}^{n} \lambda^{-3}_{N,j}$$

When N is very large and $\lambda_{N,j} = jk(\sigma)$, the summations in Eqs. (12) can be expressed as Riemann integrals by setting $y = j/N$:

$$\lim_{N\to\infty} E_{N,n}\{t\} = \int_{n/N}^{1} [dy/yk(\sigma)] = [1/k(\sigma)] \log(N/n) \tag{13a}$$

$$\lim_{N\to\infty} \Xi^2_{N,n}\{t\} = N^{-1}\int_{n/N}^{1} \frac{dy}{y^2k^2(\sigma)} = \frac{1}{k^2(\sigma)}\left(\frac{1}{n} - \frac{1}{N}\right) \tag{13b}$$

$$\lim_{N\to\infty} \mu^3{}_{N,n}\{t\} = 2N^{-2}\int_{n/N}^{1} \frac{dy}{y^3k^3(\sigma)} = \frac{1}{k^3(\sigma)}\left(\frac{1}{n^2} - \frac{1}{N^2}\right) \tag{13c}$$

Note that for large n, both $\Xi^2_{N,n}$ and $\mu^3_{N,n}$ approach zero. This implies that in the limit as $N \to \infty$ and $n \to \infty$, $n(t)$ should obey the deterministic equation

$$-\,dn(t)/dt = nk(\sigma) \tag{14}$$

with negligible fluctuations since Eq. (13a) is a solution of Eq. (14).

The lifetime of the entire ensemble is obtained by setting n equal to zero in Eq. (13a):

$$\lim_{\substack{N\to\infty \\ n\to 0}} E_{N,n}(t) \to \infty \tag{15}$$

Equation (15) states that for a given load σ, there exist fibers in the ensemble which do not break as $t \to \infty$. This result is consistent with the static tensile strength distributions (e.g., Weibull) discussed earlier, where a vanishingly small fraction of the fibers have a strength in excess of any finite σ. In the remainder of Section II, we will take both N and n to be large so that the behavior of the ensemble is adequately described by the deterministic equation (14).

We next consider the behavior of the first-order ensemble fibers under variable loads. It will be assumed that Eq. (14) is also valid for variable stresses $\sigma(t)$. We shall, however, exclude from consideration impulsive loads and oscillating loads with a periodicity of the same order of magnitude as the relaxation times for tensile motions in the fibers. Equation (14) may be rewritten as

$$-dn = n(t)k(\sigma)\ dt \tag{16}$$

where $-dn$ denotes the number of fibers breaking in the time interval between t and $t + dt$. Integrating (14), we obtain

$$n(t) = N \exp\left(-\int_0^t k(\sigma)\ dt\right) \tag{17}$$

where N is the original number of fibers. Substitution of Eq. (17) into Eq. (16) yields

$$-dn/N = k[\sigma(t_B)]\ \exp\left\{-\int_0^{t_B} k[\sigma(t)]\ dt\right\} dt_B \tag{18}$$

where t_B is the time required to break a fiber. Note that $-dn/N$ is the probability that a fiber breaks in the time interval dt_B when subjected to an arbitrary loading history $\sigma(t)$.

The average lifetime $\langle t_B \rangle_f$ under an arbitrary loading history $\sigma(t)$ is given by

$$\langle t_B \rangle_f = \int_0^\infty t(-dn/N) = \int_0^\infty tk[\sigma(t)] \times \exp\left\{-\int_0^t k[\sigma(t)]\, dt\right\} dt \tag{19}$$

On substituting for $k(\sigma)$ from Eq. (2) into Eq. (19), we obtain the following simple expression for the average lifetime under a constant stress σ:

$$\langle t_B \rangle_f = \int_0^\infty t\ell K e^{-\ell K t}\, dt = (K\ell)^{-1} \tag{20}$$

If $K(\sigma)$ is given by Eq. (3), then we obtain from Eq. (20)

$$\log\langle t_B \rangle_f = -(\log \ell - \log \xi_1 + \xi_2 \sigma) \tag{21}$$

Equation (21) suggests a simple method for determining the constants ξ_1 and ξ_2. By measuring the lifetime of the ensemble under constant stress, $\log\langle t_B \rangle_f$ may be expressed as a linear function of σ. The slope of this curve yields ξ_2; ξ_1 may then be obtained from Eq. (21) (see also Chapter 7). On the other hand, if $K(\sigma)$ is given by Eq. (4), we have from Eq. (20)

$$\log\langle t_B \rangle_f = -(\log \ell + \log \xi_3 + \xi_4 \log \sigma) \tag{22}$$

Once again, constant stress experiments suffice to determine the constants ξ_3 and ξ_4 occurring in Eq. (4).

It was previously mentioned that $-dn/N$ in Eq. (18) denotes the probability that a fiber breaks in the time interval dt_B when subjected to an arbitrary loading history $\sigma(t)$. As an example, let us consider a constant stress buildup rate. In this case,

$$\sigma(t) = \eta t \tag{23a}$$

and

$$\sigma^* = \eta t_B \tag{23b}$$

where η is a constant and σ^* is the tensile breaking strength. The distribution of breaking strengths $g(\sigma^*)$ is obtained from Eqs. (18) and (23) as

$$g(\sigma^*)\ d\sigma^* = (1/\eta)k(\sigma^*)\ \exp\left[-(1/\eta)\int_0^{\sigma^*} k(\sigma)\ d\sigma\right]d\sigma^* \tag{24}$$

When $K(\sigma)$ is given by Eq. (3), we have

$$\langle\sigma^*\rangle_f = \int_0^{\infty} \sigma^* g(\sigma^*)\ d\sigma^* = \xi_2^{-1} \log(\xi_2 \eta \xi_1 \ell^{-1}) - 0.5772\xi_2^{-1}$$

and

$$\Xi_f^2 = \langle\sigma^{*2}\rangle_f - \langle\sigma^*\rangle_f^2 = \pi^2/6\xi_2^{\ 2} \tag{25}$$

Note that $\langle\sigma^*\rangle_f$ decreases with increasing fiber length ℓ and decreasing rate of loading. However, the variance Ξ_f^2 is independent of these quantities. It is interesting to compare these results with static strength theories, where the average tensile strength is independent of the rate of loading. As shown in Chapter 7, for example, when the fibers are characterized by the Weibull distribution, the average strength decreases with increasing length but is independent of the strain rate. The variance, however, is independent of both the fiber length and the strain rate.

The theory of breaking kinetics can be shown to be consistent with the weakest-link hypothesis as follows. Let $G_{m\ell}(\sigma^*)$ and $G_\ell(\sigma^*)$ denote the cumulative tensile strength distributions of fibers of length $m\ell$ and ℓ, respectively. The weakest-link hypothesis can then be stated as

$$1 - G_{m\ell}(\sigma^*) = \left[1 - G_\ell(\sigma^*)\right]^m \tag{26a}$$

Let us assume $\sigma(t)$ to be monotonically increasing with time. Equations (2) and (18) yield

$$G_\ell(\sigma^*) = 1 - \exp[-\ell\phi(\sigma^*)]$$

$$G_{m\ell}(\sigma^*) = 1 - \exp[-m\ell\phi(\sigma^*)] \tag{26b}$$

where

$$\phi(\sigma^*) = \int_0^{\sigma^*} K(\sigma)(d\sigma/dt)^{-1}\, d\sigma$$

which are obviously solutions to Eq. (26a). The proof for the case in which the increase in $\sigma(t)$ is not monotonic with time is given in Ref. [1]. The first-order ensemble theory implies that a fiber breaks at the cross section that has the shortest lifetime under $\sigma(t)$ and is thus always compatible with the weakest-link hypothesis. The converse of this statement does not necessarily follow.

III. BUNDLE THEORY

A. Introduction

A bundle is defined to be a parallel array of fibers of equal length randomly selected from a first-order ensemble and having the following properties:

(a) All unbroken fibers support equal loads.

(b) Broken fibers support no load.

(c) All cross-sectional planes of the fiber perpendicular to its axis bear equal loads.

Consider a bundle containing N fibers subjected to a constant load L at time $t = 0$. Let $n(t)$ be the number of fibers which survive to time t. The lifetime t_B of the bundle is defined to be the time required for $n(t)$ to go to zero. If A_f denotes the initial total cross-sectional area of the fibers, then the stress σ at $t = 0$ is given by

$$\sigma = L/A_f \tag{27}$$

The definition of a bundle implies that the actual stress in an unbroken fiber at time t is $\sigma N/n(t)$. We assume that the bundle can undergo transitions of the type $\varepsilon_{N,N} \to \varepsilon_{N,N-1}, \ldots, \varepsilon_{N,0}$. The probability $\lambda_{N,n}\,dt$ of the transition $\varepsilon_{N,n} \to \varepsilon_{N,n-1}$ occurring in a small time dt is given by

$$\lambda_{N,n}\,dt = nk(\sigma N/n)\,dt \tag{28}$$

The bundle lifetime $\langle t_B\rangle_B$; the variance in t_B, Ξ_B^2; and the third moment about the mean, μ_B^3; are obtained by substituting for $\lambda_{N,n}$ from Eq. (28) into Eqs. (12):

$$\langle t_B \rangle_B = \sum_{n=N}^{1} \lambda_{N,n}^{-1} = \sum_{n=N}^{1} [nk(\sigma N/n)]^{-1} = \ell^{-1} \sum_{n=N}^{1} [nK(\sigma N/n)]^{-1}$$

$$\Xi_B^2 = \sum_{n=N}^{1} \lambda_{N,n}^{-2} = \ell^{-2} \sum_{n=N}^{1} [nK(\sigma N/n)]^{-2} \qquad (29)$$

$$\mu_B^3 = 2 \sum_{n=N}^{1} \lambda_{N,n}^{-3} = 2\ell^{-3} \sum_{n=N}^{1} [nK(\sigma N/n)]^{-3}$$

When N is very large, the summations in Eqs. (29) can be made into Riemann integrals by substituting $y = n/N$:

$$\lim_{N\to\infty} \langle t_B \rangle_B = \ell^{-1} \int_0^1 \frac{dy}{yK(\sigma/y)}$$

$$\lim_{N\to\infty} \Xi_B^2 = \ell^{-2} N^{-1} \int_0^1 \frac{dy}{y^2 K^2(\sigma/y)} \qquad (30)$$

$$\lim_{N\to\infty} \mu_B^3 = 2\ell^{-3} N^{-2} \int_0^1 \frac{dy}{y^3 K^3(\sigma/y)}$$

Since $K(\sigma/y)$ is a positive monotonic function of σ/y, we need only to postulate that

$$\lim_{y\to 0} K(\sigma/y) \to \infty$$

as some arbitrarily small positive power of σ/y, to guarantee that all of the integrals in Eqs. (30) are finite in the limit as $N \to \infty$. This implies that

(a) The average lifetime of the bundle is always finite.

(b) As $N \to \infty$, μ_B^3 becomes negligible compared to Ξ_B^3.

Therefore, we conclude that in the limit as $N \to \infty$, the distribution of $\langle t_B \rangle_B$ is asymptotically normal. It is of interest to compare this result with Daniels' static bundle theory discussed in Chapter 7. We recall that in the limit as $N \to \infty$, static bundle strength approaches a normal distribution. Equations (30) for large N justify the intuitive notion that in the limit as $N \to \infty$, $n(t)$ should obey the deterministic equation

$$- dn/dt = nk(\sigma N/n) \tag{31}$$

with negligible fluctuations.

When relationship (3) is valid, the substitution for K in Eqs. (30) yields

$$\lim_{N\to\infty} \langle t_B \rangle_B = -\xi_1 \ell^{-1} \mathrm{Ei}(-\xi_2 \sigma) \tag{32a}$$

$$\lim_{N\to\infty} \Xi^2{}_B = \left(2N\ell^2 \xi_2 \sigma\right)^{-1} \xi_1^2 \exp(-2\xi_2 \sigma) \tag{32b}$$

where the function $-\mathrm{Ei}(\xi_2 \sigma)$ is tabulated (see Ref. [8]). Numerical evaluation of Eq. (32a) yields the result that the stress required to break the bundle in a given time is approximately 90% of that required to break a single fiber in that time. The results are reproduced in Fig. 1.

For the case when the power law (4) holds, we obtain from Eqs. (30)

$$\lim_{N\to\infty} \langle t_B \rangle_B = \left(\xi_3 \ell \xi_4 \sigma^{\xi_4}\right)^{-1} \tag{33a}$$

$$\lim_{N\to\infty} \Xi_B^2 = N^{-1}\left(2\xi_4 - 1\right)^{-1}\left(\ell^2 \xi_3^2 \sigma^{2\xi_4}\right)^{-1} \tag{33b}$$

Comparison of Eqs. (33a) and (22) confirms the result that for $\xi_4 > 1$, a bundle is weaker than its component fibers.

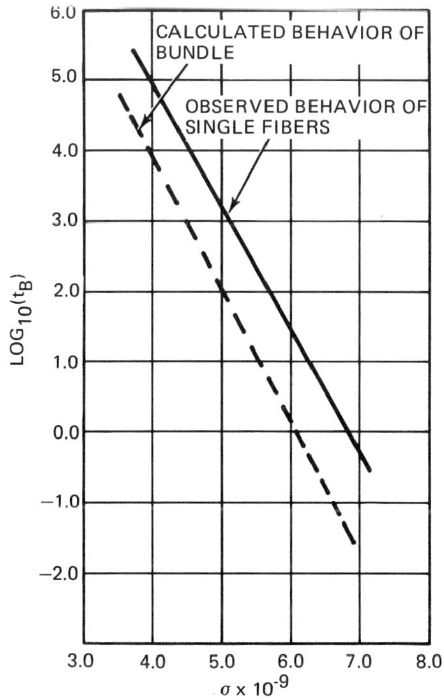

FIG. 1. Average bundle lifetime--constant load; t_B is in sec, and σ is in dynes/cm^2 (from Ref. [3]).

B. Bundle under Time-Dependent Loads

In order to investigate the behavior of a bundle under time-dependent loads, we will assume that the deterministic equation (31) is also valid for time-dependent loads $\sigma(t)$. We shall restrict the following discussion to $\sigma(t)$ linearly increasing with time, as given by Eqs. (23). When the power law (4) is valid, the deterministic equation becomes

$$-dn/dt = n\ell\xi_3(\eta tN/n)^{\xi_4} \tag{34}$$

subject to the boundary conditions

$$n(0) = N, \quad n(t_B) = 0$$

Integration of Eq. (34) yields

$$\langle\sigma^*\rangle_B = [\eta(\xi_4 + 1)/\xi_4 \ell \xi_3]^{1/(1 + \xi_4)} \tag{35}$$

For this case, the average strength of single fibers is obtained from Eq. (24):

$$\langle\sigma^*\rangle_f = \int_0^\infty \sigma^* g(\sigma^*)\, d\sigma^* = \left[\frac{(\xi_4 + 1)\eta}{\ell\xi_3}\right]^{1/(1 + \xi_4)} \times \Gamma\left(1 + \frac{1}{\xi_4 + 1}\right) \tag{36}$$

where $\Gamma(x)$ denotes the gamma function with argument x. Comparison of Eqs. (35) and (36) shows that $\langle\sigma^*\rangle_B$ is less than $\langle\sigma^*\rangle_f$ for $\xi_4 > 2$. However, for $\xi_4 = 1$, $\langle\sigma^*\rangle_B$ is greater than $\langle\sigma^*\rangle_f$. Thus there exists a value of ξ_4 between one and two which must be exceeded to have $\langle\sigma^*\rangle_B$ less than $\langle\sigma^*\rangle_f$.

In the case when the exponential law (3) holds, the deterministic equation (31) may be written as

$$-dn/dt = n(\ell/\xi_1) \exp(\xi_2 \eta N t/n) \tag{37}$$

On making the substitutions $y \equiv \xi_2 \eta N \xi_1 / n\ell$ and $\bar{t} \equiv \xi_1^{-1} \ell t$ in Eq. (37), we obtain the differential equation

$$(1/y)\, dy/d\bar{t} = e^{y\bar{t}} \tag{38}$$

subject to the boundary conditions

$$y = y_0 = \xi_2 \eta \xi_1 \ell^{-1} \quad \text{at} \quad \bar{t} = 0 \tag{39a}$$

$$y = \infty \quad \text{at} \quad \bar{t} = \bar{t}_B = \xi_1^{-1} \ell t_B \tag{39b}$$

Now let $y(\bar{t}, y_0)$ be the solution of Eq. (38) subject to the boundary condition (39a), and let $\bar{t}(y_0, y)$ be the inverse function of y. Putting

$$C(y_0) \equiv \lim_{y\to\infty} y_0\, \bar{t}(y, y_0) = y_0 \bar{t}_B$$

we obtain from Eqs. (39) and (23)

$$\xi_2 \langle\sigma^*\rangle_B = C(y_0)$$

or

$$\langle\sigma^*\rangle_B = \xi_2^{-1} C\left(\xi_2 \eta \xi_1 \ell^{-1}\right) \tag{40}$$

Thus, calculating $\langle\sigma^*\rangle_B$ amounts to finding the value of $\bar{t}_B$, as a function of y_0, for which $y(\bar{t})$ becomes infinite. This is accomplished through a numerical procedure. The details of the method may be found in Ref. [4]; the numerical results are plotted in Fig. 2. Once again, it is observed that the average bundle strength is less than that of its constituent fibers. Also, the bundle strength increases with an increase in the rate of loading. This result may be compared with Daniels' static bundle theory, where the bundle strength is independent of the rate of loading. In contradistinction to the "first-order ensemble" theory, static bundle theory considers fibers whose strength does not depend upon the rate of loading.

IV. STRENGTH OF A COMPOSITE UNDER CONSTANT LOAD

The geometric model employed by Coleman [3] to analyze composite† strength is similar to the Gücer-Gurland model

†Coleman does not use the term "composite" but considers a "tight bundle" in which fibers press against each other with a force sufficient to impede their relative motion.

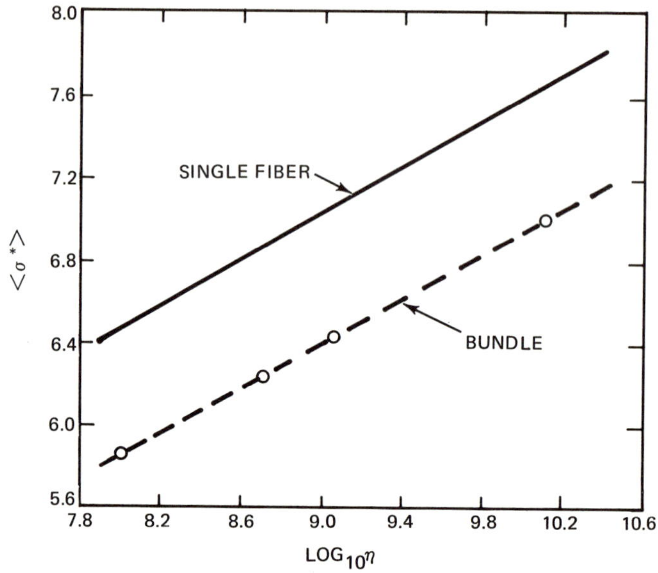

FIG. 2. Average bundle strength--linearly increasing load; σ^* and η are in C.G.S. units (from Ref. [4]).

discussed in Chapter 8. The composite is divided into M statistically independent slabs S_i $(i = 1, \ldots, M)$ of width $\delta = \ell/M$. The fracture of the composite is assumed to be contained in a particular slab S_B. The region S_i initially contains N fibers, all of which must eventually break if the composite fracture is to occur at S_i. After the constant load L is applied at $t = 0$ and as time increases, the number of unbroken fibers in each slab decreases. Let $n_i(t)$ denote the number of unbroken fibers at time t in slab S_i. The slab for which $n(t)$ first reaches zero is S_B. The time required for $n_i(t)$ to go to zero is the lifetime of S_i, and thus the lifetime of S_B is t_B.

It is now postulated that within a particular slab, the behavior of fibers roughly approximates that of a bundle with $\lambda_{N,n}$ given by

$$\lambda_{N,n} = \delta n K(\sigma N/n) \tag{41}$$

The probability that a slab S_i has a lifetime less than t when subjected to a constant load σ is $P_{N,0}(t/\sigma)$. The cumulative lifetime distribution function is given by Eq. (8) with $\lambda_{N,j}$ defined by Eq. (41). The probability that S_i has a lifetime greater than t is $1 - P_{N,0}(t/\sigma)$. Application of the "weakest-link hypothesis" leads to the following result for the cumulative lifetime distribution $Q_c(t_B)$ of the composite:

$$1 - Q_c(t_B) = [1 - P_{N,0}(t_B/\sigma)]^M \tag{42}$$

When M is very large, the right-hand side of Eq. (42) is appreciably different from zero only for t_B close to zero. The Taylor expansion of Eq. (42) along with Eq. (7) yields

$$\lim_{M\to\infty}\left[1 - Q_c(t_B)\right] = \lim_{M\to\infty}\left(1 - \Lambda_N t_B^N + \cdots\right)^M$$

$$= \exp\left(-M\Lambda_N t_B^N\right) \tag{43}$$

The probability density distribution $q_c(t_B)$ for the lifetime distribution t_B is obtained by differentiating Eq. (43) with respect to t:

$$q_c(t_B) = NM\Lambda_N t_B^{N-1} \exp\left(-M\,\Lambda_N t_B^N\right) \tag{44}$$

The mth moment for this distribution is given by

$$\langle t_B^m\rangle_c = (M\Lambda_N)^{-m/N}\Gamma(1 + m/N) \tag{45}$$

The average lifetime $\langle t_B\rangle_c$, the coefficient of variation $\Xi_c/\langle t_B\rangle_c$, and the coefficient of skewness μ_c^3/Ξ_c^3 are given by

$$\langle t_B \rangle_c = (M\Lambda_N)^{-1/N}\,\Gamma(1 + 1/N)$$

$$\frac{\Xi_c}{\langle t_B \rangle_c} = \frac{\left[\langle t_B^2 \rangle_c - \langle t_B \rangle_c^2\right]^{1/2}}{\langle t_B \rangle_c} = \left[\frac{\Gamma(1 + 2/N)}{\Gamma^2(1 + 1/N)} - 1\right]^{1/2} \tag{46}$$

$$\frac{\mu_c^3}{\Xi_c^3} = \frac{\langle t_B^3 \rangle_c - 3\langle t_B \rangle_c \langle t_B^2 \rangle_c + 2\langle t_B \rangle_c^3}{\left[\langle t_B^2 \rangle_c - \langle t_B \rangle_c^2\right]^{3/2}}$$

$$= \frac{\Gamma(1 + 3/N) - 3\Gamma(1 + 1/N)\Gamma(1 + 2/N) + 2\Gamma^3(1 + 1/N)}{[\Gamma(1 + 2/N) - \Gamma^2(1 + 1/N)]^{3/2}}$$

In the limit as N approaches infinity, Eqs. (46) reduce to

$$\lim_{N\to\infty} \langle t_B \rangle_c = (M\Lambda_N)^{-1/N}, \quad \lim_{N\to\infty} (\Xi_c / \langle t_B \rangle_c) = \pi 6^{-1/2} N^{-1}$$

$$\lim_{N\to\infty} (\mu_c^3 / \Xi_c^3) = -2.404 \times 6^{3/2} \pi^{-3} \tag{47}$$

The lifetime of the composite $\langle t_B \rangle_c$ is thus seen to depend upon M, and hence upon the length. The dependence is, however, slight. Note that the coefficient of skewness approaches a constant value. Hence the lifetime distribution is always skewed. It is of interest to compare this result with the strength distribution given by the Gücer-Gurland model of composite strength. We recall from Chapter 8 that the Gücer-Gurland model also predicts only a slight dependence of composite strength on length. Furthermore, the Gücer-Gurland composite strength distribution is also skewed.

No comparison of the results of this model with experimental data is available. However, it is readily seen that this theory suffers from the same drawbacks as the Gücer-Gurland model. To realistically predict the behavior of a composite, it is necessary to account for local stress concentrations resulting from fiber failure in a particular slab. Coleman also does not present any method for calculating δ. As a first approximation, one might employ the model for ineffective length discussed in Chapter 8.

V. GENERALIZED THEORY OF BREAKING KINETICS

In the preceding discussion, the theory of breaking kinetics for fibers drawn from a first-order ensemble was introduced. The classical form of the weakest-link hypothesis states that the fiber strength is that of its weakest cross section. In the theory of breaking kinetics, the lifetime of a fiber under an arbitrary load history $\sigma(t)$ is equal to the lifetime of that cross section of the fiber which has the shortest lifetime under $\sigma(t)$. The first-order ensemble fibers have the simplest time-dependent breakdown behavior compatible with the generalized weakest-link hypothesis, insofar as the probability that an unbroken fiber breaks in a time interval dt at t depends only on the force on the fiber at time t and is independent of the past loading history. Below, we outline a generalized theory of breaking kinetics which includes memory effects [7] and is also compatible with the weakest-link hypothesis.

If the fiber population is very large, the following deterministic equation can be written for each $\sigma(t)$:

$$Q_\ell(t_B) - 1 - [n(t_B)/N] \tag{48}$$

where $Q_\ell(t_B)$, the cumulative distribution function of lifetimes, is the probability that a given fiber breaks in the time interval $0 < t < t_B$ under the loading history $\sigma(t)$. Corresponding to the cumulative distribution function $Q_\ell(t_B)$, we can introduce a probability density function $q_\ell(t_B)$:

$$q_\ell(t_B) = \left.\frac{\partial Q_\ell(t)}{\partial t}\right|_{t=t_B} = -\left.\frac{1}{N}\frac{dn(t)}{dt}\right|_{t=t_B} \tag{49}$$

To calculate $Q_\ell(t_B)$ for the fiber ensemble, we first divide each fiber into M noninteracting slabs of width δ ($M = \ell/\delta$). Each slab has its own lifetime t_B under the load $\sigma(t)$. The weakest-link hypothesis can then be stated as

$$1 - Q_\ell(t_B) = [1 - Q_\delta(t_B)]^M \tag{50}$$

For large M, the right-hand side of Eq. (50) is appreciably different from zero only for $Q_\delta(t_B)$ close to zero. This implies that we are only interested in values of $Q_\delta(t_B)$ near $t_B \approx 0$.

A. Breakdown Rules

It is assumed that for every loading history $\sigma(t)$, there exists a scalar function $\Omega(t)$, called the breakdown function, which measures the breakdown in the slabs. The general procedure used in calculating $\Omega(t)$ is called the breakdown rule. The function $\Omega(t)$ is assumed to have the following two properties:

(a) For any $\sigma(t)$, $\Omega = 0$ for $t \leq 0$.

(b) For any $\sigma(t)$ and all $t \geq 0$, $\Omega(t)$ is a single-valued, monotonically increasing, and continuous function of t. Statement (b) has the physical significance of excluding any healing processes. We present next three classes of breakdown rules which obey requirements (a) and (b).

Class A:

$$d\Omega = \phi[\sigma(t)]\ dt \tag{51}$$

where ϕ is a continuous function of σ with $\phi(\sigma) \geq 0$ for all $\sigma \geq 0$.

Class B:

$$d\Omega/dt = \phi[\sigma(t), \Omega] \tag{52}$$

where ϕ is a continuous function of σ and Ω, with $\phi \geq 0$, $\partial\phi/\partial\sigma \geq 0$, and $\partial\phi/\partial\Omega \geq 0$ for all $\sigma \geq 0$.

Class C:

$$\frac{d\Omega}{dt} = \phi\left[\sigma, \frac{d\sigma}{dt}, \frac{d^2\sigma}{dt^2}, \ldots, \Omega, T, \frac{dT}{dt}, \frac{d^2T}{dt^2}, \ldots\right] \tag{53}$$

where T is the temperature.

In order for the breakdown function to have a precise physical meaning, the conditions for failure of a slab must be specified in terms of Ω. To do this, a death function $\psi(\Omega)$ is introduced for each fiber ensemble such that for every loading history we have

$$Q_\delta(t_B) = \psi[\Omega(t_B)] \tag{54}$$

The death function is a property of the ensemble, i.e., it does not depend upon $\sigma(t)$. It is assumed to have the following properties:

(c) $\psi(0) = 0$.

(d) For $\Omega > 0$, $\psi(\Omega)$ is a single-valued, positive, monotonically increasing function of Ω.

(e) $\lim_{\xi \to 0} [\psi(\Omega\xi)/\psi(\xi)] = \Omega^a$, where ξ is a non-negative dummy variable and a is a positive constant.

Equation (54) together with the assumptions (c)-(e) states that the probability that a particular slab of the ensemble has broken at a time less than t_B depends only on the magnitude of the breakdown function at time t_B. Once a breakdown rule has been accepted, the specification of a death function defines the ensemble of fibers. The cumulative lifetime function $Q_\ell(t_B)$ for the ensemble may then be calculated from the weakest-link hypothesis, Eq. (50).

B. Fibers of Class A

By combining Eqs. (50) and (54), we obtain

$$1 - Q_\ell(t_B) = [1 - \psi(\Omega)]^M \tag{55}$$

For large M, the right side of Eq. (55) has an appreciable value only for small ψ, which corresponds to small Ω. Postulates (c)-(e) imply that for small Ω, $\psi(\Omega)$ is of the form

$$\lim_{\Omega \to 0} \psi(\Omega) = C\Omega^a \tag{56}$$

where C and a are positive constants. Substituting Eq. (56) into Eq. (55), we obtain for large M (long fibers)

$$\lim_{M \to \infty} Q_\ell(t_B) = 1 - \exp\left[-MC\Omega^a(t_B)\right] \tag{57}$$

Equation (57) is valid regardless of the breakdown rule relating Ω and σ. For fibers obeying the class A rule, we substitute

$$K(\sigma) = (C/\delta)^{1/a}\phi(\sigma) \tag{58}$$

in Eqs. (51) and (57) to obtain

$$\lim_{M\to\infty} Q_\ell(t_B) = 1 - \exp\left\{-\ell\left[\int_0^{t_B} K[\sigma(t)]dt\right]^a\right\} \tag{59}$$

This is the general expression for long fibers of class A. Thus, under constant stress $\sigma(t) = \sigma$ = constant, the cumulative distribution becomes

$$\lim_{M\to\infty} Q_\ell(t_B) = 1 - \exp\left\{-\ell[K(\sigma)]^a t_B{}^a\right\} \tag{60}$$

The function $K(\sigma)$ is found by experimental measurements. Two possible forms of $K(\sigma)$ were discussed earlier.

Substituting Eq. (59) into Eq. (49), we obtain, on differentiating with respect to t and setting $a = 1$, the familiar first-order ensemble equation:

$$- dn/dt = n\ell K[\sigma(t)]$$

This demonstrates that only when $a = 1$ do the fibers of class A not remember their past loading history, i.e., become identical with fibers of the first-order ensemble. In other words, first-order ensemble fibers are a special case of class A fibers.

In order to investigate the behavior of class A fibers under constant stress, we require the probability density function of lifetimes, $q_\ell(t_B)$, under constant stress. Differentiating Eq. (60) with respect to t, we obtain

$$q_\ell(t_B)\ dt_B = \ell a K^a t_B^{a-1} \exp\left(-\ell\ K^a t_B^a\right) dt_B \tag{61}$$

The mth moment of this lifetime distribution is given by

$$\begin{aligned} \langle t_B{}^m\rangle_\Gamma &= \int_0^\infty t_B{}^m q_\ell(t_B)\ dt_B \\ &= \ell^{-m/a}\Gamma(1 + m/a)K^{-m} \end{aligned} \tag{62}$$

The average lifetime, the coefficient of variation in t_B, and the coefficient of skewness are

$$\langle t_B\rangle_f = \ell^{-1/a}\Gamma(1 + 1/a)K^{-1}$$

$$\frac{\Xi_f}{\langle t_B\rangle_f} = \frac{\left(\langle t_B^2\rangle_f - \langle t_B\rangle_f^2\right)^{1/2}}{\langle t_B\rangle_f} = \left[\frac{\Gamma(1 + 2/a)}{\Gamma^2(1 + 1/a)} - 1\right]^{1/2} \tag{63}$$

$$\frac{\mu_f^3}{\Xi_f^3} = \frac{\langle t_B^3\rangle_f - 3\langle t_B\rangle_f\langle t_B^2\rangle_f + 2\langle t_B\rangle_f^3}{\left[\langle t_B^2\rangle_f - \langle t_B\rangle_f^2\right]^{3/2}}$$

$$= \frac{\Gamma(1 + 3/a) - 3\Gamma(1 + 1/a)\Gamma(1 + 2/a) + 2\Gamma^3(1 + 1/a)}{[\Gamma(1 + 2/a) - \Gamma^2(1 + 1/a)]^{3/2}}$$

Note that the dimensionless quantities $\Xi_f/\langle t_B\rangle_f$ and μ_f^3/Ξ_f^3 are independent of the functional form of $K(\sigma)$. Equations (63) may be specialized to yield results for first-order ensemble fibers by substituting $a = 1$.

Let us now consider the behavior of the class A fibers under periodic loading histories $\sigma_\omega(t)$ with frequency $\omega/2\pi$ and period b $(= 2\pi/\omega)$

$$\sigma_\omega(t) = \sigma_\omega(t + mb) \geq 0 \tag{64}$$

where m denotes any positive integer. We shall denote the cumulative lifetime distribution function for periodic loads by $Q_{\ell,\omega}(t_B)$. Let us introduce $\langle K\rangle$ as the average value of $K[\sigma_\omega(t)]$ averaged over one cycle in time:

$$\langle K\rangle = (1/b)\int_0^b K[\sigma_\omega(t)]\, dt \tag{65}$$

Note that $\langle K\rangle$ is independent of ω. Coleman [7] shows that in the limit as $\omega t_B \to \infty$, we can write

$$\lim_{\omega t_B \to \infty} Q_{\ell,\omega}(t_B) = 1 - \exp\left(-\ell t_B^a \langle K \rangle^a\right) \tag{66}$$

In other words, whenever t_B is long enough or the period is short enough, such that $\omega t_B \to \infty$, the distribution function for t_B may be calculated by using Eq. (59) with $K[\sigma_\omega(t)]$ replaced by its average value over one cycle. Note that Eq. (66) has the same form as Eq. (60) for constant stress, with $K(\sigma)$ replaced by $\langle K \rangle$. This implies that Eqs. (63) also apply to this case with $K(\sigma)$ replaced by $\langle K \rangle$. Thus we conclude that whenever $\omega t_B \gg 1$, the moments of the lifetime distribution for class A fibers (hence also first-order ensemble fibers) must be independent of frequency. If an emsemble of fibers exhibits a fatigue life dependent on frequency, then the ensemble is not of Class A. The converse of this statement is not true, as will be evident from a discussion of class B fibers, which also exhibit a frequency-independent behavior under periodic loads.

The observation that class A and class B fibers exhibit frequency-independent lifetime behavior agrees with the response of polymers to periodic loads at room temperatures [7]. However, class A and class B assumptions cannot be consistently applied to cases for which the primary cause of fatigue failure is one of the following mechanisms:

(a) Abrasion.

(b) Work-hardening of plastic inclusions.

(c) A strain-dependent unidirectional propagation of the slip bands and cracks.

Such systems (e.g., steel) exhibit a fatigue behavior in which it is the number of cycles to failure, and not the actual lifetime, that is relatively independent of frequency. Materials of this type properly belong to class C, as will soon be shown.

C. Fibers of Class B

Introducing the function $\zeta(t)$,

$$\zeta(t) \equiv (C/\delta)^{1/a}\Omega(t) \tag{67}$$

Eq. (57) becomes

$$Q_\ell(t_B) = 1 - \exp\left[-\ell\zeta^a(t_B)\right] \tag{68}$$

Breakdown rules of class B are defined by Eq. (52). We now substitute, in analogous fashion to Eq. (58),

$$K[\sigma(t),\zeta] \equiv (C/\delta)^{1/a}\phi[\sigma(t), \Omega] \tag{69}$$

in Eq. (52), to obtain

$$d\zeta/dt = K[\sigma(t), \Omega] \tag{70}$$

where $\zeta(0) = 0$. It is assumed that $0 \leq \sigma < \infty$ and $0 \leq K < \infty$ for all t in some range $0 \leq t \leq t^*$. Also for constant $\zeta > 0$, and/or constant $\sigma > 0$, K is a continuous, monotonically increasing function of Ω.

Let us now consider a periodic loading history $\sigma_\omega(t)$. For each value of ω, a different $\zeta_\omega(t_B)$ appears in Eq. (68). Therefore,

$$Q_{\ell,\omega}(t_B) = 1 - \exp\left[-\ell\zeta_\omega^a(t_B)\right] \tag{71}$$

where $\zeta_\omega(t_B)$ is a solution of the equation

$$d\zeta_\omega/dt_B = K[\sigma_\omega(t), \zeta] \tag{72}$$

Now we introduce the function

$$H(\zeta) = \frac{\omega}{2\pi}\int_0^b K[\sigma_\omega(t), \zeta]\, dt \tag{73}$$

where the integration is performed with ζ held constant. Note that $H(\zeta)$ is independent of ω. It can now be proven [7] that in the range $0 \leq t < t^*$, the following limit is approached uniformly in t:

$$\lim_{\omega\to\infty} \zeta_\omega(t) = \overline{\zeta}(t) \tag{74}$$

where $\overline{\zeta}(t)$ is a solution of

$$d\overline{\zeta}/dt = H(\overline{\zeta})$$

and is thus independent of ω. Hence

$$\lim_{\omega\to\infty} Q_{\ell,\omega}(t_B) = 1 - \exp\left[-\ell\overline{\zeta}^a(t_B)\right] \tag{75}$$

The right-hand side of Eq. (75) is independent of ω. It therefore follows that for $\omega t_B >> 1$, class B fibers, like class A fibers, exhibit a frequency-independent behavior.

D. Fibers of Class C

Class C fibers have a more general breakdown rule than either class A or class B fibers, in the sense that fatigue life for class C fibers may depend upon either frequency or the number of cycles.

Consider an ensemble of class C fibers for which the breakdown rule is given by the equation

$$d\zeta/dt = \theta[\sigma(t), \zeta]\, d\sigma/dt \tag{76}$$

where θ is a continuous, monotonically increasing function in σ and ζ. Furthermore, it is assumed that $0 \leq \theta < \infty$ if both σ and ζ are finite.

Note that, in Eq. (76), if σ is constant, then $d\zeta/dt = 0$. This implies that if constant loads are applied, then the fibers do not undergo progressive damage in time.

A given fiber either breaks immediately or it does not break at all. Let us now consider the behavior of fibers obeying Eq. (76) subjected to a monotonically increasing stress $\sigma(t)$. The probability $G_\ell(\sigma^*)$ that the fiber breaks before the stress reaches the value σ^* is independent of the functional dependence of σ on t, and is given by

$$G_\ell(\sigma^*) = 1 - \exp\left[-\ell\phi^a(\sigma^*)\right] \tag{77}$$

where $\phi(\sigma^*) = \zeta(t_B)$ is the solution of the equation

$$d\phi(\sigma^*)/d\sigma^* = \theta[\sigma^*, \phi(\sigma^*)] \tag{78a}$$

$$= \psi(\sigma^*) \tag{78b}$$

with the boundary condition

$$\phi(0) = 0$$

The function ψ depends only on σ^*. This results from the fact that the solution of Eq. (78a) is unique, regardless of the dependence of σ on t, provided that $\sigma(t)$ is monotonic.

The fibers obeying Eqs. (77) and (78) are referred to as "classical fibers" insofar as they exhibit a strength behavior independent of the rate of loading (see Chapter 7). As an example, let us assume that $\psi(\sigma^*)$ is given by

$$\psi(\sigma^*) = B(\sigma^*)^D, \quad B > 0 \text{ and } D > 0 \tag{79}$$

Substitution of Eq. (79) into Eq. (78) yields

$$\phi(\sigma^*) = B(D + 1)^{-1}(\sigma^*)^{D+1} \tag{80}$$

Combining Eqs. (77) and (80), we obtain the well-known Weibull distribution (see Chapter 7)

$$G_\ell(\sigma^*) = 1 - \exp\left[-\ell\alpha(\sigma^*)^\beta\right] \tag{81}$$

where

$$\alpha = B^a(D + 1)^{-a} \quad \text{and} \quad \beta = (D + 1)a$$

Now consider the behavior of fibers obeying Eq. (76) under periodic loading histories $\sigma_\omega(t)$. We define the following new variables

$$z = \omega t/2\pi$$

$$\eta(z) = \sigma_\omega(t)$$

$$\xi(z) = \frac{d\eta}{dz} = \frac{2\pi}{\omega}\frac{d\sigma_\omega(t)}{dt}$$

$$y(z) = \zeta(t)$$

where z denotes the number of cycles that have elapsed up to time t; $\eta(z)$ and $\xi(z)$ are periodic with unit period. The function $y(z)$ obeys the differential equation

$$dy/dz = \theta[\eta(z), y]\xi(z) \tag{82}$$

Since the right-hand side of Eq. (82) is independent of ω, y is independent of frequency. Hence ζ is also independent of ω. It can be shown [7] that if θ and ξ are bounded in some interval $0 \leq t \leq t^*$, then the following limit is approached uniformly in the interval $0 \leq t \leq t^*$:

$$\lim_{\omega\to\infty} \zeta(t) = \bar{y}(z) \tag{83}$$

where $\bar{y}$ is given by the equation

$$d\bar{y}/dz = \int_0^1 \xi(z)\theta[\eta(z), \bar{y}]\, dz \tag{84}$$

The integration in Eq. (84) is performed with $\bar{y}$ held constant. Note that $\bar{y}$ is independent of ω. Let z_B be the number of complete cycles which elapse prior to failure, and let $Q_{\ell,\omega}(z_B)$ be the cumulative distribution function for z_B. Then it follows from Eqs. (71) and (83) that

$$\lim_{\omega\to\infty} Q_{\ell,\omega}(t_B) = 1 - \exp\left[-\ell\bar{y}^a(z_B)\right] \tag{85}$$

Therefore if the classical breakdown rule is valid, cycle-dependent fatigue behavior must result.

In the preceding discussion, we have indicated how certain intuitively evident *a priori* assumptions combined with the breakdown rules may be employed to deduce the strength behavior of fibers under time-dependent uniaxial tensile stresses. Detailed results were derived for fibers drawn from a first-order ensemble. The generalized theory of breaking kinetics has not been utilized thus far to any great extent to analyze composite materials. It should serve as a point of departure for analyzing composites made of fibers which exhibit a complex strain-rate-dependent behavior. A word of caution is in order, however. Fracture is not one single physical phenomenon. The same material may fail by different mechanisms under different test conditions. Thus, whenever there is a change in fracture morphology with loading history, a change in breakdown rule is to be expected. This may either involve a change in the functional form of the breakdown rule, or a complete transition from one class of breakdown to another. Therefore, it follows that in applying the foregoing analysis to cases of practical interest, one must ascertain that the breakdown rule appropriate for one loading history is applicable under another, and that there is no change in fracture morphology under the two conditions.

REFERENCES

1. B. D. Coleman, "Time Dependence of Mechanical Breakdown Phenomena," J. Appl. Phys., 27 (8), 862-866 (1956).

2. B. D. Coleman, "A Stochastic Process Model for Mechanical Breakdown," Trans. Soc. Rheology, I, 153-168 (1957).

3. B. D. Coleman, "Time Dependence of Mechanical Breakdown in Bundles of Fibers. I. Constant Total Load," J. Appl. Phys., 28 (9), 1058-1064 (1957).

4. B. D. Coleman and D. W. Marquardt, "Time Dependence of Mechanical Breakdown in Bundles of Fibers. II. The Infinite Ideal Bundle under Linearly Increasing Loads," J. Appl. Phys., 28 (9), 1065-1067 (1957).

5. B. D. Coleman, "Time Dependence of Mechanical Breakdown in Bundles of Fibers. III. The Power Law Breakdown Rule," Trans. Soc. Rheology, II, 195-218 (1958).

6. B. D. Coleman and D. W. Marquardt, "Time Dependence of Mechanical Breakdown in Bundles of Fibers. IV. Infinite Ideal Bundle under Oscillating Loads," J. Appl. Phys., 29 (7), 1091-1099 (1958).

7. B. D. Coleman, "Statistics and Time Dependence of Mechanical Breakdown in Fibers," J. Appl. Phys., 29 (6), 968-983 (1958).

8. E. Jahnke and F. Emde, Table of Functions with Formulae and Curves, Dover Publications, New York, 1945.

CHAPTER 12

INTRODUCTION TO ELASTIC WAVE PROPAGATION IN COMPOSITES

I. INTRODUCTION

In this chapter, we will first briefly introduce the elastodynamic equations governing the motion of a homogeneous, isotropic medium. These will then be specialized to deal with plane wave propagation in a number of idealized geometries representing composites with two constituents. While the constituent materials may be isotropic, the composite will in general behave anisotropically. Methods of solution for steady-state wave propagation problems will be discussed.

II. VECTOR EQUATIONS OF MOTION

The strain displacement and constitutive relations governing elastic media have been presented in Chapter 3. We begin here by deriving the equations of motion in a Cartesian coordinate system.

Consider the forces acting on the differential volume element of Fig. 1. If we neglect body forces, the x_1 component of the resultant force acting on the volume element is

$$\left(\sigma_{11} + \frac{\partial\sigma_{11}}{\partial x_1} dx_1 - \sigma_{11}\right) dA_1 + \left(\tau_{12} + \frac{\partial\tau_{12}}{\partial x_2} dx_2 - \tau_{12}\right) dA_2 + \left(\tau_{13} + \frac{\partial\tau_{13}}{\partial x_3} dx_3 - \tau_{13}\right) dA_3$$

where $dA_1 = (dx_2 \cdot dx_3)$, etc. This component may be rewritten as

$$\left(\frac{\partial\sigma_{11}}{\partial x_1} + \frac{\partial\tau_{12}}{\partial x_2} + \frac{\partial\tau_{13}}{\partial x_3}\right) dx_1\, dx_2\, dx_3$$

For the element to be in equilibrium, Newton's second law requires that the resulting force exactly balance the intertial forces acting in the x_1 direction

$$\left(\rho \cdot \frac{\partial^2 u_1}{\partial t^2}\right) dx_1\, dx_2\, dx_3$$

Here (u_1, u_2, u_3) denote the displacements in the (x_1, x_2, x_3) directions and ρ is the material density. Using a similar argument in the x_2 and x_3 directions, the equations of motion in a Cartesian coordinate system can be written as

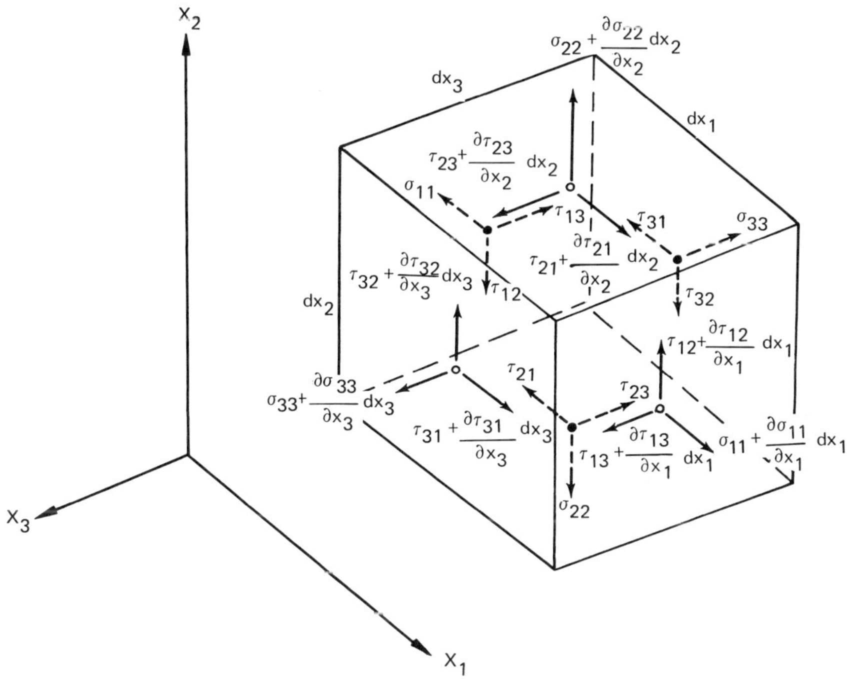

FIG. 1. Differential volume element.

$$\rho \frac{\partial^2 u_1}{\partial t^2} = \frac{\partial \sigma_{11}}{\partial x_1} + \frac{\partial \tau_{12}}{\partial x_2} + \frac{\partial \tau_{13}}{\partial x_3}$$

$$\rho \frac{\partial^2 u_2}{\partial t^2} = \frac{\partial \tau_{21}}{\partial x_1} + \frac{\partial \sigma_{22}}{\partial x_2} + \frac{\partial \tau_{23}}{\partial x_3} \qquad (1)$$

$$\rho \frac{\partial^2 u_3}{\partial t^2} = \frac{\partial \tau_{31}}{\partial x_1} + \frac{\partial \tau_{32}}{\partial x_2} + \frac{\partial \sigma_{33}}{\partial x_3}$$

These equations are valid for both isotropic and anisotropic media. Their solution requires some means of relating the stresses acting on the body to the resulting displacements. This is accomplished by means of the constitutive relations. The simplest type of such a relation is that governing the behavior of an isotropic elastic medium. In this case the stress-strain relations are

$$
\begin{aligned}
\sigma_{11} &= \lambda e + 2G\varepsilon_{11}, \quad \tau_{12} = G\gamma_{12} \\
\sigma_{22} &= \lambda e + 2G\varepsilon_{22}, \quad \tau_{23} = G\gamma_{23} \\
\sigma_{33} &= \lambda e + 2G\varepsilon_{33}, \quad \tau_{31} = G\gamma_{31}
\end{aligned} \tag{2}
$$

where $e = \varepsilon_{11} + \varepsilon_{22} + \varepsilon_{33}$, and is called the dilatation. The two Lamé constants λ and G are all that is necessary to completely define the stress-strain behavior of an isotropic elastic medium (see Chapter 3).

The strain displacement relations are

$$
\begin{aligned}
&\varepsilon_{11} = \frac{\partial u_1}{\partial x_1}, \quad \varepsilon_{22} = \frac{\partial u_2}{\partial x_2}, \quad \varepsilon_{33} = \frac{\partial u_3}{\partial x_3} \\
&\gamma_{12} = \frac{\partial u_2}{\partial x_1} + \frac{\partial u_1}{\partial x_2}, \quad \gamma_{23} = \frac{\partial u_3}{\partial x_2} + \frac{\partial u_2}{\partial x_3}, \quad \gamma_{31} = \frac{\partial u_3}{\partial x_1} + \frac{\partial u_1}{\partial x_3}
\end{aligned} \tag{3}
$$

Substitution of Eqs. (2) and (3) into Eq. (1) yields the following:

$$
\begin{aligned}
\rho \frac{\partial^2 u_1}{\partial t^2} &= (\lambda + G) \frac{\partial e}{\partial x_1} + G\nabla^2 u_1 \\
\rho \frac{\partial^2 u_2}{\partial t^2} &= (\lambda + G) \frac{\partial e}{\partial x_2} + G\nabla^2 u_2 \\
\rho \frac{\partial^2 u_3}{\partial t^2} &= (\lambda + G) \frac{\partial e}{\partial x_3} + G\nabla^2 u_3
\end{aligned} \tag{4}
$$

where the operator $\nabla^2 \equiv (\partial^2/\partial x_1^2) + (\partial^2/\partial x_2^2) + (\partial^2/\partial x_3^2)$. In vector form, Eq. (4) becomes

$$\rho\, \partial^2 \underline{S}/\partial t^2 = (\lambda + 2G)\ \text{grad}\ e - 2G\ \text{curl}\ \underline{w} \tag{5}$$

where $\underline{S}$ is the displacement vector (with components u_1, u_2, u_3), and $\underline{w}$ is the rotation vector such that $2\underline{w} = \text{curl}\ \underline{S}$. Equation (5) is the well-known Navier relation. Transformation to other coordinate systems may be affected by using the appropriate forms of the operators grad, div, and curl.

III. WAVE EQUATION IN CARTESIAN COORDINATES

It is often convenient to express displacements in terms of potential functions. The Helmholtz decomposition theorem [1] allows one to express any analytic vector field, in this case displacement, into a scalar potential ϕ and a vector potential $\underline{\psi}$ with components ψ_1, ψ_2, and ψ_3. The components of the displacement field can be written as

$$\begin{aligned} u_1 &= \frac{\partial \phi}{\partial x_1} + \frac{\partial \psi_3}{\partial x_2} - \frac{\partial \psi_2}{\partial x_3} \\ u_2 &= \frac{\partial \phi}{\partial x_2} + \frac{\partial \psi_1}{\partial x_3} - \frac{\partial \psi_3}{\partial x_1} \\ u_3 &= \frac{\partial \phi}{\partial x_3} + \frac{\partial \psi_2}{\partial x_1} - \frac{\partial \psi_1}{\partial x_2} \end{aligned} \tag{6}$$

where

$$\partial\psi_1/\partial x_1 = \partial\psi_2/\partial x_2 = \partial\psi_3/\partial x_3 = 0$$

In terms of the potential functions, the first of Eqs. (4) becomes

$$\rho \frac{\partial}{\partial x_1}\left(\frac{\partial^2 \phi}{\partial t^2}\right) + \rho \frac{\partial}{\partial x_2}\left(\frac{\partial^2 \psi_3}{\partial t^2}\right) - \rho \frac{\partial}{\partial x_3}\left(\frac{\partial^2 \psi_2}{\partial t^2}\right)$$

$$= (\lambda + G) \frac{\partial}{\partial x_1} (\nabla^2 \phi) + G \frac{\partial}{\partial x_2} (\nabla^2 \psi_3)$$

$$+ G \frac{\partial}{\partial x_1} (\nabla^2 \phi) - G \frac{\partial}{\partial x_3} (\nabla^2 \psi_2) \qquad (7)$$

Equation (7), along with similar equations in the x_2 and x_3 directions, will be satisfied if the functions ϕ and $\underline{\psi}$ are solutions of the equations

$$\nabla^2 \phi = \frac{1}{C_\ell^2} \frac{\partial^2 \phi}{\partial t^2} \qquad (8)$$

$$\nabla^2 \psi_i = \frac{1}{C_S^2} \frac{\partial^2 \psi_i}{\partial t^2}, \quad i = 1, 2, 3 \qquad (9)$$

where

$$C_\ell^2 = (\lambda + 2G)/\rho \qquad (10)$$

and

$$C_S^2 = G/\rho \qquad (11)$$

Equations (8) and (9) are wave equations. They indicate that for an unbounded medium of the kind under consideration, i.e., elastic and isotropic, disturbances may be propagated with two possible velocities, C_ℓ or C_S. In terms of the dilatation e and the rotation vector $\underline{w}$ (components w_1, w_2, w_3), Eqs. (8) and (9) become

$$\nabla^2 e = \frac{1}{C_\ell^2} \frac{\partial^2 e}{\partial t^2}, \quad \nabla^2 w_i = \frac{1}{C_S^2} \frac{\partial^2 w_i}{\partial t^2}, \quad i = 1, 2, 3 \qquad (12)$$

These equations are derived by performing the operations div and curl, respectively, on Eq. (5). Equations (12) offer us insight into the physical nature of the disturbances. A purely dilatational deformation will propagate at the velocity C_ℓ, while a purely distortional disturbance, one in which there is no accompanying volume change, will propagate at the velocity C_S. A wide variety of adjectives are often attached to these two types of waves. The C_ℓ disturbance is known alternately as a "dilatational," "irrotational," or "longitudinal" wave, while the C_S disturbance is called "equivoluminal," "transverse," "rotational," or "shear" wave. We will adopt the convention here of referring to C_ℓ as the longitudinal and C_S as the shear wave velocity.

IV. SOLUTION OF THE WAVE EQUATION

The wave equations (8) and (9) can in general be solved using the method of "separation of variables" [2]. For one-dimensional wave propagation, however, a technique known as "d'Alembert's solution" may be employed. Consider the one-dimensional wave equation

$$\partial^2\Phi/\partial t^2 = C^2\ \partial^2\Phi/\partial x_1^2 \tag{13}$$

If f is a function possessing second derivatives and C is a wave velocity dx_1/dt, then

$$\frac{\partial f(x_1 - Ct)}{\partial t} = -Cf'(x_1 - Ct), \qquad \frac{\partial f(x_1 - Ct)}{\partial x_1} = f'(x_1 - Ct)$$

$$\frac{\partial^2 (x_1 - Ct)}{\partial t^2} = C^2 f''(x_1 - Ct), \qquad \frac{\partial^2 f(x_1 - Ct)}{\partial x_1^2} = f''(x_1 - Ct)$$

where the primes denote partial differentiation with respect to $(x_1 - Ct)$. Clearly, then,

$$\Phi = f(x_1 - Ct)$$

is a solution of Eq. (13). Similarly, if g is an arbitrary, twice differentiable function, then

$$\Phi = g(x_1 + Ct)$$

is likewise a solution of Eq. (13). Now, since Eq. (13) is a linear equation of second order, it follows that the sum

$$\Phi = f(x_1 - Ct) + g(x_1 + Ct)$$

is also a solution, and can in fact be shown to be the general solution for Eq. (13) by direct substitution.

The functions f and g are arbitrary and depend upon the initial and boundary conditions of the problem. Note that f corresponds to a disturbance propagating along the positive and g to one propagating along the negative x_1 direction. For a given Φ, if the disturbance has a particular form at time t_1, it will have an identical form at time t_2 at a distance $C(t_2 - t_1)$ removed along the x_1 direction.

Consider the case of a medium subjected to the deformation field

$$u_1 = U_0 \sin[(2\pi/\Lambda)(x_1 - Ct)], \quad u_2 = u_3 = 0 \tag{14}$$

where U_0 is a constant representing the amplitude of the harmonic disturbance, Λ is its wavelength, and C is the wave velocity. In terms of potential functions, such a field corresponds to

$$\underline{\psi} = \text{const}, \quad \phi = -(U_0\Lambda/2\pi)\cos[(2\pi/\Lambda)(x_1 - Ct)]$$

Equation (14) is thus seen to be a solution of Eq. (18) if $C = C_\ell$.

The term "phase velocity" is often used to denote the velocity at which a sinusoidal disturbance propagates through an elastic medium. For the particular disturbance described by Eq. (14), both the wave velocity C and the phase velocity are seen to correspond to the longitudinal wave velocity C_ℓ of the medium. Had we chosen a disturbance such that

$$u_1 = 0, \quad u_2 = U_0 \sin[(2\pi/\Lambda)(x_1 - Ct)], \quad u_3 = 0 \qquad (15)$$

the wave and phase velocities would have assumed the values corresponding to the shear velocity C_S of the medium.

It is important to recognize the distinction between particle velocity and wave (or phase) velocity. The particle velocity $d\underline{S}/dt$ describes the displacement of an individual particle of the material with time, while the wave velocity refers to the speed at which the disturbance propagates through the medium.

Unbounded materials in which the phase and wave velocities are equal are called "simple." Had the phase velocity been dependent upon the wavelength, as it would in a viscoelastic or geometrically dispersive material, for example, the two velocities would have been equal only as the wavelength approached infinity. For a more complete treatment of elastic waves, the reader is referred to the many textbooks dealing with elasticity and wave propagation (e.g., Refs. [3,4]).

V. HARMONIC WAVES IN PERIODIC MEDIA

We begin our study of wave propagation in composite materials with an investigation of the simplest type of periodic media. Let us consider harmonic waves propagating in an unbounded laminated medium. The material is assumed to be composed of alternating layers of two homogeneous isotropic elastic layers (see Fig. 2). The treatment presented here is due to Rytov [5]. It is convenient to separate the displacement vector $\underline{S}$ of a material particle into its irrotational $\underline{U}$ and solenoidal $\underline{W}$ parts. Then

$$\underline{S} = \underline{U} + \underline{W} \tag{16}$$

where $\underline{U} = \text{grad}\ \phi$ and $\underline{W} = \text{curl}\ \underline{\psi}$. Here ϕ and $\underline{\psi}$ are the scalar and vector potentials discussed previously, and $u_1 = U_1 + W_1$ etc. The momentum equation (5) can be uncoupled to yield

$$\begin{aligned} &(\partial^2\underline{U}/\partial t^2) - C_\ell^2\ \text{grad}(\text{div}\ \underline{U}) = 0, \quad \text{curl}\ \underline{U} = 0 \\ &(\partial^2\underline{W}/\partial t^2) - C_S^2\ \text{curl curl}\ \underline{W} = 0, \quad \text{div}\ \underline{W} = 0 \end{aligned} \tag{17}$$

Consider a disturbance propagating along the x_1-axis. A shear wave with displacements $\underline{S} = (0, u_2, 0)$ parallel to the layering is autonomous, i.e., it is independent of any longitudinal wave. Conversely, a transverse displacement of the layer u_3 will in general be accompanied by a

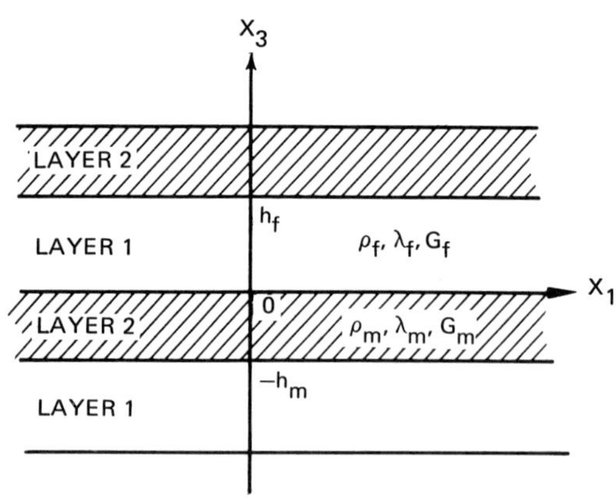

FIG. 2. Laminate geometry.

longitudinal displacement component u_1. Let us examine the latter case. The displacement vector $\underline{S}$ has the components $u_1(x_1, x_3)$, 0, $u_3(x_1, x_3)$. For harmonic waves of frequency ω, Eq. (17) becomes

$$\frac{\partial^2 \underline{U}}{\partial x_1^2} + \frac{\partial^2 \underline{U}}{\partial x_3^2} + k_\ell^2 \underline{U} = 0, \quad \frac{\partial U_3}{\partial x_1} = \frac{\partial U_1}{\partial x_3}$$

$$\frac{\partial^2 \underline{W}}{\partial x_1^2} + \frac{\partial^2 \underline{W}}{\partial x_3^2} + k_S^2 \underline{W} = 0, \quad \frac{\partial W_1}{\partial x_1} = -\frac{\partial W_3}{\partial x_3} \tag{18}$$

where

$$k_\ell^2 = \omega^2/C_\ell^2, \quad k_S^2 = \omega^2/C_S^2$$

We seek solutions to these equations which will represent traveling waves in the x_1 direction and standing (stationary) waves in the x_3 direction. The wave velocity along the x_1 axis will be denoted by C_{1j} $(j = 1,3)$ and its associated wave number* by $k^2 = \omega^2/C_{1j}^2$.

Now, still following Rytov, we will find it convenient to write the general solution to these equations in terms of specially selected partial solutions. In particular, the solution for U_{f_1} in layer 1 can be written in the form

$$U_{f_1} = P_f(x_3)e^{-ikx_1}$$

$$P_f(x_3) = A_f \cos \xi_f\left(x_3 - \tfrac{1}{2} h_f\right) + B_f \sin \xi_f\left(x_3 - \tfrac{1}{2} h_f\right)$$

$$\xi_f = k_{f\ell}^2 - k^2 \tag{19}$$

*Wave number or spatial frequency is defined in terms of wave length Λ by $k = 2\pi/\Lambda$ and is related to the phase velocity and frequency by $k = \omega/C$.

where A_f and B_f are as yet undetermined constants. Note that the expression for $P_f(x_3)$ has been separated into functions which are symmetric and antisymmetric with respect to the centerline of layer 1. From Eq. (18), then, we have

$$U_{f_3} = -[P_f'(x_3)/ik]\, e^{-ikx_1} \tag{20}$$

where the prime denotes differentiation with respect to x_3. Similarly, we find

$$W_{f_3} = Q_f(x_3)e^{-ikx_1}, \quad W_{f_1} = [Q_f'(x_3)/ik]\, e^{-ikx_1}$$

$$Q_f(x_3) = D_f \cos \zeta_f\left(x_3 - \frac{1}{2} h_f\right) + F_f \sin \zeta_f\left(x_3 - \frac{1}{2} h_f\right)$$

$$\zeta_f^2 = k_{fS}^2 - k^2 \tag{21}$$

We shall denote all quantities referring to layer 2 by a subscript m. The pertinent variables for layer 2 then become

$$U_{m_1} = P_m(x_3)e^{-ikx_1}, \quad U_{m_3} = -[P_m'(x_3)/ik]e^{-ikx_1}$$

$$W_{m_1} = [Q_m'(x_3)/ik]e^{-ikx_1}, \quad W_{m_3} = Q_m(x_3)e^{-ikx_1} \tag{22}$$

where

$$P_m(x_3) = A_m \cos \xi_m\left(x_3 + \frac{1}{2} h_m\right) + B_m \sin \xi_m\left(x_3 + \frac{1}{2} h_m\right)$$

$$\xi_m^2 = k_{m\ell}^2 - k^2$$

$$Q_m(x_3) = D_m \cos \zeta_m\left(x_3 + \frac{1}{2} h_m\right) + F_m \sin \zeta_m\left(x_3 + \frac{1}{2} h_m\right) \tag{23}$$

$$\zeta_m^2 = k_{mS}^2 - k^2$$

Using Eq. (16) and the strain displacement relations (3) for this problem, we find the components of the strain tensor in layer 2 to be

$$\varepsilon_{m_{11}} = -(ikP_m + Q_m')e^{-ikx_1}, \quad \gamma_{m_{12}} = 0$$

$$\gamma_{m_{13}} = 2\left\{P_m' + \left[\left(k^2 - \zeta_m^2\right)/2ik\right]Q_m\right\}e^{-ikx_1} \tag{24}$$

$$\varepsilon_{m_{22}} = \gamma_{m_{23}} = 0, \quad \varepsilon_{m_{33}} = \left[\left(\xi^2{}_m/ik\right)P_m + Q_m'\right]e^{-ikx_1}$$

The components of the stress tensor can now be found using the stress-strain relations (2) and Eq. (24), i.e.,

$$\tau_{m_{13}} = 2G_m\left\{P_m' + \left[\left(k^2 - \zeta_m^2\right)/2ik\right]Q_m\right\}e^{-ikx_1}$$

$$\tau_{m_{23}} = 0 \tag{25}$$

$$\sigma_{m_{33}} = \left\{\left[\left(\lambda_m k_{m\ell}^2 + 2G_m\xi_m^2\right)/ik\right]P_m + 2G_m Q_m'\right\}e^{-ikx_1}$$

Similar relations hold for layer 1 with the subscript m replaced by f in Eqs. (22), (24), and (25).

We must satisfy two sets of conditions at the laminate interfaces: those of continuity and those of periodicity. For rigid bonding between layers, continuity requires that the deformation components $u_1 = U_1 + W_1$ and $u_3 = U_3 + W_3$, and the components of the stress tensor τ_{13} and τ_{33} be continuous across the interface $x_3 = 0$, i.e.,

$$u_{f_1}(0) = u_{m_1}(0), \quad u_{f_3}(0) = u_{m_3}(0)$$

$$\tau_{f_{13}}(0) = \tau_{m_{13}}(0), \quad \sigma_{f_{33}}(0) = \sigma_{m_{33}}(0) \tag{26}$$

Further, periodicity of the structure requires that the

stresses and displacements in layer 1 at $x_3 = h_f$ be equal to the corresponding quantities in layer 2 at $x_3 = -h_m$, i.e.,

$$u_{f_1}(h_f) = u_{m_1}(-h_m), \qquad u_{f_3}(h_f) = u_{m_3}(-h_m)$$
$$\tau_{f_{13}}(h_f) = \tau_{m_{13}}(-h_m), \quad \sigma_{f_{33}}(h_f) = \sigma_{m_{33}}(-h_m) \tag{27}$$

Substitution of the appropriate stresses and displacements into Eqs. (26) and (27) yields a system of eight equations for the eight unknown constants A_f, B_f, ..., F_m.

A. Waves Which are Longitudinal in the Mean

For the case in which the coefficients $B_f = D_f = B_m = D_m = 0$, we see immediately that the longitudinal displacements u_{f_1} and u_{m_1} are symmetric with respect to their individual laminate centerlines. On the other hand, the transverse displacements u_{f_3} and u_{m_3} are antisymmetric. This means that if we are interested in displacements which are averaged over a period of the structure $(h_f + h_m)$, a wave of this type will be longitudinal, and independent of shear. If we apply the eight boundary conditions of Eqs. (26) and (27), we find four independent equations for the constants A_f, F_f, A_m, F_m:

$$A_f ik \cos\left(\frac{1}{2}\xi_f h_f\right) + F_f \zeta_f \cos\left(\frac{1}{2}\zeta_f h_f\right) = A_m ik \cos\left(\frac{1}{2}\xi_m h_m\right) + F_m \zeta_m \cos\left(\frac{1}{2}\zeta_m h_m\right)$$

$$A_f\left(\lambda_f k_{f\ell}^2 + 2G_f \xi_f^2\right) \cos\left(\frac{1}{2}\xi_f h_f\right) + F_f 2ikG_f \zeta_f \cos\left(\frac{1}{2}\zeta_f h_f\right)$$
$$= A_m\left(\lambda_m k_{m\ell}^2 + 2G_m \xi_m^2\right) \cos\left(\frac{1}{2}\xi_m h_m\right) + F_m 2ikG_m \zeta_m \cos\left(\frac{1}{2}\zeta_m h_m\right)$$

$$A_f \xi_f \sin\left(\frac{1}{2}\xi_f h_f\right) + F_f ik \sin\left(\frac{1}{2}\zeta_f h_f\right) = -A_m \xi_m \sin\left(\frac{1}{2}\xi_m h_m\right) - F_m ik \sin\left(\frac{1}{2}\zeta_m h_m\right)$$

$$A_f 2ikG_f \xi_f \sin\left(\frac{1}{2}\xi_f h_f\right) + F_f\left(\lambda_f k_{f\ell}^2 + 2G_f \xi_f^2\right) \sin\left(\frac{1}{2}\zeta_f h_f\right) = -A_m 2ikG_m \xi_m \sin\left(\frac{1}{2}\xi_m h_m\right) - F_m\left(\lambda_m k_{m\ell}^2 + 2G_m \xi_m^2\right) \sin\left(\frac{1}{2}\zeta_m h_m\right)$$

Setting the determinant of the coefficients $A_f, \ldots, F_m$ equal to zero yields the characteristic (or frequency) equation of the system

$$4(G_f - G_m)^2 X_f X_m + \omega^2 \rho_f \left[\frac{\omega^2 \rho_f}{k^2} - 4(G_f - G_m)\right] X_m \tan\frac{\zeta_f h_f}{2} + \omega^2 \rho_m \left[\frac{\omega^2 \rho_m}{k^2} + 4(G_f - G_m)\right] X_f \tan\frac{\zeta_m h_m}{2} - \frac{\omega^4 \rho_f \rho_m}{k^2}\left(Y_f \tan\frac{\zeta_m h_m}{2} + Y_m \tan\frac{\zeta_f h_f}{2}\right) = 0 \qquad (28)$$

where we have used the definitions

$$X_f = k^2 \tan\left(\frac{1}{2}\zeta_f h_f\right) + \xi_f \zeta_f \tan\left(\frac{1}{2}\xi_f h_f\right)$$

$$Y_f = k^2 \tan\left(\frac{1}{2}\zeta_f h_f\right) - \xi_m \zeta_f \tan\left(\frac{1}{2}\xi_m h_m\right)$$

$$X_m = k^2 \tan\left(\frac{1}{2}\zeta_m h_m\right) + \xi_m \zeta_m \tan\left(\frac{1}{2}\xi_m h_m\right)$$

$$Y_m = k^2 \tan\left(\frac{1}{2}\zeta_m h_m\right) - \xi_f \zeta_m \tan\left(\frac{1}{2}\xi_f h_f\right)$$

Since the phase velocity C_{11} is equal to ω/k, we see that Eq. (28) relates the propagation velocity to either frequency or wave number. This equation has multiple roots, each one corresponding to a different mode of propagation. It also implies that the sinusoidal waves of different frequency propagate at different velocities. Such a system is said to be "geometrically dispersive" (see Section IV) since a pulse made up of many frequency components would tend to spread as it propagates. This behavior is a result of the geometry of the system, and is not accompanied by any concomitant dissipation of mechanical energy, as a wave in a viscoelastic material would be, for example.

It is of interest to investigate the behavior of Eq. (28) as the frequency approaches zero. This is physically equivalent to requiring that the thickness of the laminates vanish and is accomplished by replacing all tangents by their respective arguments. On so doing, we find

$$\lim_{\omega \to 0} C_{11}^2 = \left\{(h_f + h_m)^2 + 4h_f h_m \left[(G_f - G_m)(\lambda_f + G_f - \lambda_m - G_m)\right] / \left[(\lambda_f + 2G_f)(\lambda_m + 2G_m)\right]\right\} / \left\{\rho_c (h_f + h_m)\left[h_f/(\lambda_f + 2G_f) + h_m/(\lambda_m + 2G_m)\right]\right\} \tag{29}$$

where $\rho_c = (h_f \rho_f + h_m \rho_m)/(h_f + h_m)$. Eq. (29) is thus the zero-frequency limit, and corresponds to the static elastic solution to this problem. The reader is referred to a paper by Postma [6], who reports the x_1 direction modulus of a laminated material to be equivalent to the $\rho_c C_{11}^2$ term of Eq. (29).

The case of limiting phase velocity at high frequency has been discussed by Hoffman et al. [7]. When the frequency is very high, or the corresponding wavelength small with respect to the thicknesses of the individual laminates, the tangent functions of Eq. (28) may be replaced by hyperbolic tangent functions. In the limit of infinite frequency, these hyperbolic tangent functions approach unity and the characteristic equation (28) reduces to

$$\left[(\rho_f - \rho_m)^2 - (\rho_f a_m + \rho_m a_f)(\rho_f b_m + \rho_m b_f)\right] C_{11}^4$$

$$+ \left\{4(G_f - G_m)\left[(\rho_m - \rho_f) + \rho_f a_m b_m - \rho_m a_f b_f\right]\right\} C_{11}^2$$

$$+ 4(G_f - G_m)^2 (a_m b_m - 1)(a_f b_f - 1) = 0 \qquad (30)$$

where

$$a_{f,m}^2 = 1 - (C_{11}/C_{f,m\ell})^2, \quad b_{f,m}^2 = 1 - (C_{11}/C_{f,mS})^2$$

and $C_{f,m\ell}$ and $C_{f,mS}$ are the longitudinal and shear wave velocities in the individual fiber or matrix materials, respectively. The real root of Eq. (30) agrees with the expression for the Stonely wave in seismology. This equation determines the behavior of the phase velocity C_{11} at high frequencies, and we immediately see that the laminate thicknesses h_f and h_m no longer play a part in this behavior. A Stonely wave is a surface wave, closely akin to a Rayleigh wave, which propagates along the laminate interface and decays exponentially with x_3. Its velocity is less than the shear wave velocity in either of the two laminates. The condition for the existence of a Stonely wave is that the velocities C_{fS} and C_{mS} be nearly equal [8]. In instances where this condition is not met, the high-frequency limit of the phase velocity is the lesser of C_{fS} or C_{mS}.

B. Waves Which Are Transverse in the Mean

For the case in which $A_f = F_f = A_m = F_m = 0$, the transverse displacements u_{f3} and u_{m3} will be symmetric with respect to their individual laminate centerlines, while the longitudinal displacements u_{f_1} and u_{m_1} will be antisymmetric. The longitudinal displacements and stresses will therefore vanish when averaged over the period $(h_f + h_m)$. The deformation is then predominantly shear, and the particle motion is essentially normal to the direction of wave propagation. The procedure is exactly analogous to the longitudinal case and will not be repeated. The interested reader is referred to the work of Sun et al. [9] for the complete derivation. The zero-frequency limit for the phase velocity turns out to be

$$\lim_{\omega \to 0} C_{13}^2 = \frac{h_f + h_m}{\rho_c\left[\left(\frac{h_f}{G_f}\right) + \left(\frac{h_m}{G_m}\right)\right]} \tag{31}$$

For the horizontally polarized shear wave, i.e., one propagating along the x_1 axis, in which only the u_2 component of displacement is nonzero, we have $u_2 = W_2$. In layer 1, then,

$$W_{f_2} = Q_f(x_3)e^{-ikx_1} \tag{32}$$

where

$$Q_f(x_3) = D_f \cos \zeta_f\left(x_3 - \frac{1}{2}h_f\right) + F_f \sin \zeta_f\left(x_3 - \frac{1}{2}h_f\right)$$

$$\zeta_f^2 = k_{fS}^2 - k^2$$

Similarly, in layer 2,

$$W_{m_2} = Q_m(x_3)e^{-ikx_1} \tag{33}$$

where

$$Q_m(x_3) = D_m \cos \zeta_m\left(x_3 + \frac{1}{2}h_m\right) + F_m \sin \zeta_m\left(x_3 + \frac{1}{2}h_m\right)$$

$$\zeta_m^2 = k_{mS}^2 - k^2$$

The continuity and periodicity conditions on u_2 and τ_{23} yield four equations for D_f, F_f, D_m, F_m. Setting the determinant of this system equal to zero yields the characteristic equation

$$\frac{G_m\zeta_m}{G_f\zeta_f}\left(\tan^2\frac{\zeta_f h_f}{2} + \tan^2\frac{\zeta_m h_m}{2}\right) + \left[1 + \left(\frac{G_m\zeta_m}{G_f\zeta_f}\right)^2\right]\tan\frac{\zeta_f h_f}{2}\tan\frac{\zeta_m h_m}{2} = 0 \quad (34)$$

The corresponding zero-frequency limit of the phase velocity is

$$\lim_{\omega\to 0} C_{12}^2 = (h_f G_f + h_m G_m)/[(h_f + h_m)\rho_c] \quad (35)$$

C. Waves Perpendicular to the Laminates

A longitudinal wave propagating normal to the direction of layering is not accompanied by any shear deformation. In this instance, the only nonzero component of displacement is $u_3 = U_3 = P(x_3)e^{-ikx_3}$. Similarly, a transverse wave whose only nonzero displacement component is $u_1 = W_1 = Q(x_3)e^{-ikx_3}$, parallel to the layering, may also be investigated independently. In both instances, we can conclude from Floquet's theorem [10] that the functions $P(x_3)$ and $Q(x_3)$ are periodic with a period of $(h_f + h_m)$.

For the longitudinal wave, continuity and periodicity conditions require that the stress and deformation components be continuous across $x_3 = 0$ and $x_3 = h_f, -h_m$, i.e.,

$$\begin{aligned} \sigma_{f_{33}}(0) &= \sigma_{m_{33}}(0), \quad \sigma_{f_{33}}(h_f) = \sigma_{m_{33}}(-h_m) \\ u_{f_3}(0) &= u_{m_3}(0), \quad u_{f_3}(h_f) = u_{m_3}(-h_m) \end{aligned} \tag{36}$$

The characteristic equation resulting from these conditions is

$$\begin{aligned} \cos[k(h_f + h_m)] &= \cos(k_{f\ell}h_f)\cos(k_{m\ell}h_m) \\ &\quad - \left(\frac{1+\Gamma}{2\Gamma}\right)\sin(k_{f\ell}h_f)\sin(k_{m\ell}h_m) \end{aligned} \tag{37}$$

where

$$\Gamma = [(\lambda_m + 2G_m)k_{m\ell}]/[(\lambda_f + 2G_f)k_{f\ell}]$$

The limiting value for the phase velocity as the frequency goes to zero is found by replacing the sine functions in Eq. (37) by their arguments, and limiting cosine functions to terms which are quadratic in h_f and h_m. This results in

$$\lim_{\omega \to 0} C_{33}^2 = \frac{h_f + h_m}{\rho_c\left\{[h_f/(\lambda_f + 2G_f)] + [h_m/(\lambda_m + 2G_m)]\right\}} \tag{38}$$

For the transversely polarized shear wave, the appropriate boundary conditions are

$$\begin{aligned} \tau_{f_{13}}(0) &= \tau_{m_{13}}(0), \quad \tau_{f_{13}}(h_f) = \tau_{m_{13}}(-h_m) \\ u_{f_1}(0) &= u_{m_1}(0), \quad u_{f_1}(h_f) = u_{m_1}(-h_m) \end{aligned} \tag{39}$$

The resulting characteristic equation is identical to Eq. (37), but with $k_{f\ell}$ and $k_{m\ell}$ replaced in the arguments of the sines and cosines by k_{fS} and k_{mS}, respectively, and Γ now being equal to $G_m k_{mS}/G_f k_{fS}$.

As we might have anticipated, the zero-frequency limit for the shear wave velocity C_{31} proves to be equal to C_{13} [Eq. (31)].

This concludes our brief study of harmonic wave propagation in laminated materials. In the following chapter, we will look at techniques used to study transient wave propagation in laminated and fiber-reinforced composites.

REFERENCES

1. Y. C. Fung, Foundations of Solid Mechanics, Prentice-Hall, Englewood Cliffs, N. J., 1965.
2. C. R. Wylie, Advanced Engineering Mathematics, 2nd ed., McGraw-Hill, New York, 1960.
3. H. Kolsky, Stress Waves in Solids, Dover, New York, 1963.
4. A.E.H. Love, A Treatise on the Mathematical Theory of Elasticity, Dover, New York, 1944.
5. S. M. Rytov, "Acoustical Properties of a Thinly Laminated Medium," Soviet Phys.—Acoustics, 2, 66-80, (1956).
6. G. W. Postma, "Wave Propagation in a Stratified Medium," Geophys., 20, 780-806, (1955).
7. O. Hoffman, J. A. de Runtz, D. T. Liu, A. R. Hunter, and E. L. Kyser, "Study of Advanced Filament-Reinforced Materials," Lockheed Missiles and Space Company Report LMSC-N-5M-70-2, April 1970.
8. W. M. Ewing, W. S. Jardetzky and F. Press, Elastic Waves in Layered Media, McGraw Hill, New York, 1957.
9. C. T. Sun, J. D. Achenbach, and G. Herrmann, "Time Harmonic Waves in a Stratified Medium Propagating in the Direction of Layering," J. Appl. Mech., 35, 408-411, (1968).
10. L. Brillouin, Wave Propagation in Periodic Structures, Dover, New York, 1953.

Chapter 13

APPROXIMATE ANALYSIS TECHNIQUES FOR STRESS WAVE PROPAGATION IN COMPOSITES

I. INTRODUCTION

In this chapter, we will discuss some of the analytical techniques employed to study wave propagation in composite materials. We shall primarily concern ourselves with structural (i.e., man-made) composites, but there is much carry-over of the techniques used in the study of seismic waves in geological composite media.

While we will not deal here with dynamic fracture of composites, prediction of failure is often the ultimate goal if such materials are to be successfully utilized in a structure. In addition to the dynamic failure mode of spallation (tensile failure) exhibited by homogeneous materials, the internal material boundaries in a composite often give rise to new modes of failure such as interfacial cracking and selective spalling of constituents. Such damage within the material may influence stress wave propagation. However, the difficulty in predicting dynamic failure in composites has severely limited direct analysis of its influence on wave propagation.

In order to lessen the chances of dynamic failure, structural composites are often designed to possess sufficient heterogeneity so that significant pulse dispersion will occur. Such dispersion, characterized in the case of harmonic linear elastic waves by a phase velocity dependence on wavelength can be a very effective attenuator of stress pulses. Besides this geometrically induced dispersion, there are other dispersion mechanisms which will be mentioned briefly later in this chapter.

Almost all published analytical studies have been limited to cases wherein the constituents are assumed to be linearly elastic. Wave propagation in composites with inelastic constituents is generally studied by means of finite difference or characteristic-type computer codes. One and two-dimensional finite-difference codes have been used to study waves in multimaterial inelastic media. Two-dimensional codes are quite expensive to run, however, and the results are usually limited to the specific problem being studied.

In this chapter, we will be concerned mainly with the analytical solutions. It should be emphasized, however, that the phrase "analytical solutions" is not meant to imply that a closed-form mathematical solution exists for the problem. Indeed, in almost all cases, quantitative results can only be obtained by resorting to the computer.

II. ELASTIC THEORIES

A. Pulse Propagation Normal to Laminates

An exact solution for the one-dimensional problem of square pulses propagating in a direction normal to the plane of the laminates was presented by Kinslow [1]. Figure 1 illustrates the geometry of the structure considered. The two alternating elastic laminates are assumed to extend to infinity in the x_1 and x_2 directions. Basic to Kinslow's analysis is the assumption that the longitudinal velocity (x_3-direction) remains constant through any given laminate. This assumption is made for both incident and reflected waves. The reflected wave actually traverses material through which the incident wave has recently passed, causing changes in the laminate density and elastic constants, but this effect is assumed negligible for elastic waves in most engineering materials.

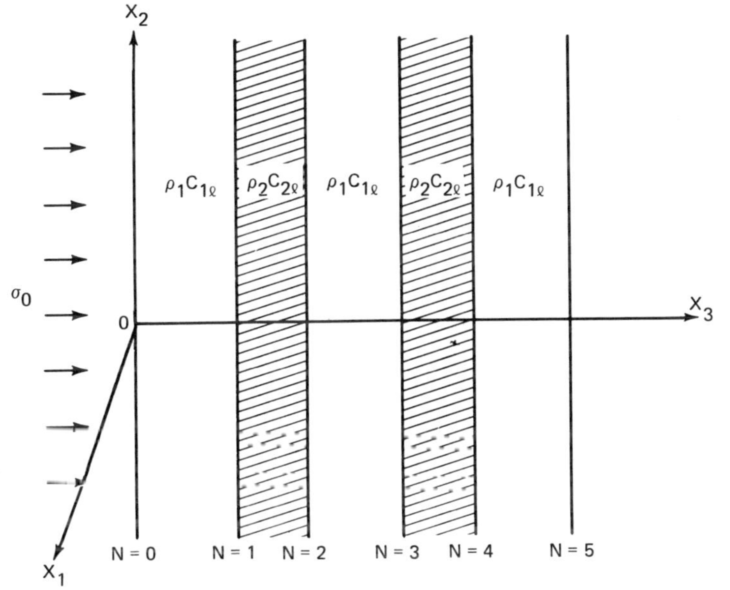

FIG. 1. Laminate geometry.

Each laminate is in a condition of uniaxial strain, and its longitudinal wave velocity is given by (see Chapter 12)

$$C_\ell = \left(\frac{\lambda + 2G}{\rho}\right)^{1/2} = \left[\frac{E(1 - \nu)}{\rho(1 + \nu)(1 - 2\nu)}\right]^{1/2}$$

where λ and G are the Lamé constants, ρ is the density, E is the modulus of elasticity, and ν is the Poisson's ratio. Across the interface between any two laminates, the conditions of continuity of stress and particle velocity must be satisfied, i.e.,

$$\sigma_O + \sigma_R = \sigma_T \tag{1}$$

and

$$v_O - v_R = v_T \tag{2}$$

where the subscripts O, R, and T refer to the original, reflected, and transmitted parts of the pulse, respectively.

Conservation of momentum requires

$$\sigma = \rho C_\ell v \tag{3}$$

Combining Eqs. (2) and (3) yields

$$\frac{\sigma_O}{\rho_O C_{O\ell}} - \frac{\sigma_R}{\rho_R C_{R\ell}} = \frac{\sigma_T}{\rho_T C_{T\ell}} \tag{4}$$

Since we are assuming that the wave velocities and densities in the individual laminates remain constant, their product, called the impedance, is also constant, i.e.,

$$\rho_O C_{O\ell} = \rho_R C_{R\ell} \tag{5}$$

Solving Eqs. (1), (4) and (5) simultaneously gives

$$\sigma_T = \left[2\rho_T C_{T\ell}/(\rho_T C_{T\ell} + \rho_O C_{O\ell})\right]\sigma_O \tag{6}$$

and

$$\sigma_R = \left[(\rho_T C_{T\ell} - \rho_O C_{O\ell})/(\rho_T C_{T\ell} + \rho_O C_{O\ell})\right]\sigma_O \tag{7}$$

Equations (6) and (7) are sometimes referred to as the "impedance mismatch" relations. If we define

$$\kappa = \rho_T C_{T\ell}/\rho_O C_{O\ell} \tag{8}$$

we find

$$\sigma_T = [2\kappa/(\kappa + 1)]\sigma_O \tag{9}$$

$$\sigma_R = [(\kappa - 1)/(\kappa + 1)]\sigma_O \tag{10}$$

and

$$\sigma_T/\sigma_R = 2\kappa/(\kappa - 1) \tag{11}$$

It is possible to draw the following conclusions from Eqs. (9)-(11):

(a) σ_T and σ_O are always of the same sign, i.e., either tensile or compressive.

(b) σ_R is of the same sign as σ_O if $\kappa > 1$ and of opposite sign if $\kappa < 1$.

(c) If $\kappa = 1$, i.e., both laminates are of equal impedance, $\sigma_T = \sigma_O$ and $\sigma_R = 0$.

The actual stress experienced by a particular laminate depends, of course, on the profile of the incident pulse as well as its magnitude σ_O and the impedance mismatch of the laminates. We will consider here rectangular input profiles only. One should recognize, however, that input profiles may profoundly influence the resulting transmitted pulse.

The above discussion was restricted to reflections at a particular interface. Let us now consider the case of multiple interfaces and reflections. Following Kinslow, we construct a two-material laminate of N interfaces joining $N + 1$ layers (Fig. 1). We apply a step pulse of magnitude σ_0 at $N = 0$. If we allow the laminates $N = 0, 2, 4, 6, \ldots$ to have impedance $\rho_1 C_{1\ell}$ and laminates $N = 1, 3, 5, 7, \ldots$ to have impedance $\rho_2 C_{2\ell}$, Eqs. (9)-(11) (with $\rho_1 C_{1\ell}$ and $\rho_2 C_{2\ell}$ replacing $\rho_0 C_{0\ell}$ and $\rho_T C_{T\ell}$) describe the magnitude of the reflected and transmitted pulses at $N = 1$. At the second interface, however, the role of the two materials is reversed, and the impedance mismatch becomes $1/\kappa$. The amplitude of the pulse transmitted across the second interface is then

$$\sigma_T\bigg|_{N=2} = \frac{2}{\kappa + 1}\,\frac{2\kappa}{\kappa + 1}\,\sigma_0 = \frac{4\kappa}{(\kappa + 1)^2}\,\sigma_0 \tag{12}$$

$$\sigma_R\bigg|_{N=2} = \frac{1 - \kappa}{1 + \kappa}\,\frac{2\kappa}{\kappa + 1}\,\sigma_0 = \frac{2\kappa(1 - \kappa)}{(\kappa + 1)^2}\,\sigma_0 \tag{13}$$

This process can be repeated indefinitely, and yields the following recursive relations:

$$\left.\begin{aligned}\sigma_T\bigg|_N &= \left(\frac{2}{1 + \kappa}\right)^N \kappa^{N/2}\sigma_0\\ \sigma_R\bigg|_N &= \frac{1 - \kappa}{2}\left(\frac{2}{1 + \kappa}\right)^N \kappa^{N/2}\sigma_0\end{aligned}\right\}\quad \text{if } N \text{ even} \tag{14}$$

or

$$\left.\begin{aligned}\sigma_T\bigg|_N &= \left(\frac{2}{1 + \kappa}\right)^N \kappa^{(N+1)/2}\sigma_0\\ \sigma_R\bigg|_N &= \frac{\kappa - 1}{2}\left(\frac{2}{1 + \kappa}\right)^N \kappa^{(N-1)/2}\sigma_0\end{aligned}\right\}\quad \text{if } N \text{ odd} \tag{15}$$

Figure 2 [1] depicts the values of σ_T in materials having values of κ ranging from 0.1 to 10 and composed of 30 laminations. Eventually, the amplitude of the transmitted pulse attenuates quite sharply, but as is evident from the figure, there exist numerous instances in which the magnitude of the transmitted pulse can actually greatly exceed that of the incident pulse. A comparison of the effect of κ on the magnitude of the transmitted wave is shown in Fig. 3.

B. Pulse Propagation Parallel to Laminates

The problem of pulse propagation parallel to the layers of a linearly elastic laminated composite was investigated by Peck and Gurtman [2]. They carried out an exact analysis of primarily compressional waves which was based upon the method of integral transforms.

The geometry considered is depicted in Fig. 2 of Chapter 12. The two layers differ in material properties and thickness and extend to infinity in the x_2 direction. The loading is a uniform normal stress σ_0 on the $x_1 = 0$ plane with a step-function dependence on time. The shear stress on the surface $x_1 = 0$ is not specified; rather, the displacements in the x_2-direction in both laminates at $x_1 = 0$ are taken to be zero. This mixed boundary condition is a familiar one in mechanical waveguide problems for rods and plates because it allows construction of solutions from sinusoidal infinite wave train modes [3-5]. One expects the far-field solution (large values of x_1) to differ little from the homogeneous boundary condition of specified normal stress and zero shear stress at $x_1 = 0$. This in fact has experimentally been shown to be the case for rods [6] as well as for laminates [7].

Since motion perpendicular to the x_1-x_3 plane is absent, the problem is two-dimensional. Peck and Gurtman obtained

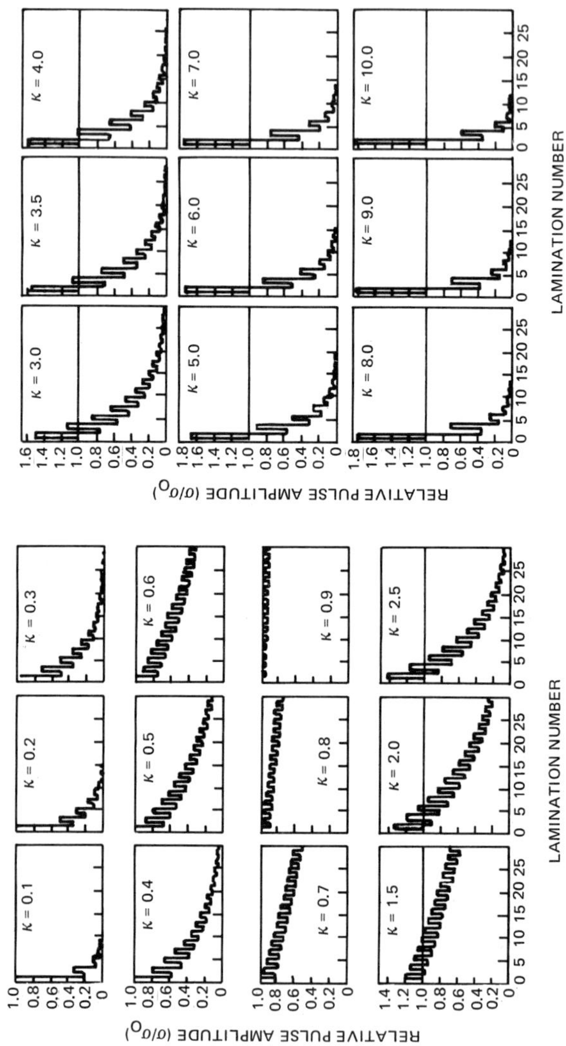

FIG. 2. Primary pulse amplitude (from Ref. [1]).

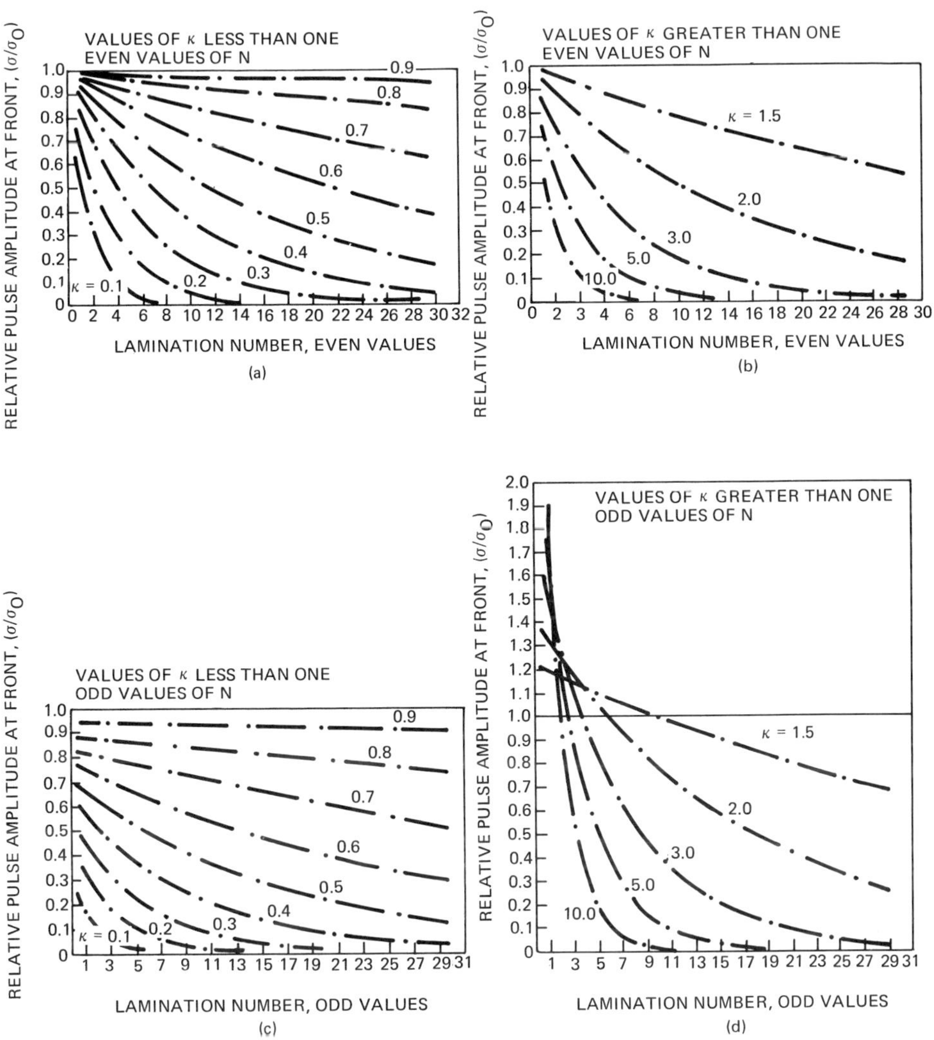

FIG. 3. Effect of impedance mismatch upon primary pulse attenuation (from Ref. [1]).

an exact solution in the form of an infinite series of integrals, from which an effective closed-form asymptotic approximation was derived. Their solution closely parallels those of Skalak [3] and Folk et al. [4], who had previously developed solutions for the closely analogous case of a homogeneous rod with transverse inertia. Beginning with the elastodynamic equations of motion in the x_1 and x_3 directions, Fourier sine and cosine transforms on x_1 are applied to u_1 and u_3 respectively, and then the Fourier exponential transform is applied on t. These transforms are defined as follows.

Sine transform:

$$f^S(k, x_3, t) = \int_0^\infty f(x_1,x_3,t) \sin(k x_1)dx_1 \tag{16}$$

Cosine transform:

$$f^C(k, x_3, t) = \int_0^\infty f(x_1,x_3,t) \cos(k x_1)dx_1 \tag{17}$$

Exponential transform:

$$f^F(x_1, x_3, \omega) = \int_0^\infty f(x_1,x_3,t)e^{i\omega t}\, dt \tag{18}$$

Application of these transforms to Eqs. (4) of Chapter 12 written in two dimensions results in

$$-\rho_j\omega^2u_{j1}^{CF} = (\lambda_j + 2G_j)\left[\frac{\sigma_0}{i\omega(\lambda_j + 2G_j)} - k^2u_{j1}^{CF}\right]$$

$$+ (\lambda_j + G_j)k\,\frac{\partial u_{j3}^{SF}}{\partial x_3} + G_j\,\frac{\partial^2 u_{j1}}{\partial x_3^2} \tag{19}$$

$$-\rho_j\omega^2 u_{j_3}^{SF} = (\lambda_j + 2G_j)\frac{\partial^2 u_{j_3}^{SF}}{\partial x_3^2} - k(\lambda_j + G_j)\frac{\partial u_{j_1}^{CF}}{\partial x_3}$$

$$- G_j k^2 u_{j_3}^{SF} \tag{20}$$

where the subscript j denotes either the fiber f or the matrix m laminate, and the boundary conditions have been applied at $x_1 = 0$.

The solution of Eqs. (19) and (20) can be written as

$$u_{j_3}^{SF} = A_j \sin(\eta_{j\ell}x_{j_3}) + B_j \cos(\eta_{j\ell}x_{j_3}) + D_j \sin(\eta_{jS}x_{j_3})$$

$$+ F_j \cos(\eta_{jS}x_{j_3}) \tag{21}$$

and

$$u_{j_1}^{CF} = -A_j\left(\frac{k}{\eta_{j\ell}}\right)\cos(\eta_{j\ell}x_{j_3}) + B_j\left(\frac{k}{\eta_{j\ell}}\right)\sin(\eta_{j\ell}x_{j_3})$$

$$+ D_j\left(\frac{\eta_{jS}}{k}\right)\cos(\eta_{jS}x_{j_3}) - F_j\left(\frac{\eta_{jS}}{k}\right)\sin(\eta_{jS}x_{j_3})$$

$$- \frac{C_{jS}^2}{C_{j\ell}^2}\frac{\sigma_0}{i\omega G_j\eta_{j\ell}} \tag{22}$$

where

$$x_{f_3} = x_3 - \frac{1}{2}h_f, \quad x_{m_3} = x_3 + \frac{1}{2}h_m$$

$$\eta_{jS}^2 = (\omega^2/C_{jS}^2) - k^2, \quad \eta_{j\ell}^2 = (\omega^2/C_{1\ell}^2) - k^2$$

The four coefficients B_j and F_j are zero by symmetry considerations, while the remaining four coefficients A_j

and D_j are determined by the application of continuity conditions at $x_3 = 0$. The denominator of the expressions for A_j and D_j is precisely the frequency equation governing the propagation of sinusoidal longitudinal wave trains in a laminated medium [Eq. (28) of Chapter 12].

Each of the infinite number of roots of this equation corresponds to a different normal mode of propagation, and since the frequency equation appears in the denominator of the integral kernels, each generates a simple pole in the transform plane.

The first of the two required inversion integrals was evaluated by means of Jordan's lemma, since the integrands approached zero uniformly over the semicircle as the radius $k \to \infty$. Cauchy's residue theorem was then invoked at each pole to yield

$$u_{j1}^{F}(x_1,x_3,\omega) = \sum_{q=1}^{\infty} \left[2i\left(\frac{\partial\phi}{\partial k}\right)^{-1} \phi\hat{u}_{j1}^{CF}(k,x_3,\omega)e^{ikx_1} \right]_{k=k_q(\omega)} \tag{23}$$

$$u_{j3}^{F}(x_1,x_3,\omega) = \sum_{q=1}^{\infty} \left[2\left(\frac{\partial\phi}{\partial k}\right)^{-1} \phi u_{j3}^{SF}(k,x_3,\omega)e^{ikx_1} \right]_{k=k_q(\omega)} \tag{24}$$

where $\hat{u}_{j1}^{CF}$ is u_{j1}^{CF} without the σ_0 term in Eq. (22), and $\phi = 0$ is the frequency equation (28) of Chapter 12. The quantities $k_q(\omega)$ are the wave numbers of sinusoidal wave propagation modes, corresponding to phase velocities $C_q(\omega)$ such that $C_q(\omega) = \omega/k_q$. The application of the second inversion integral then yields the exact solution in terms of an infinite series of integrals. The second inversion was accomplished by means of the saddle-point technique [4]. Only the first mode, and hence a single integral, was considered. This mode was the only one which was nondecaying with time, and hence it dominated the late-time, far-field solution.

Such an approximation is asymptotic in nature and becomes increasingly accurate in the vicinity of the wavefront as the pulse propagates into the material. Far behind the wavefront, or close to the source of the disturbance, the approximation breaks down. It is because of this that the technique is often called the "head of the pulse" approximation.

Peck and Gurtman found the strain in the propagation direction to be

$$\varepsilon_1(x_1, x_3, t) \sim \varepsilon_0 H(\xi) \tag{25}$$

where

$$H(\xi) = \frac{1}{3} - \int_0^{\xi} \mathrm{Ai}(\eta)\, d\eta \tag{26}$$

$$\xi = [t - (x_1/C_0)]/\overline{t} \tag{27}$$

$$\overline{t} = \left(3C_0^{-2}\alpha_1 x_1\right)^{1/3} \tag{28}$$

The α_1 term in Eq. (28) corresponds to the quadratic term coefficient in the power series expansion for the first mode longitudinal wave velocity, i.e.,

$$C_{11} = C_0 - \alpha_1 \omega^2 + \ldots \tag{29}$$

The term ε_0 is the static strain which the laminate would experience under the same load, and Ai is the Airy function. The integral of the Airy function has been tabulated [8] and the function $H(\xi)$ is shown in Fig. 4. The term ξ denotes a dimensionless time after arrival; one-third the disturbance amplitude based on unit input disturbance is seen to arrive at $t = x_1/C_0$. The quantity $\overline{t}$ may be regarded as a characteristic dispersion time of the pulse. Note that $\overline{t}$ increases as $x_1^{1/3}$ and is pro

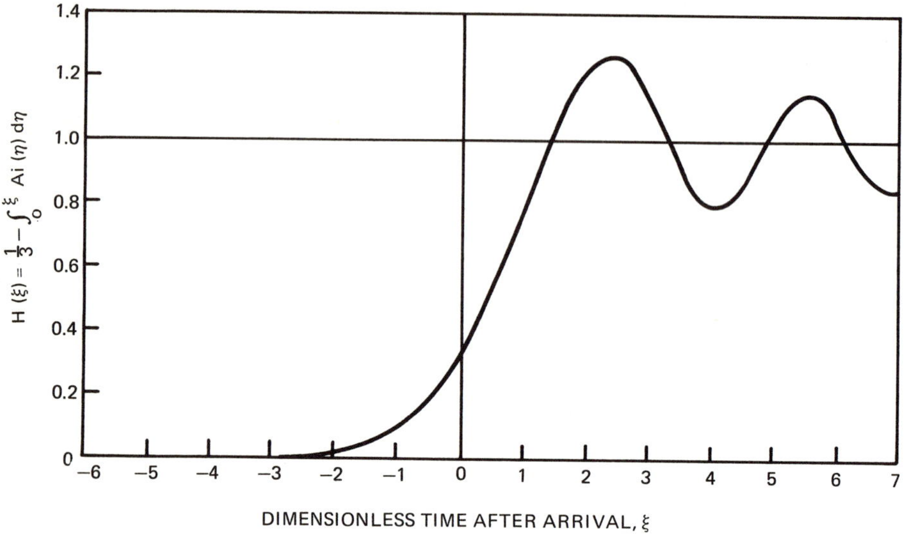

FIG. 4. The head-of-the-pulse response function.

portional to $\alpha_1^{1/3}$. The zero-frequency limit C_0 is exactly the speed one calculates from static considerations [Chapter 12, Section V(A)], so it is α_1 which actually governs the dispersivity of a particular pulse. A closed-form expression for α_1 in terms of the physical and geometric properties of the laminates is derived in Ref. [2] as

$$\alpha_1 = C_0^2\Big[4C_0^{-3}(G_f - G_m)^2 + 2C_0^{-1}(G_f - G_m)(\rho_m - \rho_f)$$

$$- C_0^{-5}(G_f - G_m)^2\zeta_{fs}^{-2}\zeta_{ms}^{-2}(\zeta_{fs}^2 + \zeta_{ms}^2)$$

$$-(1/4)C_0^{-1}\zeta_{fs}^{-2}\zeta_{ms}^{-2}(\zeta_{fs}^2 + \zeta_{ms}^2)(\rho_m - \rho_f)^2$$

$$- C_0^{-3}\zeta_{fs}^{-2}\zeta_{ms}^{-2}(G_f - G_m)(\rho_m - \rho_f)(\zeta_{fs}^2 + \zeta_{ms}^2)$$

$$- C_0^{-3}\zeta_{ms}^{-2}(G_f - G_m)^2(\zeta_{f\ell}^2 + \zeta_{ms}^2)$$

$$-(1/4)\zeta_{ms}^{-2}C_0\rho_m^2(\zeta_{f\ell}^2 + \zeta_{ms}^2) - C_0^{-1}\zeta_{ms}^{-2}\rho_m(G_f - G_m)(\zeta_{f\ell}^2 + \zeta_{ms}^2)$$

$$+ 2C_0^{-1}\zeta_{f\ell}^2(G_f - G_m)^2 - (1/2)\zeta_{f\ell}^2 C_0^3 \rho_m^2$$

$$- C_0^{-3}\zeta_{fs}^{-2}(G_f - G_m)^2(\zeta_{fs}^2 + \zeta_{f\ell}^2)$$

$$-(1/4)\zeta_{fs}^{-2}\rho_m^2 C_0(\zeta_{fs}^2 + \zeta_{f\ell}^2) - C_0^{-1}\zeta_{fs}^{-2}\rho_m(G_f - G_m)(\zeta_{fs}^2 + \zeta_{f\ell}^2)$$

$$+ 2C_0^{-1}\zeta_{m\ell}^2(G_f - G_m)^2 - (1/2)\zeta_{m\ell}^2 C_0^3 \rho_f^2$$

$$- \zeta_{ms}^{-2}C_0^{-3}(G_f - G_m)^2(\zeta_{ms}^2 + \zeta_{m\ell}^2)$$

$$-(1/4)\zeta_{ms}^{-2}\rho_f^2 C_0(\zeta_{ms}^2 + \zeta_{m\ell}^2) + C_0^{-1}\zeta_{ms}^{-2}\rho_f(G_f - G_m)(\zeta_{ms}^2 + \zeta_{m\ell}^2)$$

$$- C_0^{-3}\zeta_{fs}^{-2}(\zeta_{m\ell}^2 + \zeta_{fs}^2)(G_f - G_m)^2 - (1/4)\zeta_{fs}^{-2}C_0\rho_f^2(\zeta_{m\ell}^2 + \zeta_{fs}^2)$$

$$+ C_0^{-1}\zeta_{fs}^{-2}\rho_f(G_f - G_m)(\zeta_{m\ell}^2 + \zeta_{fs}^2) - 2C_0^{-1}(\zeta_{f\ell}^2 + \zeta_{m\ell}^2)(G_f - G_m)^2$$

$$-(1/4)\zeta_{ms}^{-2}\rho_f\rho_m C_0(\zeta_{ms}^2 + \zeta_{m\ell}^2)(h_m/h_f) - (1/2)C_0^3\rho_f\rho_m\zeta_{m\ell}^2(h_m/h_f)$$

$$-(1/4)\zeta_{fs}^{-2}\rho_f\rho_m C_0(\zeta_{fs}^2 + \zeta_{m\ell}^2)(h_m/h_f)$$

$$-(1/4)\zeta_{fs}^{-2}\rho_f\rho_m C_0(\zeta_{f\ell}^2 + \zeta_{fs}^2)(h_f/h_m)$$

$$-(1/2)C_0^3\rho_f\rho_m\zeta_{f\ell}^2(h_f/h_m) - (1/4)\zeta_{ms}^{-2}\rho_f\rho_m C_0(\zeta_{ms}^2 + \zeta_{f\ell}^2)(h_f/h_m)$$

$$- C_0^{-1}\zeta_{fs}^{-2}\zeta_{ms}^{-2}(G_f - G_m)^2\zeta_{m\ell}^2(\zeta_{fs}^2 + \zeta_{ms}^2)\Big]^{-1}$$

$$\times \Big\{ \Big[-(1/12)C_0^{-4}(G_f - G_m)^2 - (1/48)(\rho_m - \rho_f)^2$$

$$-(1/12)C_0^{-2}(G_f - G_m)(\rho_m - \rho_f)\Big](\zeta_{fs}^2 h_f^2 + \zeta_{ms}^2 h_m^2)$$

$$+ \Big[-(1/12)C_0^{-2}\zeta_{f\ell}^2(G_f - G_m)^2 - (1/48)\zeta_{f\ell}^2\rho_m^2 C_0^2$$

$$-(1/12)\zeta_{f\ell}^2\rho_m(G_f - G_m)\Big](h_m^2\zeta_{ms}^2 + h_f^2\zeta_{f\ell}^2)$$

$$+\Big[-(1/12)C_0^{-2}\zeta_{m\ell}^2(G_f - G_m)^2 - (1/48)\zeta_{m\ell}^2\rho_f^2 C_0^2$$

$$+(1/12)\rho_f\zeta_{m\ell}^2(G_f - G_m)\Big](\zeta_{fs}^2 h_f^2 + \zeta_{m\ell}^2 h_m^2)$$

$$-(1/12)(G_f - G_m)^2\zeta_{f\ell}^2\zeta_{m\ell}^2(\zeta_{f\ell}^2 h_f^2 + \zeta_{m\ell}^2 h_m^2)$$

$$-(1/48)\zeta_{m\ell}^2\rho_f\rho_m C_0^2(\zeta_{m\ell}^2 + \zeta_{fs}^2)(h_m^3/h_f)$$

$$-(1/48)\zeta_{f\ell}^2\rho_f\rho_m C_0^2(\zeta_{f\ell}^2 + \zeta_{fs}^2)(h_f^3/h_m)\Big\} \qquad (30)$$

where

$$\zeta_{i\ell}^2 \equiv C_{i\ell}^{-2} - C_0^{-2}$$

and

$$\zeta_{iS}^2 \equiv C_{iS}^{-2} - C_0^{-2}, \quad i = f,m$$

This expression is obviously very complicated in its dependence on most of the composite parameters, and is best studied by numerical means.

Equation (25) implies that the strain in the x_1-direction is uniform over the cross section of the composite. The stress in the x_1-direction is in general, therefore, highly nonuniform since the moduli of the laminates may differ arbitrarily. For the average stress over the cross section $\bar{\sigma}(x_1, t)$, one has

$$\bar{\sigma}(x_1,t) \sim \sigma_0 H(\xi) \qquad (31)$$

For the unit step pulse, we see from Fig. 4 that dispersion induces an overshoot in the composite response of 28%. For finite-width pulses, however, dispersion will often cause the initial disturbance to attenuate. It is a straightforward matter to construct the response of the composite to a rectangular pulse of duration t_0 by simply taking the difference of two separate step loads of opposite sign and applying them a time t_0 apart. Peck presents the results of such a calculation for $t_0/\bar{t} = 1$ in Ref. [9]. His results are depicted in Fig. 5.

C. Pulse Propagation Parallel to Fibers

The head-of-the-pulse technique used in determining the transient response of laminated composites is also applicable to transient pulse propagation in a fiber-reinforced composite. Essentially, the problem hinges on the determination of the α_1 term in Eq. (29). In fact, for the approximate far-field solution, it is the α_1 term

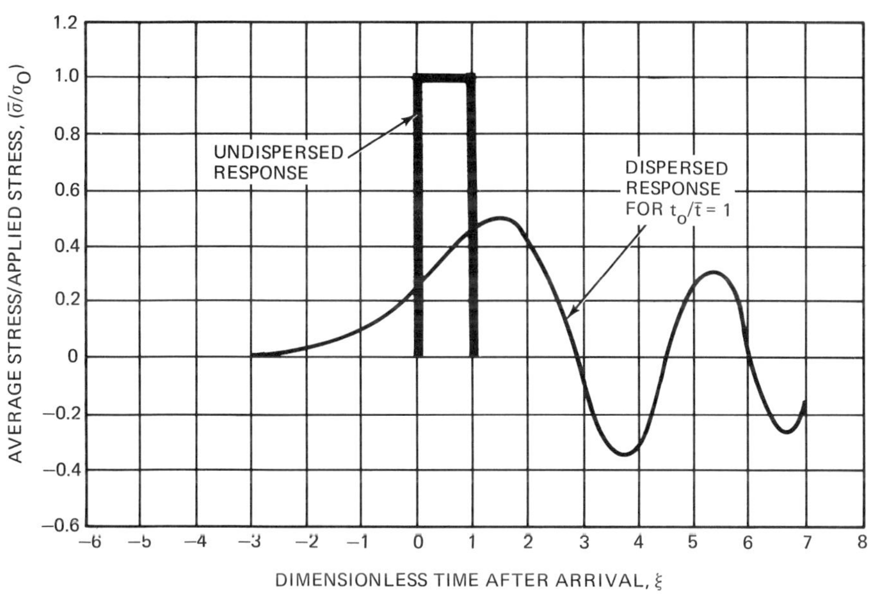

FIG. 5. An attenuated pulse according to the head-of-the-pulse approximation.

which contains all of the composite's dispersion information. Once the dispersion relation is determined, the specific geometry of the composite is no longer a factor in the head-of-the-pulse solution. As was discussed in Section II-B, α_1 corresponds to the quadratic term coefficient in the power series expansion for the first-mode longitudinal wave velocity in the vicinity of the origin $(\omega \to 0)$. To study transient response in the fiber-reinforced composites using this technique, then, we must first determine the dispersion relation.

We might reasonably expect to model a hexagonally packed fiber-reinforced composite as a single cylinder surrounded by an annular sheath (Fig. 17, Chapter 3). The annulus is subjected to the constraints of zero radial displacement and shear stress at its outermost radius (r_m).

An exact solution for this problem has been obtained by Hoffman, et al. [10]. Their solution follows closely that derived by Rytov [11] for wave propagation in the direction of laminates. The determination of α_1 from the solution of Ref. [10] involves expanding a 6×6 determinant whose arguments are in turn functions of k and ω. This obviously is an extremely tedious procedure [witness the algebraic complexity of α_1 for the laminates, Eq. (30), which was determined in this manner].

It is of interest, therefore, to pursue an approximate solution developed by Jones [12] which is based on variational principles. This solution enables one to extract the desired α_1 quite easily from a simplified dispersion relation.

Jones assumes the fiber and matrix to be rigidly bonded together, and hypothesizes a state of deformation in the two materials subject to the constraint that the fibers are considerably stiffer than the matrix material. The last statement implies that the fiber follows the assumptions made in simple rod theory, i.e., (1) plane sections remain plane and (2) transverse motion may be ignored. In light of this, the displacement field is assumed to be parabolic, such that shear stresses vanish at the boundary r_m. The geometry is depicted in Fig. 6. For this problem, then, the strain components are

$$\varepsilon_{11} = \partial u_1/\partial x_1, \quad \varepsilon_{r1} = \partial u_1/\partial r$$

where all other $\varepsilon_{ij} = 0$, $i,j = r,\theta,1$. For each material, the kinetic energy T is

$$T = (\rho/2)\int_0^{\ell}\int_A (\partial u_1/\partial t)^2 \, dx_1 \, dA \tag{32}$$

and the strain energy U is

$$U = \frac{1}{2}\int_0^{\ell}\int_A \left[(\lambda + 2G)(\partial u_1/\partial x_1)^2 + G(\partial u_1/\partial r)^2\right]dx_1 \, dA \tag{33}$$

where the interval $(0,\ell)$ is an arbitrary length in the x_1-direction, and A denotes the cross sectional area. The displacement of the fiber is assumed to be independent of r since the fiber is considerably stiffer than the matrix material:

$$u_{f_1}(r,x_1,t) = u_1(x_1,t) \tag{34}$$

Substitution of Eq. (34) into the expressions for the kinetic and strain energies of the fiber yields

$$\begin{aligned} T_f &= (\rho_f A_f/2)\int_0^{\ell} (\partial u_1/\partial t)^2 \, dx_1 \\ U_f &= (E_f A_f/2)\int_0^{\ell} (\partial u_1/\partial x_1)^2 \, dx_1 \end{aligned} \tag{35}$$

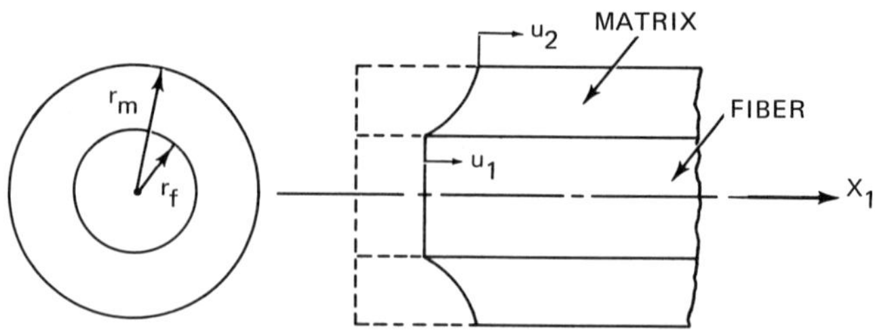

FIG. 6. Fiber geometry.

where we have replaced $\lambda_f + 2G_f$ by the modulus of elasticity E_f for the fiber. Here ρ_f and A_f are the fiber density and cross sectional area, respectively.

The use of E_f for the modulus of the fiber implies that its outer boundary r_f is completely unconstrained. While this is not the actual condition experienced by the fiber, the error introduced by this simplification should be small in cases in which the stiffness of the fiber greatly exceeds that of the matrix.

In the matrix annulus, the deformation is assumed to be given by

$$u_{m_1} = u_1(x_1, t) + u_2(x_1, t) f(r) \tag{36}$$

where $f(r)$ is a function of r only. The choice of $f(r)$ is arbitrary, but it must satisfy the continuity condition that u_2 vanish at $r = r_f$ and the symmetry condition $df/dr = 0$ at $r = r_m$ to ensure that the shear stress τ_{r1} vanish at $r = r_m$. A parabola satisfying these conditions is

$$f(r) = 2(r - r_f)/h - (r - r_f)^2/h^2 \tag{37}$$

where $h = r_m - r_f$ is the thickness of the annulus. Substitution of Eqs. (36) and (37) into the expressions for the kinetic and potential energies of the annulus (denoted by subscript m) yields

$$T_m = \frac{\rho_m A_m}{2} \int_0^{\ell} \left[\left(\frac{\partial u_1}{\partial t} \right)^2 + \frac{4}{3} \beta_1 \frac{\partial u_1}{\partial t} \frac{\partial u_2}{\partial t} + \frac{8}{15} \beta_2 \left(\frac{\partial u_2}{\partial t} \right)^2 \right] dx_1$$

$$U_m = \frac{A_m}{2} \int_0^{\ell} \left\{ (\lambda_m + 2G_m) \left[\left(\frac{\partial u_1}{\partial x_1} \right)^2 + \frac{4}{3} \beta_1 \frac{\partial u_1}{\partial x_1} \frac{\partial u_2}{\partial x_1} \right. \right.$$

$$\left. \left. + \frac{8}{15} \beta_2 \left(\frac{\partial u_2}{\partial x_1} \right)^2 \right] + G_m \frac{4}{3} \beta_3 (u_2)^2 \right\} dx_1 \tag{38}$$

where

$$\beta_1 = \frac{1 + (5/8a)}{1 + (a/2)}, \quad \beta_2 = \frac{1 + (11/16a)}{1 + (a/2)}$$

$$\beta_3 = \frac{1 + (a/4)}{1 + (a/2)}, \quad a = h/r_f$$

The Lagrangian $\mathcal{L}$ is now defined as

$$\mathcal{L} = T_m + T_f - U_m - U_f \tag{39}$$

The equations of motion are obtained from Hamilton's principle,

$$\delta \int_{t_0}^{t} \mathcal{L} \, dt = 0 \tag{40}$$

where δ denotes the first variation.

Performing the indicated variation in Eq. (40) and using Eqs. (35) and (38), we obtain

$$\left[E_c \frac{\partial^2 u_1}{\partial x_1^2} - \rho_c \frac{\partial^2 u_1}{\partial t^2} \right] + \frac{2}{3} V_m \beta_1 \left[(\lambda_m + 2G_m) \frac{\partial^2 u_2}{\partial x_1^2} \right.$$

$$\left. - \rho_m \frac{\partial^2 u_2}{\partial t^2} \right] = 0$$

$$\beta_1 \left[(\lambda_m + 2G_m) \frac{\partial^2 u_1}{\partial x_1^2} - \rho_f \frac{\partial^2 u_1}{\partial t^2} \right] + \frac{4}{5} \beta_2 \left[(\lambda_m + 2G_m) \frac{\partial^2 u_2}{\partial x_1^2} \right.$$

$$\left. - \rho_m \frac{\partial^2 u_2}{\partial t^2} - \frac{5}{2} \beta_4 \frac{G_m}{h^2} u_2 \right] = 0 \tag{41}$$

where

$$V_m = A_m/(A_f + A_m)$$

and

$$\beta_4 = \frac{1 + (a/4)}{1 + (11/16a)}$$

The terms E_c and ρ_c are so-called "law-of-mixtures" quantities defined by

$$E_c = (\lambda_m + 2G_m)V_m + E_f(1 - V_m), \quad \rho_c = \rho_m V_m + \rho_f(1 - V_m)$$

We next assume harmonic forms for the displacements u_1 and u_2 of the type

$$u_1 = Ae^{i(kx-\omega t)}, \quad u_2 = Be^{i(kx-\omega t)} \tag{42}$$

Substituting Eqs. (42) into Eq. (41) and requiring that the system of equations have nontrivial solutions, we obtain the dimensionless frequency equation

$$(1 - \overline{C}^2)\left(1 - \frac{C_0^2}{C_m^2}\overline{C}^2 + \frac{5}{2}\frac{G_m}{\lambda_m + 2G_m}\frac{\beta_4}{\overline{K}^2}\right)$$

$$- \frac{5}{6} b_0 \frac{\beta_1^2}{\beta_2}\left(1 - \frac{C_0^2}{C_m^2}\overline{C}^2\right)^2 = 0 \tag{43}$$

where

$$C_0^2 = E_c/\rho_c \ , \quad C_m^2 = (\lambda_m + 2G_m)/\rho_m \ , \quad \overline{k} = kh \ , \quad \overline{\omega} = \omega h/C_0$$

and

$$\overline{C} = \overline{\omega}/\overline{k} = C/C_0 \ , \quad b_0 = \left[(\lambda_m + 2G_m)/E_c\right]V_m$$

Equation (43) is quadratic in C^2 and has, therefore, two roots. These roots correspond to the first two modes of longitudinal wave propagation. Solutions to Eq. (43) are compared with exact elasticity solutions in Fig. 7. As can be seen, agreement between the approximate and exact theories is excellent for the first mode, but the agreement for the second mode is rather poor.

According to the head-of-the-pulse theory, the averaged transient response of a fiber-reinforced composite to a uniform step pressure pulse σ_0 is therefore

$$\overline{\sigma}(x_1,t)/\sigma_0 = \frac{1}{3} - \int_0^{\xi} Ai(\eta)\, d\eta \qquad (31')$$

where

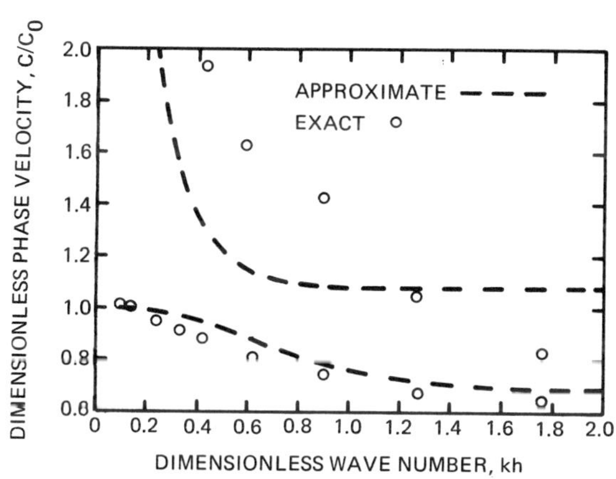

FIG. 7. Dimensionless phase velocity versus dimensionless wave number (from Ref. [12]).

$$\xi = \left[t - (\overline{x}_1/C_0)\right]/\overline{t} \tag{27'}$$

and

$$\overline{t} = \left(3C_0^{-2}\alpha_1 x_1\right)^{1/3} \tag{28'}$$

Since the integral of the Airy function is tabulated, one need determine only the quantities C_0 and α_1 to determine the composite response.

In Jones' solution to the problem, C_0 corresponds roughly to the zero-frequency limit of the first-mode longitudinal wave velocity.* The real advantage of Jones' solution to the problem lies in the ease in which the α_1 term can be determined. Writing Eq. (43) in the form

$$(1 - \overline{C}^2)\left(1 - \frac{C_0^2}{C_m^2}\overline{C}^2 + \frac{5}{2}\beta_4 \frac{G_m}{\lambda_m + 2G_m}\frac{\overline{C}^2}{\overline{\omega}^2}\right) - \frac{5}{6} b_0 \frac{\beta_1^2}{\beta_2}\left(1 - \frac{C_0^2}{C_m^2}\overline{C}^2\right)^2 = 0 \tag{44}$$

we search for a solution of the form $\overline{C} = 1 - \overline{\alpha}_1\overline{\omega}^2 + \ldots$ as $\overline{\omega} \to 0$. This is easily found to be

$$\overline{\alpha}_1 = \frac{1}{6}\frac{\beta_1^2}{\beta_2\beta_4} b_0 \frac{\lambda_m + 2G_m}{G_m}\left(1 - \frac{C_0^2}{C_m^2}\right)^2 \tag{45}$$

In dimensional form, we have

*An "exact" elastic prediction of C_0 as given by Hoffman, et al. [10] is

$$C_0^2 = \rho_c^{-1}\left[(\lambda_m + 2G_m)V_m + (\lambda_f + 2G_f)(1 - V_m) - \frac{(\lambda_f - \lambda_m)^2 V_m(1 - V_m)}{(\lambda_f + G_f)V_m + (\lambda_m + G_m)(1 - V_m) + G_f}\right]$$

$$\alpha_1 = \bar{\alpha}_1 h^2/C_0^2 \tag{46}$$

Using these values of α_1 and C_0, transient solutions of stress or strain similar to those of Figs. 4 and 5 may be found for the fiber-reinforced composite.

III. CONTINUUM THEORIES

From the point of view of mathematical analysis, it would be particularly advantageous to be able to represent a heterogeneous composite possessing discontinuous material properties as an equivalent homogeneous continuum whose material properties are continuous, albeit anisotropic. Such a model obviously cannot be used to determine local stress or strain fields at discontinuities, but if interest centers on the gross response of the composite, i.e., the dispersion and attenuation of stress waves propagating through such a medium, continuum theories can be extremely useful.

So-called "effective modulus" theories, based on elementary averaging procedures [13, 14] are generally useful only in static analyses since such theories cannot adequately account for the geometrically dispersive nature of elastic waves in composites. In two papers dealing with sound wave propagation in laminated and fibrous composites, Behrens [15, 16] attempts to partially alleviate this shortcoming. He derives approximate phase velocity relations for the propagation of elastic waves that are valid for wavelengths which are long compared with the intercomponent spacings. This approach yields averaged elastic constants for the composite in terms of geometric data and the elastic constants of the constituents. While this technique does not properly model the general dispersive behavior of composites, it does reveal the conditions under which the composite can be adequately described as a quasihomogeneous material. If the wavelength is short compared with the intercomponent spacings, the wave distinguishes each constituent separately by being reflected at each interface. For long wavelengths, however, the wave acts as if it is traversing a quasihomogeneous medium, and the frequency-independent phase velocity determines the averaging rule for calculating composite elastic constants.

A continuum approach to the study of elastic waves in heterogeneous materials has also been proposed by Ben-Amoz [17]. He substitutes a homogeneous medium endowed with dispersion characteristics for the given heterogeneous material. The new medium is characterized by (1) average, directionally dependent elastic moduli, (2) influence coefficients which connect the displacement fields in the two media, and (3) a set of parameters which relate the average wave velocities of the composite to that of the continuum. In another paper, Ben-Amoz [18] applied extremum principles to establish upper and lower bounds on the effective elastic moduli and wave velocities in heterogeneous materials.

Khaikovich and Khalfin [19] derived effective dynamic parameters for an acoustic medium containing regularly spaced spherical acoustic scatterers. The effective parameters with which they chose to characterize the original composite are phase velocity, density, and specific heats. These parameters may take on complex values and it is this property which allows the homogeneous media to behave in a manner at least phenomenologically similar to the actual composite. A similar continuum approach has been suggested by Mok [20, 21] in a study of plane sinusoidal waves propagating through a fiber-reinforced elastic material in a direction normal to the uniaxially aligned fibers. Noting that the effective modulus of a composite material is frequency dependent due to dispersion and that composites often appear to be dissipative because of the transformation of energy between longitudinal and shear modes of wave propagation, Mok proposed comparing the gross composite properties with those of a homogeneous, linear viscoelastic solid. The derivation consists in obtaining the scattered waves from a single fiber and integrating them to obtain the resultant transmitted and reflected waves for a layer of the composite. These expressions are then compared with equivalent ones obtained for a homogeneous viscoelastic plate and relations are derived for the effective dynamic moduli of the composite. These moduli have complex values and imply that the composite material is both dispersive and dissipative. While the latter result is seemingly undesirable for elastic wave propagation, the close resemblence of phenomenological response for the actual composite and viscoelastic plate model appears to justify the use of the theory for certain types of problems.

Yet another type of treatment is the utilization of the so-called "extended" or nonclassical theories of continuum mechanics. Such theories hypothesize continua possessing microstructure, i.e., the medium does not merely consist of assemblages of continuous points, but also includes the orientation of each infinitesimal rigid mass element as an independent kinematic quantity. Cosserat and Cosserat [22] constructed a theory of elasticity based on this notion and introduced the concept of couple stresses. More recently, Mindlin [23] extended their theory using the idea of a micromedium unit cell which is itself free to deform at each point within the macromedium. This model possesses the dispersive wave behavior exhibited by composite materials on the macroscale. An attempt to apply this type of theory to composites was made by Habip [24]. He studied the propagation of plane harmonic waves in an unbounded elastic particulate solid. Habip derived expressions for the strain and kinetic energies of the medium. From these, explicit dispersion relations governing longitudinal wave propagation in terms of the relative properties of the constituent materials were obtained.

Tauchert et al. [25] formulated a continuum theory for composites based on a lattice concept. The matrix and inclusions of the composite are idealized as a gridwork of thermoelastic bars attached to small, rigid masses. This formulation leads to a system of equations which are identical with those of the linear couple stress theory.

Extended continuum theories have also been employed by Achenbach, Herrmann and Sun to model the dynamic behavior of composites. They have called their approach the "effective stiffness" theory and have applied it to a wide variety of composite geometries and modes of wave propagation. In some of their earlier papers [26-29], Achenbach, Herrmann and Sun used representative elastic moduli for the matrix material and combined the elastic and geometric properties of the fibers or laminates into "effective stiffnesses". An undesirable result of this approach was that the representative elastic moduli of the matrix were not the actual elastic constants of the material, but were instead fictitious moduli which were computed from the zero-frequency limit of the phase velocity.

In later papers, a more accurate system of displacement equations of motion was derived. This later procedure [30] involves expressing the displacements in both matrix and reinforcing (or fiber) laminates as linear expansions about their respective centerlines. It is assumed that the dimensions of the laminates are small relative to the overall dimensions of the body and the characteristic wavelength of the disturbance. The authors consider a medium as depicted in Fig. 8, consisting of two alternating, homogeneous, isotropic laminates. Two local coordinate systems $(x_1, x_2, \bar{x}_3^m)$ and $(x_1, x_2, \bar{x}_3^f)$ are introduced

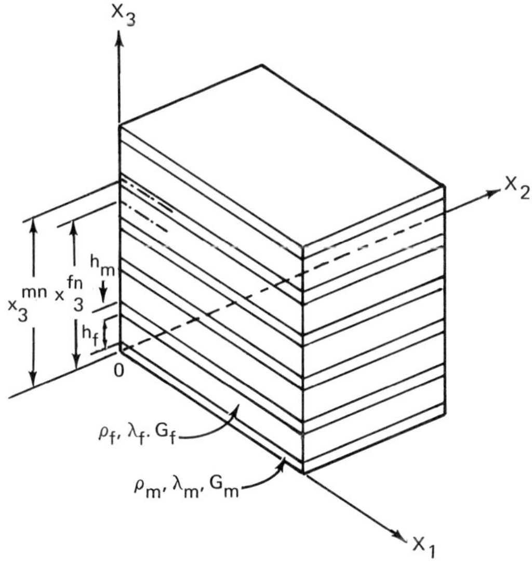

FIG. 8. Laminated medium.

for the nth pair of laminates (see Fig. 9). The midplane of the matrix laminate is denoted by x_3^{mn} and that of the fiber laminate by x_3^{fn}. The displacements at these midpoints are denoted by $u_{0i}^{mn}(x_1, x_2, x_3^{mn}, t)$ and $u_{0i}^{fn}(x_1, x_2, x_3^{fn}, t)$.

The displacements throughout the nth matrix and fiber layers, u_i^{mn} and u_i^{fn} respectively, are expressed as an infinite series of Legendre polynomials. For the fiber layer, we have

$$u_i^{fn} = \sum_{J=0}^{\infty} P_J\left(\bar{x}_3^f/h_f\right)u_{Ji}^{fn}\left(x_1, x_2, x_3^{fn}, t\right) \tag{47}$$

where h_f is the fiber laminate thickness and $P_J\left(\bar{x}_3^f/h_f\right)$ are the Legendre polynomials

$$\begin{aligned} P_0\left(\bar{x}_3^f/h_f\right) &= 1, \quad P_1\left(\bar{x}_3^f/h_f\right) = \bar{x}_3^f/h_f \\ P_2\left(\bar{x}_3^f/h_f\right) &= \left[3\left(\bar{x}_3^f/h_f\right)^2 - 1\right]/2 \end{aligned} \tag{48}$$

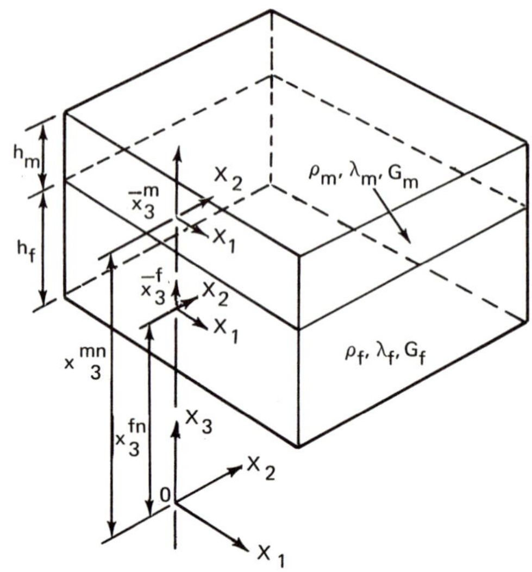

FIG. 9. Pair of reinforcing and matrix layers.

Similar expressions may be written for $u_i^{mn}(x_1, x_2, x_3^{mn}, \bar{x}_3^m, t)$. Since the laminate thicknesses are small with respect to a characteristic disturbance wavelength, the polynomial series are truncated and only the linear terms in Eq. (47) are retained. The displacements thus become

$$u_i^{fn} = u_{0i}^{fn}\left(x_1, x_2, x_3^{fn}, t\right) + \bar{x}_3^f \psi_{3i}^{fn}\left(x_1, x_2, x_3^{fn}, t\right) \tag{49}$$

and

$$u_i^{mn} = u_{0i}^{mn}\left(x_1, x_2, x_3^{mn}, t\right) + \bar{x}_3^m \psi_{3i}^{mn}\left(x_1, x_2, x_3^{mn}, t\right) \tag{50}$$

In Eq. (49), u_{0i}^{fn} is the displacement at the midplane of the nth fiber layer, ψ_{31}^{fn} and ψ_{32}^{fn} represent antisymmetric thickness shear deformations, and ψ_{33}^{fn} represents symmetric thickness stretch deformation. Continuity of displacements across the interface between the nth matrix and fiber layers provides the relation

$$u_{0i}^{mn}\left(x_1, x_2, x_3^{mn}, t\right) - u_{0i}^{fn}\left(x_1, x_2, x_3^{fn}, t\right)$$

$$= \frac{1}{2} h_f \psi_{3i}^{fn}\left(x_1, x_2, x_3^{fn}, t\right) + \frac{1}{2} h_m \psi_{3i}^{mn}\left(x_1, x_2, x_3^{mn}, t\right) \tag{51}$$

For an isotropic, linearly elastic material, the strain energy density U can be written as

$$U = \frac{1}{2}\lambda\varepsilon_{ii}\varepsilon_{jj} + G\varepsilon_{ij}\varepsilon_{ij} \tag{52}$$

where λ and G are the Lamé constants and ε_{ij} is the tensorial strain. An approximate expression for the strain energy density in the nth fiber layer may then be found by substituting Eq. (49) into the strain displacement equations (3) of Chapter 12 and then using Eq. (52). Integration over the thickness h_f yields the strain energy per unit surface area stored in the nth fiber laminate, i.e.,

$$\begin{aligned} U^{fn} = {} & \frac{1}{2} h_f(\lambda_f + 2G_f)\left[\left(\frac{\partial u_{01}^{fn}}{\partial x_1}\right)^2 + \left(\frac{\partial u_{02}^{fn}}{\partial x_2}\right)^2 + \left(\psi_{33}^{fn}\right)^2\right] \\ & + h_f\lambda_f\left[\frac{\partial u_{01}^{fn}}{\partial x_1}\frac{\partial u_{02}^{fn}}{\partial x_2} + \frac{\partial u_{01}^{fn}}{\partial x_1}\psi_{33}^{fn} + \frac{\partial u_{02}^{fn}}{\partial x_2}\psi_{33}^{fn}\right] \\ & + \frac{1}{2} h_f G_f\left[\left(\frac{\partial u_{03}^{fn}}{\partial x_1} + \psi_{31}^{fn}\right)^2 + \left(\frac{\partial u_{03}^{fn}}{\partial x_2} + \psi_{32}^{fn}\right)^2\right. \\ & \left. + \left(\frac{\partial u_{01}^{fn}}{\partial x_2} + \frac{\partial u_{02}^{fn}}{\partial x_1}\right)^2\right] \end{aligned} \tag{53}$$

A similar expression can be derived for the matrix layer.

The kinetic energy density T for a homogeneous material may be written as

$$T = \sum_{i=1}^{3} \frac{1}{2} \rho(\partial u_i/\partial t)^2 \tag{54}$$

Substituting the displacement expansion (49) into Eq. (54) yields the kinetic energy per unit surface area for the nth fiber layer

$$T^{fn} = \sum_{i=1}^{3} \left\{ \frac{1}{2}(h_f + h_m)V_f\rho_f\left(\frac{\partial u_{0i}^{fn}}{\partial t}\right)^2 + \frac{1}{2}(h_f + h_m)I_f\left(\frac{\partial \psi_{3i}^{fn}}{\partial t}\right)^2 \right\} \tag{55}$$

where ρ_f is the fiber laminate density and

$$V_f = h_f/(h_f + h_m) \tag{56}$$

$$I_f = (1/12)h_f^2V_f\rho_f \tag{57}$$

The kinetic energy per unit surface area of the nth matrix laminate is obtained by changing the subscripts and superscripts f in Eq. (55) to m, replacing V_f by $(1 - V_f)$, and noting that

$$I_m = (1/12)h_m^2(1 - V_f)\rho_m \tag{58}$$

In a given parallelepiped of the laminate of sides 1, 1, and ℓ in the x_1, x_2, and x_3 directions respectively, the total strain and kinetic energies are

$$U_\ell = \sum_{n=1}^{N} (U^{fn} + U^{mn}) \tag{59}$$

$$T_\ell = \sum_{n=1}^{N} (T^{fn} + T^{mn}) \tag{60}$$

where

$$N = \ell/(h_f + h_m)$$

Sun et al. now effect the transition from the original laminate to an equivalent homogeneous continuum by means of the following procedure. The summations of Eqs. (59) and (60) over 2N discrete points x_3^{fn} and x_3^{mn} are approximated by weighted integrations over x_3. This "smoothing operation" is defined by

$$U_\ell = \sum_{n=1}^{N} (U^{fn} + U^{mn}) \simeq \int_\ell \frac{1}{h_f + h_m} (U^f + U^m)\, dx_3 \tag{61}$$

$$T_\ell = \sum_{n=1}^{N} (T^{fn} + T^{mn}) \simeq \int_\ell \frac{1}{h_f + h_m} (T^f + T^m)\, dx_3 \tag{62}$$

The superscript n on the right-hand side of Eqs. (61) and (62) has been removed since U^f, U^m, T^f and T^m are now defined for all x_3. Thus, these new strain and kinetic energy densities are continuous functions of x_1, x_2, x_3 and t, i.e.,

$$U = \left[U^f(x_i,t) + U^m(x_i,t)\right]/(h_f + h_m) \tag{63}$$

$$T = \left[T^f(x_i,t) + T^m(x_i,t)\right]/(h_f + h_m) \tag{64}$$

Hamilton's principle is now invoked to obtain the equations of motion. Thus

$$\delta \int_{t_0}^{t_1} (T - U)\, dx_1\, dx_2\, dx_3\, dt = 0 \tag{65}$$

where δ denotes the first variation. The variation of the integral yields the Euler-Lagrange equations of motion in terms of the displacements of the composite, i.e.,

$$\sum_{j=1}^{4} \frac{\partial}{\partial q_j} \left[\frac{\partial (T - U)}{\partial(\partial \Gamma_s/\partial q_j)}\right] - \frac{\partial (T - U)}{\partial \Gamma_s} = 0 \tag{66}$$

where f_s are the dependent variables (u_i, ψ^f_{3i}, ψ^m_{3i}) and q_j are the spatial and time coordinates x_i and t. The coefficients in these equations are functions of the geometric and material properties of the original laminates. The resulting equations are very similar to those obtained by Mindlin in his previously mentioned paper dealing with higher-order continuum theories. The relationship between the "effective stiffness" and higher-order continuum theories is explained more fully by Herrmann and Achenbach in Ref. [31].

Assuming harmonic forms for the displacement u_i, the equations of motion were used to study the propagation of plane harmonic waves both in the direction of and normal to, the direction of layering (see also Chapter 12, Section IV). Results of longitudinal phase velocity versus wave number for propagation in the direction of layering are depicted in Fig. 10. The solid curve represents the numerical

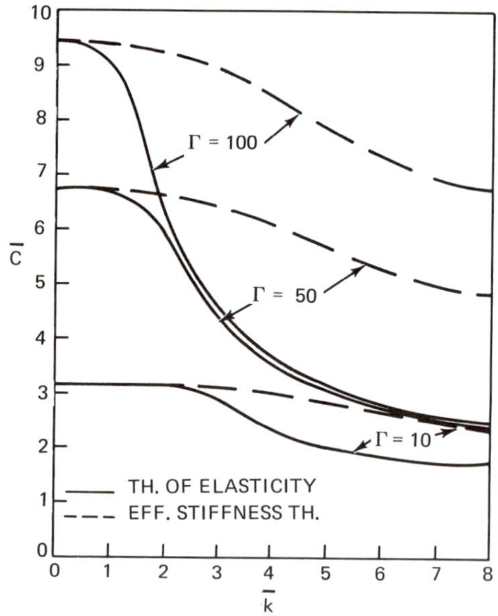

FIG. 10. Lowest symmetric mode propagating in the direction of the layering (from Ref. [30]).

solution to Rytov's frequency equation as given in Ref. [30]. The dashed curve represents the "effective stiffness prediction". Figure 11 shows both the exact and effective stiffness results for longitudinal waves propagating normal to the layers. In both these plots, the dimensionless phase velocity $\overline{C}$ is defined as

$$\overline{C} = C/(G_m/\rho_m)^{1/2} \tag{67}$$

where C is the phase velocity for the particular mode of propagation being studied, and the subscript m refers to the matrix laminate material properties. The dimensionless wave number $\overline{k}$ is defined as

$$\overline{k} = kh_f \tag{68}$$

where k is the dimensional wave number. The parameter Γ is the ratio of the shear moduli for the two laminates, i.e.,

$$\Gamma = G_f/G_m \tag{69}$$

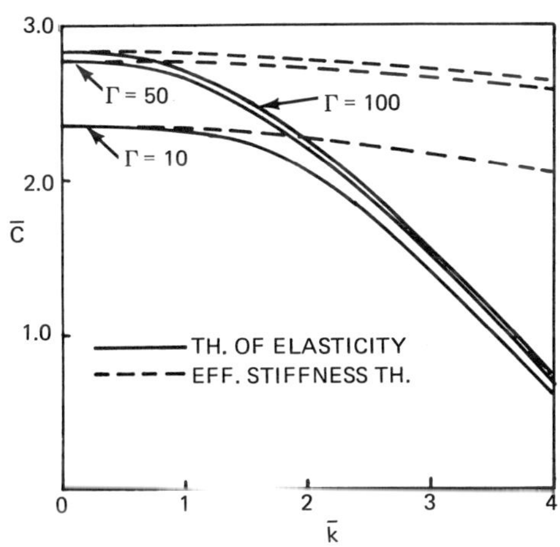

FIG. 11. Lowest longitudinal mode propagating normal to the layering (from Ref. [30]).

The calculations were carried out for three different values of Γ using a density ratio of $\rho_f/\rho_m = 3$, a fiber volume fraction $V_f = 0.8$, and Poisson's ratios of $\nu_f = 0.3$ and $\nu_m = 0.35$.

In all cases in which a zero-frequency limit of phase velocity existed, the "effective stiffness" and elasticity solutions gave essentially identical results. This is to be expected, since this limit corresponds to the static elastic solution, and the procedure developed by Sun et al. guarantees that this solution is identically satisfied. Differences between the "effective stiffness" and elasticity solutions become gradually more pronounced as the wave number (or frequency) increases. As discussed in Ref. [31], accuracy of the effective stiffness theory is limited by the particular displacement fields assumed in each laminate. Thus, improvement in the theory could be achieved by the inclusion of nonlinear terms in the displacement expansions [Eqs. (49) and (50)].

REFERENCES

1. R. Kinslow, "Stress Waves in Laminated Materials," AIAA Paper No. 67-140, AIAA 5th Aerospace Sciences Meeting, January 1967.
2. J. C. Peck and G. A. Gurtman, "Dispersive Pulse Propagation Parallel to the Interfaces of a Laminated Composite," J. Appl. Mech., 36, 479-484 (1969).
3. R. Skalak, "Longitudinal Impact of a Circular Elastic Rod," J. Appl. Mech., 24, 59-64 (1957).
4. R. Folk, G. Fox, C. A. Shook, and C. W. Curtis, "Elastic Strain Produced by Sudden Application of Pressure to One End of a Cylindrical Bar. I. Theory," J. Acoust. Soc. Am., 30, 552-558 (1958).

5. J. Miklowitz, "Transient Compressional Waves in an Infinite Plate or Elastic Layer Overlaying a Rigid Half-Space," J. Appl. Mech., 29, 53-60 (1962).

6. G. Fox and C. W. Curtis, "Elastic Strain Produced by Sudden Application of Pressure to One End of a Cylindrical Bar. II. Experimental Observations," J. Acoust. Soc. Am., 30, 559-563 (1958).

7. J. S. Whittier and J. C. Peck, "Experiments on Dispersive Pulse Propagation in Laminated Composites and Comparison with Theory," J. Appl. Mech., 36, 485-490 (1969).

8. M. Abramowitz and I. A. Stegun (eds.), Handbook of Mathematical Functions, NBS Applied Mathematics Series, U. S. Government Printing Office, Washington, D. C., 1965.

9. J. C. Peck, "Pulse Attenuation in Composites," in Shock Waves and the Mechanical Properties of Solids, J. J. Burke and V. Weiss, eds., Proc. 17th Army Materials Research Conf., Syracuse Univ. Press, 1971, pp. 155-184.

10. O. Hoffman, J. A. de Runtz, D. T. Liu, A. R. Hunter, and E. L. Kyser, "Study of Advanced Filament-Reinforced Materials," Lockheed Missiles and Space Company Report LMSC-N-5M-70-2, April 1970.

11. S. M. Rytov, "Acoustical Properties of a Thinly Laminated Medium," Soviet Phys.—Acoustics, 2, 68-80 (1956).

12. J. P. Jones, "Stress Wave Dispersion in Rod-Reinforced Composites," Aerospace Corp. Report No. TR-0066 (5240-30)-3, January 1970.

13. G. W. Postma, "Wave Propagation in a Stratified Medium," Geophys., 20, 780-806 (1955).

14. J. E. White and F. A. Angona, "Elastic Wave Velocities in Laminated Media," J. Acoust. Soc. Am., 27 (2), 311-317 (1955).

15. E. Behrens, "Sound Propagation in Lamellar Composite Materials and Averaged Elastic Constants," J. Acoust. Soc. Am., 42, 378-383 (1967).

16. E. Behrens, "Elastic Constants of Filamentary Composites with Rectangular Symmetry," J. Acoust. Soc. Am., 42, 367-377 (1967).

17. M. Ben-Amoz, "A Continuum Approach to the Theory of Elastic Waves in Heterogeneous Materials," Int. J. Engng. Sci., 6, 209-218 (1968).

18. M. Ben-Amoz, "On Some Extremum Principles in Linear Elastodynamics," Int. J. Engng. Sci., 5, 869-877 (1967).

19. I. M. Khaikovich and L. A. Khalfin, "The Effective Dynamic Properties of Sound Propagation," Soviet Phys.—Acoustics, 4 (3), 280-286 (1958).

20. C. H. Mok, "Wave Propagation in a Fiber-Reinforced Composite Material and Effective Moduli," General Electric Co. Technical Report 067-03, June 1967.

21. C. H. Mok, "The Effective Dynamic Properties of a Fiber-Reinforced Material and the Propagation of Sinusoidal Waves," General Electric Co. Technical Report R67-SD66, November 1967.

22. E. Cosserat and F. Cosserat, Théorie des Corps Déformables, A. Hermann et Fils, Paris, 1909.

23. R. D. Mindlin, "Micro-Structure in Linear Elasticity," Arch. Rat. Mech. Anal., 16, 51-78 (1964).

24. L. M. Habip, "A Continuum Theory for an Elastic Solid with Elastic Micro-Inclusions," NASA Report CR-1321, March 1969.

25. T. R. Tauchert, T. Y. Chang, and A. C. Eringen, "A Lattice Model for Composite Materials," NASA Report CR-1322, April 1969.

26. J. D. Achenbach, "Propagation of Harmonic Waves in a Laminated Composite," ASCE Engineering Mechanics Specialty Conference Proceedings, 1967, pp. 44-48.

27. G. Herrmann and J. D. Achenbach, "On Dynamic Theories of Fiber-Reinforced Composites," AIAA/ASME 8th Structures, Structural Dynamics and Materials Conference, March 1967.

28. J. D. Achenbach and G. Herrmann, "Dispersion of Free Harmonic Waves in Fiber-Reinforced Composites," AIAA J., 6 (10), 1832-1836 (1968).

29. C. T. Sun, J. D. Achenbach, and G. Herrmann, "Time Harmonic Waves in a Stratified Medium Propagating in the Direction of Layering," J. Appl. Mech., 35, 408-411 (1968).

30. C. T. Sun, J. D. Achenbach, and G. Herrmann, "Continuum Theory for a Laminated Medium," J. Appl. Mech., 35, 467-475 (1968).

31. G. Herrmann and J. D. Achenbach, "Applications of Theories of Generalized Cosserat Continua to the Dynamics of Composite Materials," Northwestern University Report 67-7, October 1967.

CHAPTER 14

APPLICATION OF CONTINUUM MIXTURE THEORIES TO THE STUDY OF ELASTIC WAVE PROPAGATION IN COMPOSITE MATERIALS

I. INTRODUCTION

In Chapter 13, Section III, we briefly discussed a number of different continuum or mixture models which various investigators have proposed for the study of elastic waves in composites. In this chapter, we will explore in detail a specific type of dynamic mixture theory. In particular, we will consider a theory in which the composite constituents are assumed to coexist in space but are allowed to interact with one another in a dynamic manner throughout the course of the composite's response.

A number of such "interacting-continua" theories have been utilized in the past to study wave propagation in particulate or geologic media but their application to structural composites is a relatively recent innovation.

Both Lempriere [1] and Bedford and Stern [2] have proposed interacting-continua-based models of the elastodynamics of composites. In their theories, the composite constituents are superimposed in space but are allowed to undergo individual deformations. The microstructure of the actual composite is then simulated by specifying the interactions between the constituents. For the case of a one-dimensional (on the average) wave propagating in a two-component mixture, Lempriere hypothesizes the nature of these interactions as being dependent upon either the displacement or the velocity difference in the two constituents. For laminates running in the direction of wave propagation, Bedford and Stern offer an analytical procedure to determine these interactions. Their procedure involves a quasistatic analysis of the composite.

In this chapter, we will present an alternate analytical procedure for the development of binary mixture theories for the modeling of wave propagation in laminated and fiber-reinforced waveguide-type (waves propagating in the direction of laminates/fibers) composites. The underlying physical assumption made in the analysis is that the characteristic wavelength of the particular disturbance being studied is long when compared to the microdimension (i.e., center-to-center spacing of the reinforcement) of the composite. The following development closely parallels that presented by Hegemier et al. [3].

II. ANALYSIS OF LAMINATES

A. Introduction

Consider a periodic array of two linearly elastic, isotropic, and homogeneous laminates bonded at their interfaces. A state of plane strain will be assumed in the x_2 direction ($u_2 = 0$), as well as wave motions yielding symmetric u_1 and antisymmetric u_3 distributions with respect to the midplane of each laminate. Consequently, motion of the composite, when averaged over both laminates, exists only in the x_1 direction. It is therefore sufficient to consider a typical bilaminate as depicted in Fig. 1. The basic equations for the laminates ($\alpha = 1$ or 2) are as follows.

(i) Equations of motion:

$$\frac{\partial \overset{(\alpha)}{\sigma}_{11}}{\partial x_1} + \frac{\partial \overset{(\alpha)}{\tau}_{13}}{\partial x_3} = \overset{(\alpha)}{\rho} \frac{\partial^2 \overset{(\alpha)}{u}_1}{\partial t^2} \tag{1a}$$

$$\frac{\partial \overset{(\alpha)}{\sigma}_{33}}{\partial x_3} + \frac{\partial \overset{(\alpha)}{\tau}_{13}}{\partial x_1} = \overset{(\alpha)}{\rho} \frac{\partial^2 \overset{(\alpha)}{u}_3}{\partial t^2} \tag{1b}$$

(ii) Constitutive relations:

$$\overset{(\alpha)}{\sigma}_{11} = \left(\overset{(\alpha)}{\lambda} + 2\overset{(\alpha)}{G}\right)\frac{\partial \overset{(\alpha)}{u}_1}{\partial x_1} + \overset{(\alpha)}{\lambda}\frac{\partial \overset{(\alpha)}{u}_3}{\partial x_3} \tag{2a}$$

$$\overset{(\alpha)}{\sigma}_{33} = \left(\overset{(\alpha)}{\lambda} + 2\overset{(\alpha)}{G}\right)\frac{\partial \overset{(\alpha)}{u}_3}{\partial x_3} + \overset{(\alpha)}{\lambda}\frac{\partial \overset{(\alpha)}{u}_1}{\partial x_1} \tag{2b}$$

$$\overset{(\alpha)}{\tau}_{13} = \overset{(\alpha)}{G}\left(\frac{\partial \overset{(\alpha)}{u}_1}{\partial x_3} + \frac{\partial \overset{(\alpha)}{u}_3}{\partial x_1}\right) \tag{2c}$$

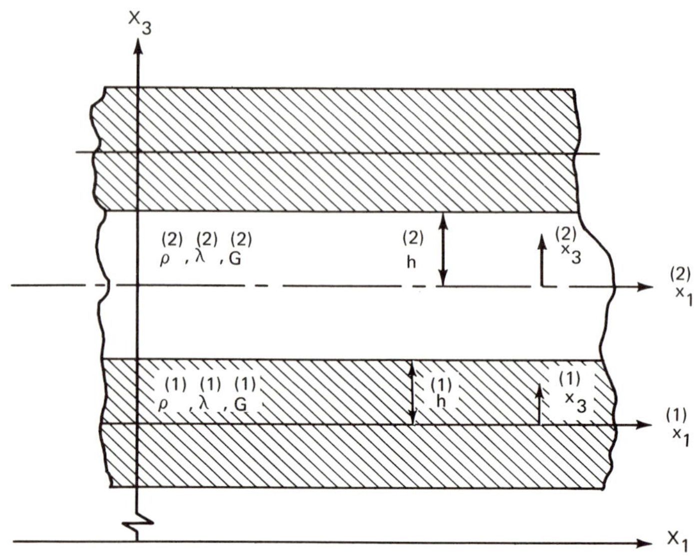

FIG. 1. Laminated composite geometry and coordinate system.

In addition to Eqs. (1) and (2), the complete specification of motion requires

(iii) Symmetry conditions:

$$\overset{(\alpha)}{u}_3 = 0, \quad \overset{(\alpha)}{\tau}_{13} = 0 \quad \text{at} \quad \overset{(\alpha)}{x}_3 = 0 \tag{3}$$

(iv) Interface conditions:

$$\overset{(1)}{u}_1 = \overset{(2)}{u}_1, \quad \overset{(1)}{u}_3 = \overset{(2)}{u}_3, \quad \overset{(1)}{\tau}_{13} = \overset{(2)}{\tau}_{13}, \quad \overset{(1)}{\sigma}_{33} = \overset{(2)}{\sigma}_{33}$$

at

$$\overset{(1)}{x}_3 = \overset{(1)}{h}, \quad \overset{(2)}{x}_3 = -\overset{(2)}{h} \tag{4}$$

(v) Initial conditions at $t = 0$, and appropriate boundary data at $x_1 = 0, \ell$, where $0 \le x_1 \le \ell$. Here ℓ denotes the length of the laminates.

B. Mixture Equations of Motion

If Eq. (1a) is integrated with respect to $\overset{(\alpha)}{x}_3$ from 0 to $\overset{(\alpha)}{h}$ and average or "effective" stresses and displacement are defined according to

$$\left(\overset{(\alpha)}{\ldots}{}^{e}\right) = \left(1/\overset{(\alpha)}{h}\right)\int_0^{\overset{(\alpha)}{h}} \left(\overset{(\alpha)}{\ldots}\right) d\overset{(\alpha)}{x}_3 \tag{5}$$

where $\overset{(\alpha)}{h}$ denotes the half-thickness of laminate (α), then Eq. (1a) can be written as

$$\overset{(\alpha)}{h}\left(\partial \overset{(\alpha)}{\sigma}{}_{11}^{e}/\partial x_1\right) - \overset{(\alpha)}{h}\,\overset{(\alpha)}{\rho}\left(\partial^2 \overset{(\alpha)}{u}{}_1^{e}/\partial t^2\right) = -\overset{(\alpha)}{\tau}_{13}\left(x_1, \overset{(\alpha)}{h}, t\right) \tag{6}$$

Since $\overset{(\alpha)}{\tau}_{13}$ must be continuous across laminate interfaces, and is antisymmetric in $\overset{(\alpha)}{x}_3$, we have

$$-\overset{(1)}{\tau}_{13}\left(x_1, \overset{(1)}{h}, t\right) = \overset{(2)}{\tau}_{13}\left(x_1, \overset{(2)}{h}, t\right) = \tau^*_{13}\left(x_1, t\right) \tag{7}$$

We now define so-called "partial" stresses and densities as

$$\overset{(\alpha)}{\sigma}{}_{11}^{p} = \overset{(\alpha)}{V}\,\overset{(\alpha)}{\sigma}{}_{11}^{e}, \quad \overset{(\alpha)}{\rho}{}^{p} = \overset{(\alpha)}{V}\,\overset{(\alpha)}{\rho} \tag{8}$$

where the volume fractions $\overset{(\alpha)}{V}$ are

$$\overset{(\alpha)}{V} = \overset{(\alpha)}{h}/\left(\overset{(1)}{h} + \overset{(2)}{h}\right) \tag{9}$$

Substituting Eqs. (7) and (8) into Eq. (6) results in the following

$$\left(\partial \overset{(1)}{\sigma}{}_{11}^{p}/\partial x_1\right) - \overset{(1)}{\rho}{}^{p}\left(\partial^2 \overset{(1)}{u}{}_1^{p}/\partial t^2\right) = \rho_c \beta$$

$$\left(\partial\, {}^{(2)}\sigma_{11}^{p}/\partial x_1\right) - {}^{(2)}\rho^{p}\left(\partial^2\, {}^{(2)}u_1^{p}/\partial t^2\right) = -\rho_c\beta \tag{10}$$

where

$$\rho_c\beta = \tau^*_{13}/\left({}^{(1)}h + {}^{(2)}h\right) \tag{11}$$

and

$${}^{(\alpha)}u_1^{p} = {}^{(\alpha)}u_1^{e}$$

The interaction term $\rho_c\beta$, reflecting momentum transfer from one constituent to another, corresponds to a body force equivalent to shear interaction at the laminate interfaces. The problem now hinges on the determination of the interaction term $\rho_c\beta$ and constitutive relations for the composite constituents. The major differences between the various approaches [1-3] is the manner in which these relations are derived. Lempriere hypothesizes forms for the interaction terms which depend on effective displacement or velocity differences between the two laminates. He also develops constitutive relations for the partial stresses in terms of the effective displacement gradients but does not analytically determine the coefficients of these equations. Bedford and Stern derive expressions for the interaction terms and constitutive relation coefficients based on a quasistatic analysis of laminates. Hegemier et al. develop the required expressions through recourse to asymptotic expansions. This technique appears to be the most general, particularly in its application to other geometries (see Refs. [4] and [5]) and is the one followed here.

The procedure entails assuming a power series expansion for stresses and displacements about the midplane of each laminate. Thus we have

$$\overset{(\alpha)}{g}\left(x_1, \overset{(\alpha)}{x_3}, t\right) = \overset{(\alpha)}{g}_{(0)}(x_1, t) + \overset{(\alpha)}{g}_{(1)}(x_1,t)\overset{(\alpha)}{x_3}$$

$$+ \dots + \overset{(\alpha)}{g}_{(n)}(x_1,t)\left\{\left[\overset{(\alpha)}{x_3}\right]^n/n!\right\} + \dots \tag{12}$$

where $\overset{(\alpha)}{g}$ denotes either a stress or a displacement component in the α-constituent. The series (12) need not be convergent, but only asymptotic in a parameter ε representing the ratio of composite micro to macro dimension as $\varepsilon \to 0$. This point is discussed in detail in Ref. [4].

Substituting Eq. (12) into Eqs. (1)-(2) and equating terms of similar order in $\overset{(\alpha)}{x_3}$, one obtains differential-recurrence relations for the coefficients of the expansion (12), i.e.,

$$\left(\partial \overset{(\alpha)}{\sigma}_{11(n)}/\partial x_1\right) + \overset{(\alpha)}{\tau}_{13(n+1)} = \overset{(\alpha)}{\rho}\ \partial^2 \overset{(\alpha)}{u}_{1(n)}/\partial t^2 \tag{13a}$$

$$\overset{(\alpha)}{\sigma}_{33(n+1)} + \left(\partial \overset{(\alpha)}{\tau}_{13(n)}/\partial x_1\right) = \overset{(\alpha)}{\rho}\ \partial^2 \overset{(\alpha)}{u}_{3(n)}/\partial t^2 \tag{13b}$$

$$\overset{(\alpha)}{\sigma}_{11(n)} = \left(\overset{(\alpha)}{\lambda} + 2\overset{(\alpha)}{G}\right)\left(\partial \overset{(\alpha)}{u}_{1(n)}/\partial x_1\right) + \overset{(\alpha)}{\lambda}\ \overset{(\alpha)}{u}_{3(n+1)} \tag{14a}$$

$$\overset{(\alpha)}{\sigma}_{33(n)} = \left(\overset{(\alpha)}{\lambda} + 2\overset{(\alpha)}{G}\right)\overset{(\alpha)}{u}_{3(n+1)} + \overset{(\alpha)}{\lambda}\left(\partial \overset{(\alpha)}{u}_{1(n)}/\partial x_1\right) \tag{14b}$$

$$\overset{(\alpha)}{\tau}_{13(n)} = \overset{(\alpha)}{G}\left[\overset{(\alpha)}{u}_{1(n+1)} + \left(\partial \overset{(\alpha)}{u}_{3(n)}/\partial x_1\right)\right] \tag{14c}$$

Equations (13a), (14a), and (14b) apply for $n = 0,2,4,\dots$ whereas Eqs. (13b) and (14c) are valid for $n = 1,3,5,\dots$. We note that from symmetry or asymmetry of the appropriate stresses or displacements,

$$\overset{(\alpha)}{\tau}_{13(n)} = \overset{(\alpha)}{u}_{3(n)} = 0 \quad \text{for} \quad n = 0, 2, 4, \ldots$$

$$\overset{(\alpha)}{\sigma}_{11(n)} = \overset{(\alpha)}{\sigma}_{33(n)} = \overset{(\alpha)}{u}_{1(n)} = 0 \text{ for } n = 1, 3, 5, \ldots \tag{14d}$$

Using Eqs. (13)-(14), all stresses and displacements can be written in terms of the dependent variables of the mixture momentum equations (10) as follows

$$\overset{(\alpha)}{g}\left(x_1, \overset{(\alpha)}{x}_3, t\right) = \sum_n \left(\overset{(\alpha)}{\mathscr{L}}_{(n)}(x_1, t) \frac{\left[\overset{(\alpha)}{x}_3\right]^n}{n!} \right) \overset{(\alpha)}{u}{}_1^e + \sum_n \left(\overset{(\alpha)}{\mathscr{M}}_{(n)}(x_1, t) \frac{\left[\overset{(\alpha)}{x}_3\right]^n}{n!} \right) \overset{(\alpha)}{\sigma}{}_{11}^p \tag{15}$$

where $\overset{(\alpha)}{\mathscr{L}}_{(n)}$ and $\overset{(\alpha)}{\mathscr{M}}_{(n)}$ are differential operators with respect to x_1 and t. These operators form the basis for the construction of an entire hierarchy of higher-order continuum theories [4]. In the present exposition, however, only the first term or two of the aforementioned operators will be retained.

C. Mixture Constitutive Relations

If we average Eqs. (2a) and (2b) according to Eq. (5) and invoke continuity of $\overset{(\alpha)}{u}_3$ across laminate interfaces, we obtain

$$\overset{(1)}{\mathbf{V}} \left[\left(\overset{(1)}{\sigma}{}_{11}^e / \overset{(1)}{\lambda} \right) - \left(\overset{(1)}{E} / \overset{(1)}{\lambda} \right) \left(\partial \overset{(1)}{u}{}_1^e / \partial x_1 \right) \right] = -S \tag{16a}$$

$$\overset{(2)}{V} \left[\left(\overset{(2)}{\sigma}{}_{11}^e / \overset{(2)}{\lambda} \right) - \left(\overset{(2)}{E} / \overset{(2)}{\lambda} \right) \left(\partial \overset{(2)}{u}{}_1^e / \partial x_1 \right) \right] = S \tag{16b}$$

$$\overset{(1)}{V}\left[\left(\overset{(1)}{\sigma}{}^{e}_{33}/\overset{(1)}{E}\right) - \left(\overset{(1)}{\lambda}/\overset{(1)}{E}\right)\left(\partial\overset{(1)}{u}{}^{e}_{1}/\partial x_{1}\right)\right] = -S \tag{16c}$$

$$\overset{(2)}{V}\left[\left(\overset{(2)}{\sigma}{}^{e}_{33}/\overset{(2)}{E}\right) - \left(\overset{(2)}{\lambda}/\overset{(2)}{E}\right)\left(\partial\overset{(2)}{u}{}^{e}_{1}/\partial x_{1}\right)\right] = S \tag{16d}$$

where

$$S = u_3^*(x_1, t)/\left(\overset{(1)}{h} + \overset{(2)}{h}\right) \tag{17a}$$

and

$$u_3^* = \overset{(2)}{u}_3\left(x_1, \overset{(2)}{h}, t\right) = \overset{(1)}{u}_3\left(x_1, -\overset{(1)}{h}, t\right) = -\overset{(1)}{u}_3\left(x_1, \overset{(1)}{h}, t\right) \tag{17b}$$

$$\overset{(\alpha)}{E} = \overset{(\alpha)}{\lambda} + 2\overset{(\alpha)}{G} \tag{17c}$$

Now let the characteristic dominant signal wavelength of a disturbance propagating in the composite be ℓ, and the typical composite microdimension be $\overset{(1)}{h} + \overset{(2)}{h}$. We introduce the dimensionless variables

$$\xi = x_1/\ell, \quad \overset{(\alpha)}{\zeta} = \overset{(\alpha)}{x}_3/\left(\overset{(1)}{h} + \overset{(2)}{h}\right)$$

$$\bar{t} = tC_0/\ell, \quad \varepsilon = \left(\overset{(1)}{h} + \overset{(2)}{h}\right)/\ell \tag{18a}$$

where C_0 denotes a representative "mixture" velocity which we will define later. If $0 \le x_1 \le \ell$ and $0 \le t \le \ell/C_0$ we have $0 \le \xi \le 1$, $0 \le \bar{t} \le 1$, i.e., the typical macrodimension is of $O(1)$, whereas the typical microdimension is of $O(\varepsilon)$. With the aid of Eqs. (13)-(14), the dimensionless variables (18a), and the assumption that ℓ may be selected such that

$$\partial(\ldots)/\partial\xi = O(1), \quad \partial(\ldots)/\partial\bar{t} = O(1)$$

where (...) represents $\overset{(\alpha)}{u}{}_1^e$ or $\overset{(\alpha)}{\sigma}{}_{11}^p$, the stress boundary condition

$$\overset{(1)}{\sigma}_{33}\left(x_1, \overset{(1)}{h}, t\right) = \overset{(2)}{\sigma}_{33}\left(x_1, -\overset{(2)}{h}, t\right)$$

can be written as

$$\left[1 + O(\varepsilon^2)\right]\overset{(1)}{\sigma}{}_{33}^e = \left[1 + O(\varepsilon^2)\right]\overset{(2)}{\sigma}{}_{33}^e \tag{18b}$$

In what follows, it will be assumed that $\varepsilon << 1$. We will therefore neglect terms of $O(\varepsilon^2)$, and adopt the approximation

$$\overset{(1)}{\sigma}{}_{33}^e \simeq \overset{(2)}{\sigma}{}_{33}^e \tag{18c}$$

whereupon Eqs. (16c) and (16d) yield

$$S = (1/E^*)\left[\overset{(1)}{\lambda}\left(\partial \overset{(1)}{u}{}_1^e/\partial x_1\right) - \overset{(2)}{\lambda}\left(\partial \overset{(2)}{u}{}_1^e/\partial x_1\right)\right] \tag{19a}$$

with

$$E^* = \left(\overset{(1)}{E}/\overset{(1)}{V}\right) + \left(\overset{(2)}{E}/\overset{(2)}{V}\right) \tag{19b}$$

Substitution of Eqs. (19) into Eqs. (16a) and (16b) results in

$$\begin{aligned} \overset{(1)}{\sigma}{}_{11}^p &\simeq c_{11}\left(\partial \overset{(1)}{u}{}_1^e/\partial x_1\right) + c_{12}\left(\partial \overset{(2)}{u}{}_1^e/\partial x_1\right) \\ \overset{(2)}{\sigma}{}_{11}^p &\simeq c_{12}\left(\partial \overset{(1)}{u}{}_1^e/\partial x_1\right) + c_{22}\left(\partial \overset{(2)}{u}{}_1^e/\partial x_1\right) \end{aligned} \tag{20}$$

where

$$c_{\alpha\alpha} = \overset{(\alpha)}{V}\overset{(\alpha)}{E} - \left[(\overset{(\alpha)}{\lambda})^2/E^*\right], \quad c_{\alpha\beta} = \overset{(\alpha)}{\lambda}\overset{(\beta)}{\lambda}/E^*$$

$$\alpha, \beta = 1, 2; \alpha \neq \beta \tag{21}$$

Equations (20) are the desired mixture constitutive relations, subject to the approximation (18c). As such, they relate the stress in each composite constituent to the displacement gradient in both constituents. When the displacement gradients are the same in both the constituents, as they would be for a static deformation field, these constitutive relations reduce to those of Ref. [2].

D. Laminate Interaction Term

To solve for $\rho_c \beta$, we multiply Eq. (2c) by $\overset{(\alpha)}{x}_3$ and integrate over each laminate; thus

$$\left(1/\overset{(\alpha)}{h}\right)\int_0^{\overset{(\alpha)}{h}} \overset{(\alpha)}{x}_3\left[\left(\partial \overset{(\alpha)}{u}_1/\partial \overset{(\alpha)}{x}_3\right) + \left(\partial \overset{(\alpha)}{u}_3/\partial x_1\right) - \left(\overset{(\alpha)}{\tau}_{13}/\overset{(\alpha)}{G}\right)\right] d\overset{(\alpha)}{x}_3 = 0 \tag{22}$$

If we now expand $\left(\partial \overset{(\alpha)}{u}_3/\partial x_1\right) - \left(\overset{(\alpha)}{\tau}_{13}/\overset{(\alpha)}{G}\right)$ in powers of $\overset{(\alpha)}{x}_3$ according to Eqs. (12)-(15), and subsequently integrate by parts, we obtain

$$\overset{(\alpha)}{u}_1\left(x_1, \overset{(\alpha)}{h}, t\right) - \overset{(\alpha)}{u}_1{}^{e} + \left[\left(\overset{(\alpha)}{h}\right)^2/3\right]\left[\left(\partial \overset{(\alpha)}{u}_{3(1)}/\partial x_1\right) - \left(\partial \overset{(\alpha)}{\tau}_{13(1)}/\overset{(\alpha)}{G}\right)\right] + \ldots = 0 \tag{23a}$$

Using Eqs. (12) and the relations (13) and (14), the higher-order terms in Eq. (23a) may be grouped in the form

$$\overset{(\alpha)}{u}_1\left(x_1, \overset{(\alpha)}{h}, t\right) - \overset{(\alpha)}{u}_1{}^{e} + \frac{1}{3}\overset{(\alpha)}{h}\left[\left(1 + \overset{(\alpha)}{\eta}_1 \varepsilon^2 + \ldots\right)\right.$$

$$\cdot \left(\partial \overset{(\alpha)}{u}_3/\partial x_1\right)\left(x_1, \overset{(\alpha)}{h}, t\right) - \left(1/\overset{(\alpha)}{G}\right)\left(1 + \overset{(\alpha)}{\eta}_2 \varepsilon^2 + \ldots\right)$$

$$\cdot \overset{(\alpha)}{\tau}_{13}\left(x_1, \overset{(\alpha)}{h}, t\right)\Bigg] = 0 \tag{23b}$$

where $\overset{(\alpha)}{\eta}_\beta$ are second-order differential operators in ξ and $\bar{t}$. As before, since $\varepsilon << 1$, we neglect terms of $O(\varepsilon^2)$. Using Eqs. (7) and (17b), then, we may write

$$\overset{(1)}{u}_1\left(x_1, \overset{(1)}{h}, t\right) - \overset{(1)}{u}_1{}^e + \frac{1}{3}\overset{(1)}{h}\left[\left(1/\overset{(1)}{G}\right)\tau^*_{13} - \left(\partial u^*_3/\partial x_1\right)\right] = 0$$

$$\overset{(2)}{u}_1\left(x_1, \overset{(2)}{h}, t\right) - \overset{(2)}{u}_1{}^e + \frac{1}{3}\overset{(2)}{h}\left[\left(\partial u^*_3/\partial x_1\right)\right.$$

$$\left. - \left(1/\overset{(2)}{G}\right)\tau^*_{13}\right] = 0 \tag{24}$$

Since

$$\overset{(1)}{u}_1\left(x_1, \overset{(1)}{h}, t\right) = \overset{(1)}{u}_1\left(x_1, -\overset{(1)}{h}, t\right) = \overset{(2)}{u}_1\left(x_1, \overset{(2)}{h}, t\right) = u^*_1 \tag{25}$$

equation (24) can be solved for τ^*_{13}, i.e.,

$$\tau^*_{13} = \left[3\overset{(1)}{G}\,\overset{(2)}{G}/\left(\overset{(1)}{h}\,\overset{(2)}{G} + \overset{(2)}{h}\,\overset{(1)}{G}\right)\right]\Bigg\{\left[1 + \left(\overset{(1)}{\lambda}\varepsilon^2/3E^*\right)\right.$$

$$\left.\cdot \left(\partial^2/\partial\xi^2\right)\right]\overset{(1)}{u}_1{}^e - \left[1 + \left(\overset{(2)}{\lambda}\varepsilon^2/3E^*\right)\left(\partial^2/\partial\xi^2\right)\right]\overset{(2)}{u}_1{}^e\Bigg\} \tag{26a}$$

If $\varepsilon << 1$ and we neglect terms of $O(\varepsilon^2)$, the interaction term $\rho_c\beta$ becomes

$$\rho_c\beta \simeq a\left(\overset{(1)}{u}_1{}^e - \overset{(2)}{u}_1{}^e\right)/\left(\overset{(1)}{h} + \overset{(2)}{h}\right)^2 \tag{26b}$$

where

$$a = 3\overset{(1)}{G}\,\overset{(2)}{G}/\left(\overset{(1)}{V}\,\overset{(2)}{G} + \overset{(2)}{V}\,\overset{(1)}{G}\right) \tag{26c}$$

Although derived in a somewhat different manner, Eq. (26b) is identical with the interaction term found by Bedford and Stern [2].

E. Phase Velocity Spectrum

To determine the accuracy of the foregoing mixture continuum model, we now investigate the longitudinal phase velocity spectrum, which will then be compared with the exact elasticity solution (Chapter 12, Section V.A). Equations (10), (20), and (26b) form a complete system which may be reduced to a single equation for either partial stresses or effective displacements of either constituent; i.e.,

$$\left\{\left[R/\left(\overset{(1)}{h}+\overset{(2)}{h}\right)^2\right]\left[C_0^2\left(\partial^2/\partial x_1^2\right)-\left(\partial^2/\partial t^2\right)\right]+\left(C_1^2C_2^2-C_3^4\right)\left(\partial^4/\partial x_1^4\right)\right.$$
$$\left.-\left(C_1^2+C_2^2\right)\left(\partial^2/\partial x_1^2\right)\left(\partial^2/\partial t^2\right)+\left(\partial^4/\partial t^4\right)\right\}\left\{\begin{matrix}\overset{(\alpha)}{u}{}_1^e\\ \overset{(\alpha)}{\sigma}{}_{11}^p\end{matrix}\right\}=0 \qquad (27)$$

where

$$C_0^2=(c_{11}+c_{22}+2c_{12})/\rho_C,\quad C_1^2=c_{11}/\overset{(1)}{\rho}{}^p$$

$$C_2^2=c_{22}/\overset{(2)}{\rho}{}^p,\quad C_3^4=c_{12}^2/\overset{(1)}{\rho}{}^p\overset{(2)}{\rho}{}^p$$

$$R-a\left[\rho_C/\overset{(1)}{\rho}{}^p\overset{(2)}{\rho}{}^p\right] \qquad (28)$$

and ρ_C is the "composite density" defined by $\rho_C = \overset{(1)}{\rho}{}^p + \overset{(2)}{\rho}{}^p$. The longitudinal phase velocity is obtained by seeking solutions of Eq. (27) of the form

$$\begin{Bmatrix} {}^{(\alpha)}u_1^{e} \\ {}^{(\alpha)}\sigma_{11}^{p} \end{Bmatrix} = \begin{Bmatrix} {}^{(\alpha)}u_1^{*e} \\ {}^{(\alpha)}\sigma_{11}^{*p} \end{Bmatrix} e^{i(kx_1 - \omega t)} \tag{29}$$

where k, ω, $\begin{Bmatrix} {}^{(\alpha)}u_1^{*e} \\ {}^{(\alpha)}\sigma_{11}^{*p} \end{Bmatrix}$ denote wave number, circular frequency, and peak wave amplitude, respectively. Substitution of Eq. (29) into Eq. (27) yields the characteristic equation relating phase velocity C_{11} $(= \omega/k)$ and frequency ω:

$$\omega = \left\{ \left[R/\left({}^{(1)}h + {}^{(2)}h \right)^2 \right] \left(C_{11}^2 - C_0^2 \right) C_{11}^2 / \left[\left(C_-^2 - C_{11}^2 \right) \left(C_+^2 - C_{11}^2 \right) \right] \right\}^{1/2} \tag{30a}$$

where

$$C_{\pm}^2 = \frac{1}{2}\left\{ C_1^2 + C_2^2 \pm \left[\left(C_1^2 - C_2^2 \right) + 4C_3^4 \right]^{1/2} \right\} \tag{30b}$$

Equation (30) is identical in form with the one derived by Lempriere [1].

The basic character of Eq. (30) is worthy of note. If the material properties of both constituents are the same, then we have $C_{11} = C_0$. This together with Eq. (30a) implies that the longitudinal wave velocity in the medium is independent of frequency, i.e., nondispersive.

In the general case, it is convenient to divide the solution into four parts.

I. $0 \leq C_{11} \leq C_-$, no real solution.

II. $C_- \leq C_{11} \leq C_0$, real solution
$\lim_{\omega \to 0} C_{11} = C_0$, $\lim_{\omega \to \infty} C_{11} = C_-$.

III. $C_0 \leq C_{11} \leq C_+$, no real solution.

IV. $C_{11} \geq C_+$, real solution, $\lim_{\omega \to \infty} C_{11} = C_+$,

$\lim_{C_{11} \to \infty} \omega = \left[\sqrt{R} / \left(\overset{(1)}{h} + \overset{(2)}{h} \right) \right]$ (cutoff frequency).

The qualitative behavior of the solution is illustrated in Fig. 2. The solution in region II corresponds to the primary mode of propagation, while that in region IV is a secondary mode. The zero-frequency limit of the primary mode, C_0, can be shown to agree identically with the

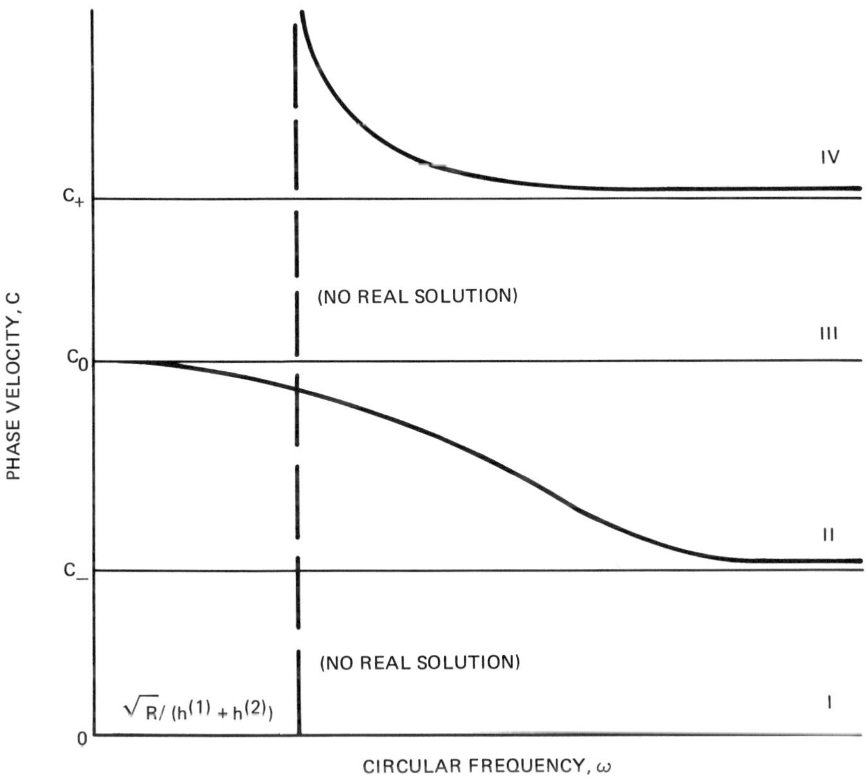

FIG. 2. Qualitative behavior of mixture continuum theory, prediction of longitudinal phase velocity.

elasticity solution of Rytov [see Eq. (29) of Chapter 12]. This result is to be expected since the zero-frequency limit corresponds to the static elastic solution and the continuum relations were derived in such a way as to guarantee satisfaction of the static solution. The mixture continuum model does, however, diverge increasingly from the exact solution with increasing frequency, becoming asymptotic to velocity C_- rather than the shear, or in some cases, the Stoneley wave velocity predicted by elasticity theory [6].

Typical quantitative behavior of the first mode is illustrated in Figs. 3a and 3b. Here the continuum model is compared with the exact elasticity solution derived by Rytov [Eq. (28) of Chapter 12] and experimental data reported by Whittier and Peck [7], for thornel- or boron-reinforced carbon phenolic laminates. Material properties for the two materials studied are given in Tables 1 and 2. As can be observed, agreement between the exact and approximate theories is good for wavelengths Λ greater than the typical composite microdimension. Fortunately, in cases involving sufficiently long input pulses, it is the low-frequency components (wavelengths greater than the typical microdimension) which appear to dominate the overall character of the transient solution [8].

F. Pulse Propagation in Laminates

The utility of the mixture continuum models is perhaps best demonstrated by comparison of theoretical pulse propagation results with experimental data. To this end, the momentum equations (10), the interaction term (26b), and the constitutive relations (20) were incorporated by Hegemier, et al. into a numerical finite-difference computer code and the results of the calculations were compared with the experimental data of Whittier and Peck [7]. The test materials were composed of boron or thornel layers reinforcing a carbon phenolic matrix.

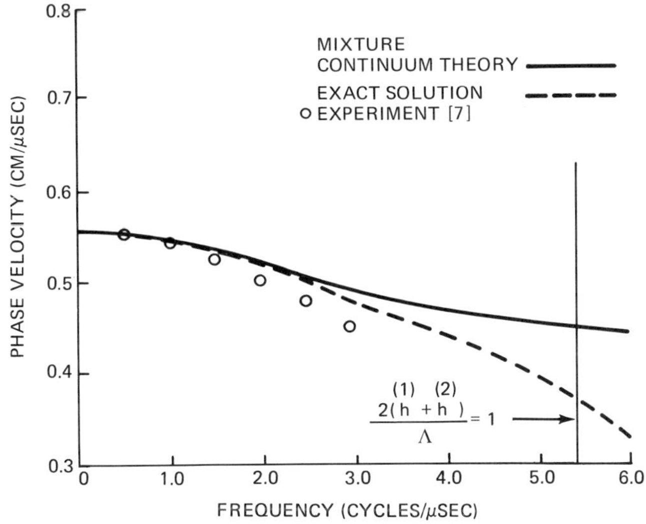

FIG. 3a. Phase velocity versus frequency for thornel-reinforced carbon-phenolic laminate.

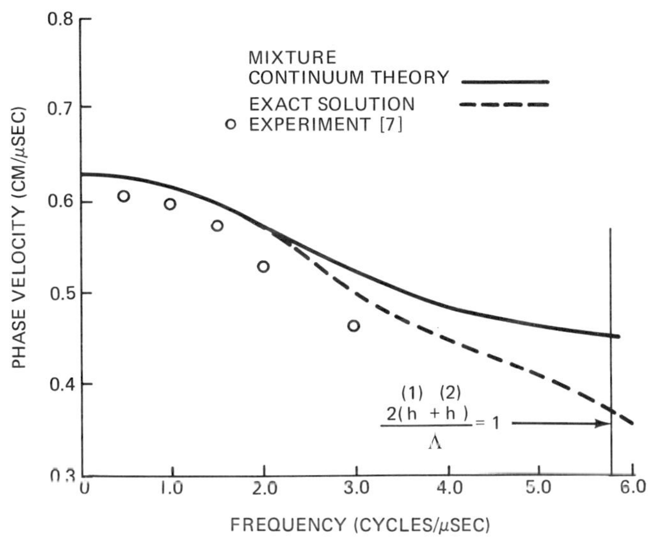

FIG. 3b. Phase velocity versus frequency for boron-reinforced carbon-phenolic laminate.

TABLE 1

Thornel-Carbon Phenolic Laminate

	Thornel laminate	Carbon phenolic laminate
Density (g/cc)	1.47	1.42
Shear modulus $G^{(\alpha)}$ (dynes/cm^2)	7.56×10^{11}	6.62×10^{10}
Lamé constant $\lambda^{(\alpha)}$ (dynes/cm^2)	7.56×10^{11}	11.4×10^{10}
Laminate Thickness $2h^{(\alpha)}$ (cm)	0.0064	0.0558
Volume fraction $V^{(\alpha)}$	0.104	0.896

TABLE 2

Boron-Carbon Phenolic Laminate

	Boron laminate	Carbon phenolic laminate
Density (g/cc)	2.37	1.42
Shear modulus $G^{(\alpha)}$ (dynes/cm^2)	9.51×10^{11}	6.62×10^{10}
Lamé constant $\lambda^{(\alpha)}$ (dynes/cm^2)	8.06×10^{11}	11.4×10^{10}
Laminate Thickness $2h^{(\alpha)}$ (cm)	0.0104	0.0521
Volume fraction $V^{(\alpha)}$	0.166	0.834

Specimens of 1/4-in. thickness were subjected to a uniform pressure at one boundary, with a step-function time dependence induced by a gas-dynamic shock wave of approximately 70 psi. Rear surface velocity, averaged across a 3/8-in.-diameter area of the specimen, was measured by means of a capacitance gage.

The calculations, as reported in Ref. [9], were initiated by imparting a step function velocity of 7.786 cm/sec to both constituents at the boundary $x_1 = 0$. While this condition did not correspond precisely to that of the experiment, which applied uniform stress at the boundary, the authors felt that the error introduced would be negligible at the rear surface of the specimens.

Figures 4 and 5 present the comparison between experimental and theoretical results for averaged rear surface velocity normalized with respect to the impact velocity $v_0 = 7.786$ cm/sec. The physical data for the

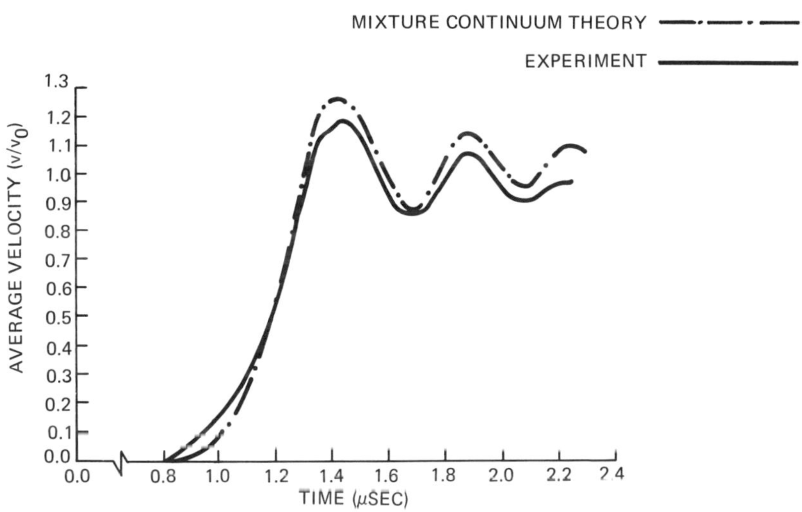

FIG. 4. Comparison of mixture continuum theory with experimental rear surface velocity results for thornel-reinforced carbon-phenolic laminates (from Ref. [9]).

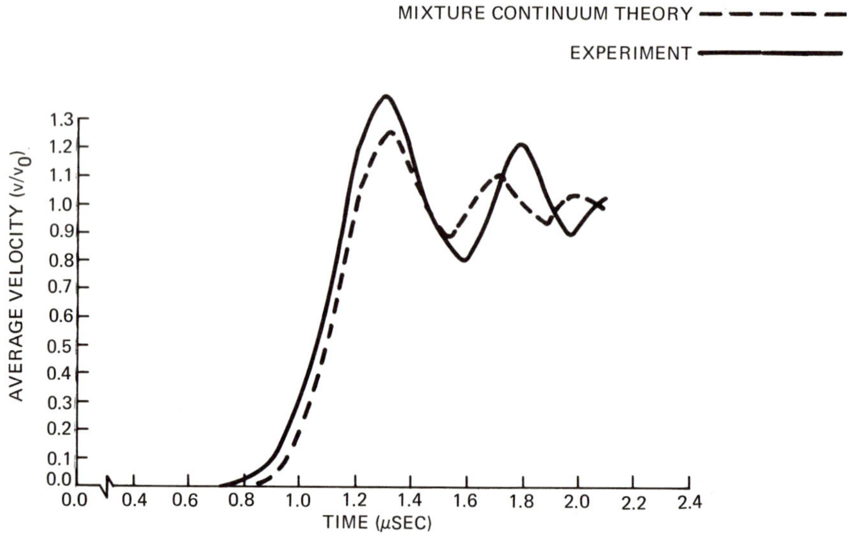

FIG. 5. Comparison of mixture continuum theory with experimental rear surface velocity results for boron-reinforced carbon-phenolic laminates (from Ref. [9]).

composites are presented in Tables 1 and 2. Since absolute times were not measured in these tests, theoretical and experimental results were matched at their respective first peak arrivals. The times shown in the figures were those predicted by the mixture continuum model.

The results for the thornel composite agree somewhat better with the experiments than do those of the boron composite. As noted by Whittier and Peck, this may be due to the fact that the thornel reinforcement more nearly approximates a homogeneous layer. Considering this, as well as the other differences between the laminated medium idealization and the actual test specimens, correlation is quite good, particularly at the head of the wave.

Typical theoretical stress histories of the individual constituents, as well as the area-averaged composite stress, are illustrated in Figs. 6 and 7 for the rear face of the specimens.

III. ANALYSIS OF FIBER REINFORCED COMPOSITES

A. INTRODUCTION

With minor modifications, the procedure outlined in Section II can be applied to wave propagation parallel to the fibers of a hexagonally arrayed, unidirectionally reinforced composite. We begin by approximating a hexagonal array by concentric, linearly elastic cylinders

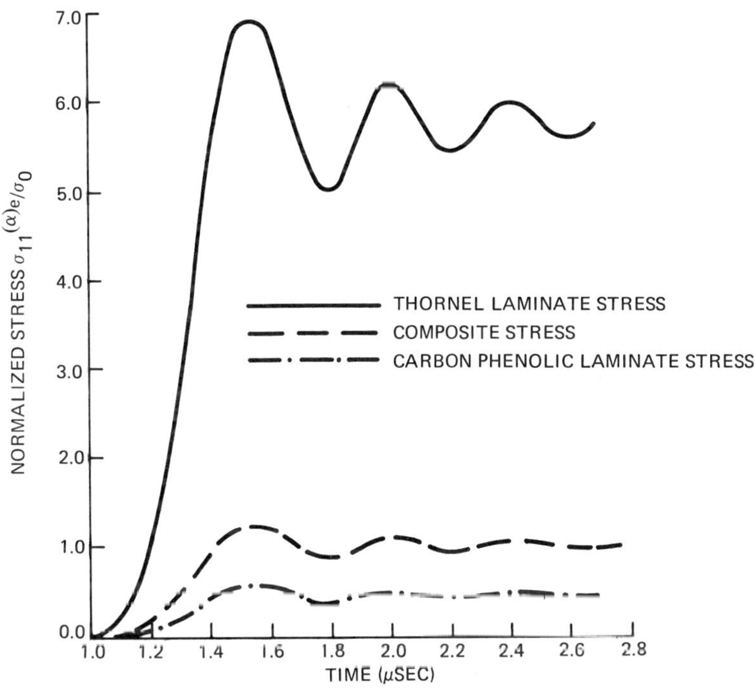

FIG. 6. Finite-difference code predictions of normalized thornel, carbon-phenolic, and composite stress histories (from Ref. [9]).

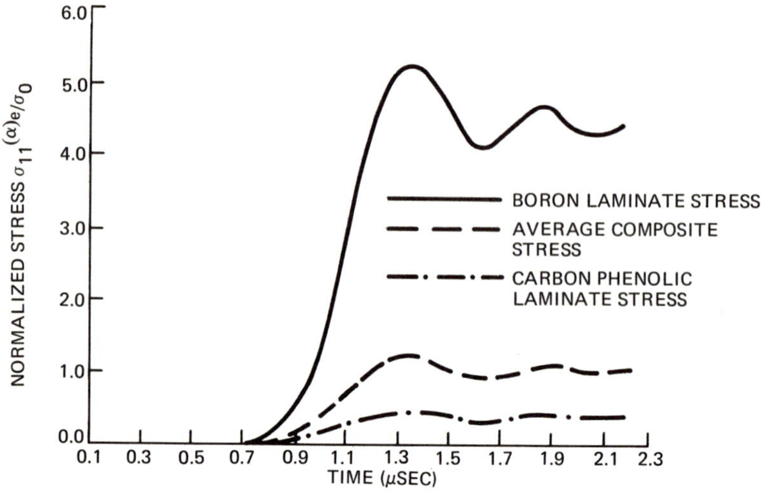

FIG. 7. Finite-difference code predictions of normalized boron, carbon-phenolic, and composite stress histories (from Ref. [9]).

with rigid bonds between interfaces and subject to vanishing shear stress and radial displacements on the outer boundaries, as illustrated in Fig. 17 of Chapter 3. The relevant equations (in polar coordinates x_1, r, θ) are as follows

(i) Equations of motion:

$$\left(\partial \overset{(\alpha)}{\sigma}_{11}/\partial x_1\right) + (1/r)(\partial/\partial r)\left(r \overset{(\alpha)}{\tau}_{1r}\right) = \overset{(\alpha)}{\rho}\ \partial^2 \overset{(\alpha)}{u}_1/\partial t^2 \tag{31a}$$

$$\left(\partial \overset{(\alpha)}{\sigma}_{rr}/\partial r\right) + \left(\partial \overset{(\alpha)}{\tau}_{r1}/\partial x_1\right) + \left[\left(\overset{(\alpha)}{\sigma}_{rr} - \overset{(\alpha)}{\sigma}_{\theta\theta}\right)/r\right]$$

$$= \overset{(\alpha)}{\rho}\ \partial^2 \overset{(\alpha)}{u}_r/\partial t^2 \tag{31b}$$

(ii) Constitutive relations:

$$\overset{(\alpha)}{\sigma}_{11} = \left(\overset{(\alpha)}{\lambda}/r\right)(\partial/\partial r)\left(r\ \overset{(\alpha)}{u}_r\right) + \left(\overset{(\alpha)}{\lambda} + 2\overset{(\alpha)}{G}\right)\left(\partial \overset{(\alpha)}{u}_1/\partial x_1\right) \tag{32a}$$

$$\overset{(\alpha)}{\sigma}_{rr} = \overset{(\alpha)}{\lambda}\left[\left(\overset{(\alpha)}{u}_r/r\right) + \left(\partial \overset{(\alpha)}{u}_1/\partial x_1\right)\right] + \left(\overset{(\alpha)}{\lambda} + 2\overset{(\alpha)}{G}\right) \cdot \left(\partial \overset{(\alpha)}{u}_r/\partial r\right) \quad (32b)$$

$$\overset{(\alpha)}{\sigma}_{\theta\theta} = \overset{(\alpha)}{\lambda}\left[\left(\partial \overset{(\alpha)}{u}_r/\partial r\right) + \left(\partial \overset{(\alpha)}{u}_1/\partial x_1\right)\right] + \left(\overset{(\alpha)}{\lambda} + 2\overset{(\alpha)}{G}\right) \cdot \left(\overset{(\alpha)}{u}_r/r\right) \quad (32c)$$

$$\overset{(\alpha)}{\tau}_{r1} = \overset{(\alpha)}{G}\left[\left(\partial \overset{(\alpha)}{u}_r/\partial x_1\right) + \left(\partial \overset{(\alpha)}{u}_1/\partial r\right)\right] \quad (32d)$$

Here, $\alpha = 1$ denotes the fiber and $\alpha = 2$ the matrix.

(iii) Symmetry conditions:

$$\overset{(2)}{u}_r\left(x_1, \overset{(2)}{r}, t\right) = \overset{(2)}{\tau}_{r1}\left(x_1, \overset{(2)}{r}, t\right) = 0 \quad (33)$$

(iv) Interface conditions:

$$\overset{(1)}{u}_r = \overset{(2)}{u}_r,\quad \overset{(1)}{u}_1 = \overset{(2)}{u}_1,\quad \overset{(1)}{\sigma}_{rr} = \overset{(2)}{\sigma}_{rr},\quad \overset{(1)}{\tau}_{r1} = \overset{(2)}{\tau}_{r1}$$

$$\text{at}\quad r = \overset{(1)}{r} \quad (34)$$

(v) Appropriate initial data at time $t = 0$ and boundary conditions on $x_1 = 0, \ell$ $(0 \le x_1 \le \ell)$ also need to be specified.

In Eqs. (33) and (34), $\overset{(1)}{r}$ and $\overset{(2)}{r}$ (r_f and r_m in Fig. 17 of Chapter 3) denote the fiber and the matrix outer radii, respectively.

If Eq. (31) is multiplied by $2\pi r$, integrated from $r = 0$ to $r = \overset{(1)}{r}$ for $\alpha = 1$ and from $r = \overset{(1)}{r}$ to $r = \overset{(2)}{r}$ for $\alpha = 2$, and if continuity of shear stress across $r = \overset{(1)}{r}$ is invoked, the result can be written in the form of Eq. (10), where, for this case,

$$\rho_c \beta = \left(2 \overset{(1)}{V} / \overset{(1)}{r}\right) \tau_{r1}\left(x_1, \overset{(1)}{r}, t\right)$$

$$\overset{(1)}{V} = \left(\overset{(1)}{r}\right)^2 / \left(\overset{(2)}{r}\right)^2, \quad \overset{(2)}{V} = \left[\left(\overset{(2)}{r}\right)^2 - \left(\overset{(1)}{r}\right)^2\right] / \left(\overset{(2)}{r}\right)^2$$

$$\overset{(1)}{(\ldots)}{}^e = \left[1/\pi\left(\overset{(1)}{r}\right)^2\right] \int_0^{\overset{(1)}{r}} \overset{(1)}{(\ldots)} 2\pi r \, dr \tag{35}$$

$$\overset{(2)}{(\ldots)}{}^e = \left\{1/\pi\left[\left(\overset{(2)}{r}\right)^2 - \left(\overset{(1)}{r}\right)^2\right]\right\} \int_{\overset{(1)}{r}}^{\overset{(2)}{r}} \overset{(2)}{(\ldots)} 2\pi r \, dr$$

To derive recurrence relations for the present problem, it is convenient to expand the stresses and displacement in the fiber and the matrix in the form

$$\overset{(\alpha)}{g}\left(x_1, \overset{(\alpha)}{r}{}^*, t\right) \equiv \overset{(\alpha)}{g}_{(0)}\left(x_1, t\right) + \overset{(\alpha)}{g}_{(1)}\left(x_1, t\right) \overset{(\alpha)}{r}{}^*$$
$$+ \ldots + \overset{(\alpha)}{g}_{(n)}\left(x_1, t\right)\left\{\left[\overset{(\alpha)}{r}{}^*\right]^n / n!\right\} + \ldots \tag{36}$$

where

$$\overset{(1)}{r}{}^* \equiv r, \quad \overset{(2)}{r}{}^* \equiv r - \overset{(2)}{r} \tag{37}$$

and $\overset{(1)}{g}\left(\overset{(2)}{g}\right)$ denotes a stress or a displacement component in the fiber (matrix).

Substitution of Eq. (36) into Eqs. (31) and (32) yields a system of differential recurrence relations similar to Eqs. (13) and (14).

B. Mixture Constitutive Relations

If Eq. (31a) is averaged according to Eq. (35) and the continuity of $\overset{(\alpha)}{u}_r$ at $r = \overset{(1)}{r}$ is invoked, the result is

$$\begin{aligned} \overset{(1)}{a}\,\overset{(1)}{\sigma}{}^{p}_{11} - \overset{(1)}{b}\,\overset{(1)}{V}\left(\partial\,\overset{(1)}{u}{}^{e}_{1}/\partial x_1\right) &= u^*_r \\ \overset{(2)}{a}\,\overset{(2)}{\sigma}{}^{p}_{11} - \overset{(2)}{b}\,\overset{(2)}{V}\left(\partial\,\overset{(1)}{u}{}^{e}_{1}/\partial x_1\right) &= u^*_r \end{aligned} \tag{38}$$

where

$$\begin{aligned} \overset{(\alpha)}{a} = 1/\overset{(\alpha)}{\lambda}, \quad \overset{(\alpha)}{b} &= \left(\overset{(\alpha)}{\lambda} + 2\,\overset{(\alpha)}{G}\right)/\overset{(\alpha)}{\lambda} \\ u^*_r &= \left(2\,\overset{(1)}{V}/\overset{(1)}{r}\right)\overset{(1)}{u}_r\left(x_1, \overset{(1)}{r}, t\right) \end{aligned} \tag{39}$$

Retaining only the first two terms of the expression in Eq. (36) and averaging Eq. (32b) according to Eq. (35) yields

$$\overset{(1)}{\sigma}{}^{e}_{rr} = \left(1/\overset{(1)}{a}\right)\left(\partial\,\overset{(1)}{u}{}^{e}_{1}/\partial x_1\right) + \left(\overset{(1)}{d}/\overset{(1)}{V}\right)u^*_r \tag{40}$$

$$\overset{(2)}{\sigma}{}^{e}_{rr} = \left(1/\overset{(2)}{a}\right)\left(\partial\,\overset{(2)}{u}{}^{e}_{1}/\partial x_1\right) - \left(1/\overset{(2)}{V}\right)\left[\overset{(2)}{d} + \overset{(2)}{G}\left(\overset{(2)}{r}/\overset{(1)}{r}\right)\right]u^*_r \tag{41}$$

where

$$\overset{(\alpha)}{d} = \left(\overset{(\alpha)}{\lambda} + \overset{(\alpha)}{G}\right)$$

If the $\overset{(\alpha)}{\sigma}{}^{e}_{11}$ are subjected to a condition similar to Eq. (18c), then solving Eqs. (40) and (41) for u^*_r gives

$$Eu^*_r = \left(1/\overset{(2)}{a}\right)\left(\partial\,\overset{(2)}{u}{}^{e}_{1}/\partial x_1\right) - \left(1/\overset{(1)}{a}\right)\left(\partial\,\overset{(1)}{u}{}^{e}_{1}/\partial x_1\right) \tag{42a}$$

where

$$E = \left(1/\overset{(1)}{V}\,\overset{(2)}{V}\right)\left\{\left(\overset{(1)}{\lambda} + \overset{(1)}{G}\right)\overset{(1)}{V} + \left(\overset{(2)}{\lambda} + \overset{(2)}{G}\right)\overset{(2)}{V} + \overset{(2)}{G}\left[\overset{(1)}{V}\right]^{1/2}\right\} \tag{42b}$$

Finally, substituting Eq. (42a) into Eq. (39), we find the mixture constitutive relations satisfy Eq. (20) with $c_{\alpha\beta}$ redefined as

$$c_{\alpha\alpha} = \left(\overset{(\alpha)}{\lambda} + 2\,\overset{(\alpha)}{G}\right)\overset{(\alpha)}{V} + \left\{\left[\overset{(\alpha)}{\lambda}\right]^2/E\right\}$$

$$c_{\alpha\beta} = \overset{(\alpha)}{\lambda}\,\overset{(\beta)}{\lambda}/E \quad (\alpha \neq \beta) \tag{43}$$

In contrast to the laminate analysis, the first few terms of the fiber expansion (36) do not contain the static elasticity solution. Consequently, the resultant modulus $(c_{\alpha\alpha} + c_{\beta\beta} + 2c_{\alpha\beta})$ differs from the exact static composite modulus (see footnote, Chapter 13, Section II.C). The exact modulus may be obtained by replacing the $\left[\overset{(1)}{V}\right]^{1/2}$ term in Eq. (42b) by 1.

C. Fiber Interaction Term

With the use of the first two terms of Eq. (36), Eq. (32d) can be written as

$$\partial \overset{(1)}{u}_1/\partial r = \left[\left(1/\overset{(1)}{G}\right)\overset{(1)}{\tau}_{r1(1)} - \left(\partial\, \overset{(1)}{u}_{r(1)}/\partial x_1\right)\right] r + O(r^2) \tag{44}$$

If we multiply Eq. (44) by r and integrate by parts with respect to r from $r = 0$ to $r = \overset{(1)}{r}$, we obtain

$$\overset{(1)}{u}_1\left(x_1, \overset{(1)}{r}, t\right) - \overset{(1)}{u}{}_1^{e} \simeq \frac{1}{2}\overset{(1)}{r}{}^2\left[\left(1/\overset{(1)}{G}\right)\overset{(1)}{\tau}_{r1(1)} - \left(\partial\, \overset{(1)}{u}_{r(1)}/\partial x_1\right)\right] \tag{45}$$

Similarly, in Eq. (32d) with $\alpha = 2$, we first multiply by $\left(r^2 - \overset{(1)}{r}{}^2\right)/\left(\overset{(2)}{r}{}^2 - \overset{(1)}{r}{}^2\right)$, then expand $\overset{(2)}{u}_r$ and $\overset{(2)}{\tau}_{r1}$ in powers of $\overset{(2)}{r}{}^*$, and finally integrate by parts to obtain

$$\overset{(2)}{u}_1\left(x_1, \overset{(1)}{r}, t\right) - \overset{(2)}{u}{}_1^e \simeq \left[\left(1/\overset{(2)}{G}\right)\overset{(2)}{\tau}_{r1(1)} - \left(\partial \overset{(2)}{u}_{r(1)}/\partial r\right)\right]Q \tag{46}$$

where

$$Q = \left[\left(\overset{(2)}{r} - \overset{(1)}{r}\right)^3 \Big/ \left(\overset{(1)}{r} + \overset{(2)}{r}\right)\right]\left\{\frac{2}{3}\left[\overset{(2)}{r}\Big/\left(\overset{(2)}{r} - \overset{(1)}{r}\right)\right] - \frac{1}{4}\right\} \tag{47}$$

Since $\overset{(1)}{u}_1\left(x_1, \overset{(1)}{r}, t\right) = \overset{(2)}{u}_1\left(x_1, \overset{(1)}{r}, t\right)$, Eqs. (45) and (46) give

$$\overset{(1)}{u}{}_1^e - \overset{(2)}{u}{}_1^e = \left\{\left[Q\Big/\overset{(2)}{G}\left(\overset{(1)}{r} - \overset{(2)}{r}\right)\right] - \left(\overset{(1)}{r}\Big/4\overset{(1)}{G}\right)\right\}\tau^*_{r1} - \left\{\left[Q\Big/\left(\overset{(2)}{r} - \overset{(1)}{r}\right)\right] + \left(\overset{(1)}{r}/4\right)\right\}\left(\partial u^*_r/\partial x_1\right) \tag{48}$$

where

$$\tau^*_{r1} \equiv \overset{(1)}{\tau}_{r1}\left(x_1, \overset{(1)}{r}, t\right) = \overset{(2)}{\tau}_{r1}\left(x_1, \overset{(2)}{r}, t\right)$$

However, from Eq. (42a) and following the discussion concerning the laminate interaction term, the last term in Eq. (48) can be shown to be of higher order than the first with respect to the parameter $\varepsilon = \overset{(2)}{r}/\ell$. The interaction term, therefore, takes the form of Eq. (26b) with $a/\left(\overset{(1)}{h} + \overset{(2)}{h}\right)^2$ redefined as

$$a/\left(\overset{(1)}{h} + \overset{(2)}{h}\right)^2 = 8\overset{(1)}{V}\,\overset{(1)}{G}\,\overset{(2)}{G}\Big/\overset{(1)}{r}\left\{\left[4\overset{(1)}{G}Q\Big/\left(\overset{(2)}{r} - \overset{(1)}{r}\right)\right] + \overset{(1)}{r}\,\overset{(2)}{G}\right\} \tag{49}$$

D. Phase Velocity Spectrum

The form of the fiber equations of motion and constitutive relations are similar to those of the laminates. The two differ only in the geometric dependence of their constants. Consequently Eqs. (27), (30), and (36) are applicable to the fiber geometry.

In Fig. 8, a comparison of the continuum model (first mode) dispersion relation is made with the exact elasticity solution derived by Hoffman et al. [6]. The properties selected are representative of a unidirectional fibrous quartz phenolic and are given in Table 3. Once again, agreement between the continuum and exact results is good

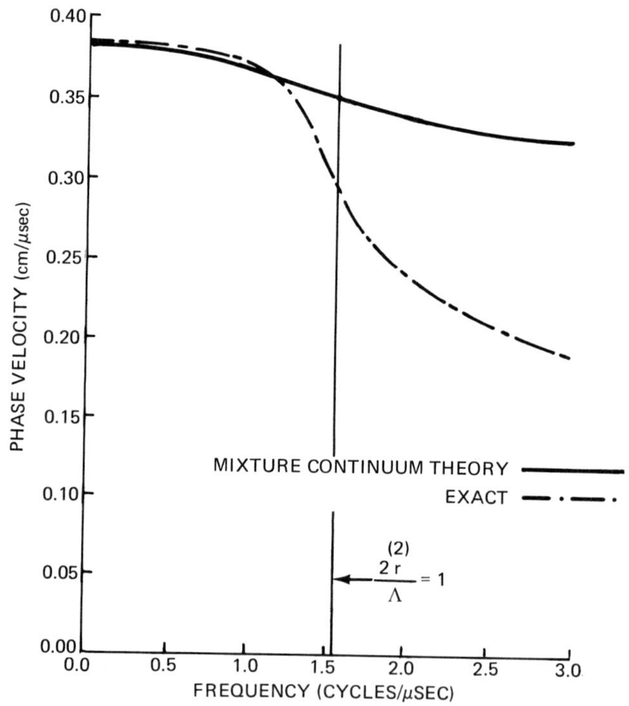

FIG. 8. Longitudinal phase velocity versus frequency for quartz-fiber-reinforced phenolic (from Ref. [3]).

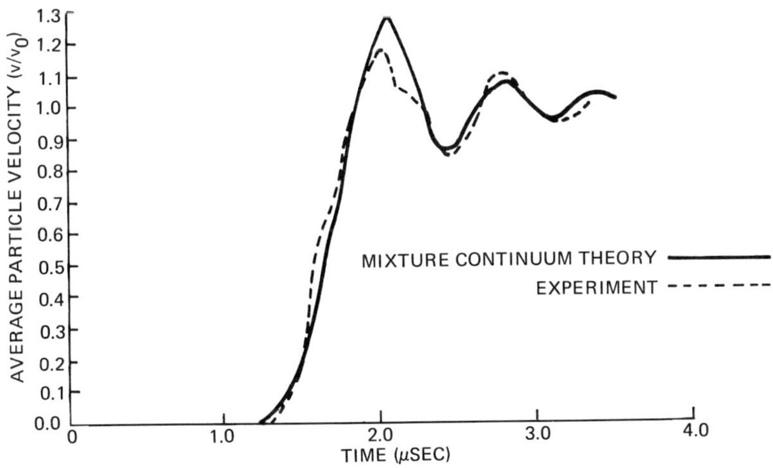

FIG. 9. Comparison of mixture continuum theory with experimental rear surface velocity results for quartz-fiber-reinforced phenolic.

for wavelengths greater than the microdimension of the composite and becomes progressively poorer as frequency increases.

E. Pulse Propagation

In Fig. 9, typical results of transient pulse data as predicted by the continuum theory are compared with experimental data for the unidirectional quartz phenolic fibrous composite described previously. The experiments were performed by the Aerospace Corp. [9]. Specimens were subjected to a 70-psi step function in pressure in a shock tube. The experimental apparatus was identical to that described in Ref. [7]. Rear surface velocities, averaged across a 1/2-in.-diameter circle, were measured on a specimen of 0.633 cm thickness. Constituent data input was as given in Table 3. Figure 9 depicts the comparison between the experimental and finite-difference code predictions of the area-averaged rear surface velocities. The calculation was begun by imparting an equal step-function velocity to both constituents at $x_1 = 0$. As was pointed out previously, the condition differs from that of uniform stress on the boundary but probably does not significantly influence the results at the rear surface of the test specimens.

As in the case of the laminate tests, absolute arrival times of the stress waves were not measured experimentally, and theoretical and experimental results were matched at

TABLE 3

Quartz-Fiber-Reinforced Phenolic

	Quartz fiber	Phenolic
Density (g/cc)	1.85	1.29
Shear modulus $G^{(\alpha)}$ (dynes/cm^2)	106×10^9	30.3×10^9
Lamé constant $\lambda^{(\alpha)}$ (dynes/cm^2)	327×10^9	75.8×10^9
Fiber radius $r^{(1)}$ (cm)	0.0508	
Outer matrix radius $r^{(2)}$ (cm)		0.0975
Volume fractions $V^{(\alpha)}$	0.272	0.728

approximately their first peak arrivals. The times shown in the figure correspond to those predicted by theory. Once again, the correlation between theory and experiment is observed to be quite good.

A continuum theory very similar to that discussed

here has been applied to a number of other geometries. The propagation of shear waves in laminated wave guides has been investigated by Nayfeh and Gurtman [10] and longitudinal wave propagation normal to the laminates has been studied by Hegemier and Nayfeh [5]. Wave propagation normal to fibers is discussed in Hegemier's general survey of continuum type theories [4].

REFERENCES

1. B. Lempriere, "On the Practicability of Analyzing Waves in Composites by the Theory of Mixtures," Lockheed Missiles and Space Company Report, LMSC-6-78-69-21, April 1969.

2. A. Bedford and M. Stern, "Toward a Diffusing Continuum Theory of Composite Materials," J. Appl. Mech., 38 (1), 8-14 (1971).

3. G. A. Hegemier, G. A. Gurtman, and A. H. Nayfeh, "A Continuum Mixture Theory of Wave Propagation in Laminated and Fiber-Reinforced Composites," Int. J. Solids and Structures, 9, 395-414 (1973).

4. G. A. Hegemier, "On a Theory of Interacting Continua for Wave Propagation in Composites," in Dynamics of Composite Materials, ASME, New York, 1972, pp. 70-121.

5. G. A. Hegemier and A. H. Nayfeh, "A Continuum Theory for Wave Propagation in Laminated Composites, Case I: Propagation Normal to the Laminates," J. Appl. Mech., 40 (2), 503-510 (1973).

6. O. Hoffman, J. A. deRuntz, D. T. Liu, A. R. Hunter, and E. L. Kyser, "Study of Advanced Filament-Reinforced Materials," Lockheed Missiles and Space Company Report No. LMSC-N-5M-70-2, April 1970.

7. J. S. Whittier and J. C. Peck, "Experiments on Dispersive Pulse Propagation in Laminated Composites and Comparisons with Theory," J. Appl. Mech., 36 (3), 485-490 (1969).

8. J. C. Peck and G. A. Gurtman, "Dispersive Pulse Propagation Parallel to the Interfaces of a Laminated Composite," J. Appl. Mech., 36 (3), 479-484 (1969).

9. G. A. Gurtman, A. H. Nayfeh, and G. A. Hegemier, "A Continuum Theory Applied to Wave Propagation in Structural Composite Materials," Air Force Weapons Laboratory Report AFWL-TR-71-167, May 1972.

10. A. H. Nayfeh and G. A. Gurtman, "A Continuum Approach to the Propagation of Shear Waves in Laminated Wave Guides," ASME Paper No. 73-APMW-12, 1973.

Chapter 15

SHOCK WAVES IN COMPOSITE MATERIALS

I. INTRODUCTION

Shock wave propagation in multiphase materials is of interest in many areas of science and engineering. The utilization of composite materials in structures designed to encounter intense dynamic loads has been an impetus for numerous investigators to study the problem of hydrodynamic shock wave propagation in such materials [1-10]. A parallel effort has been expended to analyze shock wave propagation in geologic composites. In this area, the effects of water saturation have come under increased scrutiny [11-15]. In this chapter, concepts from the theory of interacting continua [16-18] are utilized to derive a generalized set of conservation equations (i.e., the Hugoniot relations) across a steadily propagating disturbance. The primary advantage of this approach is that the work of various investigators can be compared in a rational manner, and thereby a unified description of the problem can be provided.

The regime of interest is limited to that where the constituent materials have no shear strength and may be treated as compressible fluids. Steady shock waves are, *a priori*, assumed to exist. That is, a definite propagation velocity is assigned to a (shock) disturbance. The shock retains a constant thickness as it moves through the composite. Chemical changes in the constituents are not considered. The interactions between the constituents are modeled as body forces and energy transfer processes influencing their motion. It is through these interactions that the final state behind a shock front is achieved. In most formulations of the problem, the interaction terms are implicitly specified by the conditions imposed on the mixture behind the shock. The differences between the various Hugoniot mixture models are clarified in the present analysis. For convenience, binary composites are treated in detail. The basic limitation in the theory is that it considers only those materials where the volume fraction of a constituent is the same as the surface area fraction of that component (in the direction of wave propagation). This includes laminated composites in which the direction of wave propagation is parallel to the laminates, as well as the case of macroscopically isotropic mixtures (particulate composites).

II. ANALYSIS

In the theory of interacting continua, it is assumed that every small volume of the composite body is occupied by particles of each constituent $\overset{(\alpha)}{s}$ (α = 1,..., r). The geometric arrangement of the various constituents is not explicitly considered but is implicit in the selection of the interaction terms (see Chapter 14). Furthermore, each constituent $\overset{(\alpha)}{s}$ has a velocity field $\overset{(\alpha)}{\underline{v}}(\underline{x}, t)$ through the composite, where $\underline{x}$ denotes the position vector in space (with respect to a fixed Newtonian frame) and t denotes time. The mass of constituent $\overset{(\alpha)}{s}$ per unit volume of composite is called its partial density* $\overset{(\alpha)}{\rho}(\underline{x},t)$ and the total mass per unit volume of composite $\rho(\underline{x},t)$ is given by

$$\rho = \sum_{\alpha=1}^{r} \overset{(\alpha)}{\rho} \tag{1}$$

Similarly, the total stress tensor $\underset{\sim}{\sigma}$ associated with a unit area of the composite may be decomposed into partial stresses $\underset{\sim}{\overset{(\alpha)}{\sigma}}$ associated with each component $\overset{(\alpha)}{s}$:

$$\underset{\sim}{\sigma} = \sum_{\alpha=1}^{r} \underset{\sim}{\overset{(\alpha)}{\sigma}} \tag{2}$$

General relations expressing conservation of mass, momentum, and energy may be written for each $\overset{(\alpha)}{s}$. Assuming that external body forces are absent and that there is no mass transfer between the constituents due to chemical interaction, the conservation relations for $\overset{(\alpha)}{s}$ may be written for a given control volume V (with surface area S) as follows.

Continuity:

$$\int_V \left(\partial \overset{(\alpha)}{\rho}/\partial t\right) dV + \int_S \overset{(\alpha)}{\rho}\,\overset{(\alpha)}{v}_j n_j \, dS = 0 \tag{3}$$

---

*For sake of convenience, we will drop the superscript p used to denote the partial quantities in Chapter 14.

Momentum:

$$\int_V (\partial/\partial t)\left(\overset{(\alpha)}{\rho}\,\overset{(\alpha)}{v_i}\right) dV + \int_S \overset{(\alpha)}{\rho}\,\overset{(\alpha)}{v_i}\,\overset{(\alpha)}{v_j} n_j \, dS$$

$$= \int_S \overset{(\alpha)}{\sigma_{ij}} n_j \, dS + \int_V \rho \overset{(\alpha)}{\beta_i} \, dV \tag{4}$$

Energy:

$$\int_V (\partial/\partial t)\left[\overset{(\alpha)}{\rho}\left(\overset{(\alpha)}{E} + \frac{1}{2}\overset{(\alpha)}{v_j}\,\overset{(\alpha)}{v_j}\right)\right] dV + \int_S \overset{(\alpha)}{\rho}\left(\overset{(\alpha)}{E}\right.$$

$$\left. + \frac{1}{2}\overset{(\alpha)}{v_j}\,\overset{(\alpha)}{v_j}\right)\overset{(\alpha)}{v_i}\, n_i \, dS = \int_S \overset{(\alpha)}{\sigma_{ij}}\,\overset{(\alpha)}{v_i} n_j \, dS - \int_S \overset{(\alpha)}{q_j} n_j \, dS$$

$$+ \int_V \left(\rho \overset{(\alpha)}{\beta_i}\,\overset{(\alpha)}{v_i} + \rho \overset{(\alpha)}{\psi}\right) dV \tag{5}$$

where n_j is the normal (in the outward direction) to surface S, $\alpha = 1, 2, \ldots, r$, and $i, j = 1, 2, 3$. Here, $\overset{(\alpha)}{v_i}$ is the velocity of material $\overset{(\alpha)}{s}$, $\overset{(\alpha)}{\rho}$ is the partial density of $\overset{(\alpha)}{s}$, $\rho \overset{(\alpha)}{\beta_i}$ is the momentum supply in the ith direction to $\overset{(\alpha)}{s}$ due to interaction with the remaining $\alpha - 1$ constituents, $\rho \overset{(\alpha)}{\psi}$ is the internal energy supply to $\overset{(\alpha)}{s}$ due to interaction processes, $\overset{(\alpha)}{E}$ is the internal energy per unit mass of material $\overset{(\alpha)}{s}$, and $\overset{(\alpha)}{q_i}$ is the heat flux vector into $\overset{(\alpha)}{s}$ from the remaining $\alpha - 1$ constituents.

The requirements that the total momentum and energy contributions of the internal material forces be zero may be written as

$$\sum_{\alpha=1}^{r} \rho \overset{(\alpha)}{\beta_i} = 0 \tag{6}$$

$$\sum_{\alpha=1}^{r}\left[\rho\left(\overset{(\alpha)}{\beta}_i \overset{(\alpha)}{v}_i + \overset{(\alpha)}{\psi}\right) - \left(\partial \overset{(\alpha)}{q}_i/\partial x_i\right)\right] = 0 \qquad (7)$$

In writing the above conservation relations, no reference is made to the actual mean volume $\overset{(\alpha)}{V}$ occupied by $\overset{(\alpha)}{s}$ per unit volume of composite. If the interaction terms and constitutive relations for the composite medium are to be expressed in terms of the behavior of the isolated constituents, however, reference must be made to the actual constituents. Effective densities $\overset{(\alpha)}{\rho}{}^{e}$ and effective stress tensors $\overset{(\alpha)}{\underset{\sim}{\sigma}}{}^{e}$ may be defined in terms of partial densities and partial stresses by the scaled relations

$$\overset{(\alpha)}{\rho} = \overset{(\alpha)}{V}\,\overset{(\alpha)}{\rho}{}^{e} \qquad (8)$$

$$\overset{(\alpha)}{\underset{\sim}{\sigma}} - \overset{(\alpha)}{V}\,\overset{(\alpha)}{\underset{\sim}{\sigma}}{}^{e} \qquad (9)$$

where (see also Chapter 14)

$$\sum_{\alpha=1}^{r} \overset{(\alpha)}{V} = 1 \qquad (10)$$

These definitions imply that the composite is isotropic, so that each plane through the medium intersects the same area fraction of $\overset{(\alpha)}{s}$. Then the area and the volume fractions are the same for $\overset{(\alpha)}{s}$ and a single scaling fraction $\overset{(\alpha)}{V}$ occurs in both. Note that although laminated composites are not isotropic, Eqs. (8)-(10) still hold for wave propagation in the direction of lamination.

III. INTERACTION TERMS

The following discussion is restricted to plane wave propagation in the x_1 direction. For this case, the conservation relations (3)-(5) may be written in differential form as

$$\left(\partial \overset{(\alpha)}{\rho} / \partial t\right) + \overset{(\alpha)}{v}_1 \left(\partial \overset{(\alpha)}{\rho} / \partial x_1\right) + \overset{(\alpha)}{\rho} \left(\partial \overset{(\alpha)}{v}_1 / \partial x_1\right) = 0 \tag{11}$$

$$\overset{(\alpha)}{\rho} \left[\left(\partial \overset{(\alpha)}{v}_1 / \partial t\right) + \overset{(\alpha)}{v}_1 \left(\partial \overset{(\alpha)}{v}_1 / \partial x_1\right)\right] = \rho \overset{(\alpha)}{\beta}_1 - \left(\partial \overset{(\alpha)}{p} / \partial x_1\right) \tag{12}$$

$$\overset{(\alpha)}{\rho} \left[\left(\partial \overset{(\alpha)}{E} / \partial t\right) + \overset{(\alpha)}{v}_1 \left(\partial \overset{(\alpha)}{E} / \partial x_1\right)\right] = - \overset{(\alpha)}{p} \left(\partial \overset{(\alpha)}{v}_1 / \partial x_1\right) + \rho \overset{(\alpha)}{\psi} - \left(\partial \overset{(\alpha)}{q}_1 / \partial x_1\right) \tag{13}$$

where $\overset{(\alpha)}{p} = - \overset{(\alpha)}{\sigma}_{ii}/3$ denotes the partial fluid pressure.

The momentum interaction term $\rho \overset{(\alpha)}{\beta}_1$ incorporates terms to represent both dilatational $\rho \overset{(\alpha)}{\beta}_d$ and diffusive (shear) $\rho_0 \overset{(\alpha)}{\eta}$ forces.

$$\rho \overset{(\alpha)}{\beta}_1 = \rho \overset{(\alpha)}{\beta}_d + \rho_0 \overset{(\alpha)}{\eta} \tag{14}$$

To derive an expression for $\rho \overset{(\alpha)}{\beta}_1$, we consider the usual momentum conservation equation for fluid flow with friction through a tube of variable cross section:

$$\overset{(\alpha)}{V} \overset{(\alpha)}{\rho}{}^e \left(\frac{\partial \overset{(\alpha)}{v}_1}{\partial t} + \overset{(\alpha)}{v}_1 \frac{\partial \overset{(\alpha)}{v}_1}{\partial x_1}\right) = + \rho_0 \overset{(\alpha)}{\eta} - \overset{(\alpha)}{V} \frac{\partial \overset{(\alpha)}{p}{}^e}{\partial x_1} \tag{15}$$

Here $\rho_0 \overset{(\alpha)}{\eta}$ denotes the drag force experienced by the $\overset{(\alpha)}{s}$ component. Combining Eqs. (12), (14), and (15), there follows

$$\overset{(\alpha)}{\rho}\,\overset{(\alpha)}{\beta}_1 = \overset{(\alpha)}{p}{}^{e}\left(\partial \overset{(\alpha)}{V}/\partial x_1\right) + \rho_0 \overset{(\alpha)}{\eta} \tag{16}$$

This relation is strictly applicable only to fluids. For a binary mixture, if the effective pressures are equal $\left(\overset{(1)}{p}{}^{e} = \overset{(2)}{p}{}^{e} = p^{e}\right)$, then there follows from Eqs. (6) and (16)

$$\overset{(1)}{\eta} = -\overset{(2)}{\eta} = \eta \tag{17}$$

The shear (diffusive) term $\rho_0\eta$ may be thought of as that which results when a fluid moves through a porous material. The viscous effects may be lumped into a constitutive relation. The specification of the interaction term (16) does not limit the present theory to any particular geometry. It simply distinguishes between the normal and shear components in the interaction without attempting to further specify the shear term $\overset{(\alpha)}{\eta}$. The basic restrictions of the theory are denoted by Eqs. (8)-(10).

We also have to prescribe the constitutive relations for $\overset{(\alpha)}{\psi}$ and $\overset{(\alpha)}{q}$. It is assumed that the energy transfer resulting from heat flow can be decoupled from the internal energy supply due to the interaction forces. This yields

$$\sum_{\alpha} \partial \overset{(\alpha)}{q}_1/\partial x_1 = 0 \tag{18a}$$

$$\sum_{\alpha} \rho\left(\overset{(\alpha)}{\beta}_1 \overset{(\alpha)}{v}_1 + \overset{(\alpha)}{\psi}\right) = 0 \tag{18b}$$

For a binary mixture, since $\rho\overset{(1)}{\beta}_1 = -\rho\overset{(2)}{\beta}_1 = \rho\beta$, Eq. (18b) may be rewritten as

$$\rho\beta\left(\overset{(1)}{v}_1 - \overset{(2)}{v}_1\right) + \rho\left(\overset{(1)}{\psi} + \overset{(2)}{\psi}\right) = 0 \tag{18c}$$

The interaction energy term $\overset{(\alpha)}{\rho\psi}$ contains both dilatational $\rho\overset{(\alpha)}{\psi}_d$ and diffusive (frictional or viscous) $\rho\overset{(\alpha)}{\psi}_s$ contributions:

$$\rho\overset{(\alpha)}{\psi} = \rho\overset{(\alpha)}{\psi}_d + \rho\overset{(\alpha)}{\psi}_s \tag{19}$$

Also, we have

$$\overset{(\alpha)}{E}{}^{e} \equiv \overset{(\alpha)}{E} \tag{20}$$

To derive an expression for $\rho\overset{(\alpha)}{\psi}$, we consider the usual internal energy balance equation for fluid flow with friction:

$$\overset{(\alpha)}{V}\,\overset{(\alpha)}{\rho}{}^{e}\left[\left(\partial\overset{(\alpha)}{E}/\partial t\right) + \overset{(\alpha)}{v}_1\left(\partial\overset{(\alpha)}{E}/\partial x_1\right)\right]$$

$$= -\overset{(\alpha)}{V}\,\overset{(\alpha)}{p}{}^{e}\left(\overset{(\alpha)}{\varepsilon}{}^{e}\right)^{\cdot} - \left(\partial\overset{(\alpha)}{q}_1/\partial x_1\right) + \rho\overset{(\alpha)}{\psi}_s \tag{21}$$

Here $\left(\overset{(\alpha)}{\varepsilon}{}^{e}\right)^{\cdot}$ denotes the effective volumetric strain rate. One has from mass conservation

$$\left(\overset{(\alpha)}{\varepsilon}{}^{e}\right)^{\cdot} = -\left(\overset{(\alpha)}{D}\,\overset{(\alpha)}{\rho}{}^{e}/Dt\right)/\overset{(\alpha)}{\rho}{}^{e}$$

$$= -\left[\left(\overset{(\alpha)}{D}\,\overset{(\alpha)}{\rho}/Dt\right)/\overset{(\alpha)}{\rho}\right] + \left(1/\overset{(\alpha)}{V}\right)\left(\overset{(\alpha)}{D}\,\overset{(\alpha)}{V}/Dt\right)$$

$$= \left(\partial\overset{(\alpha)}{v}_1/\partial x_1\right) + \left(1/\overset{(\alpha)}{V}\right)\left(\overset{(\alpha)}{D}\,\overset{(\alpha)}{V}/Dt\right) \tag{22}$$

where

$$\overset{(\alpha)}{D}/Dt \equiv (\partial/\partial t) + \overset{(\alpha)}{v}_1 \cdot \partial/\partial x_1$$

Substituting from Eq. (22) into Eq. (21) and utilizing Eqs. (13), (18) and (19), we obtain

$$\rho \overset{(\alpha)}{\psi}_d = -\overset{(\alpha)}{p}{}^{e}\left[\left(\partial \overset{(\alpha)}{V}/\partial t\right) + \overset{(\alpha)}{v}_1 \cdot \partial \overset{(\alpha)}{V}/\partial x_1\right] \tag{23}$$

For a binary mixture with equal effective pressures $\left(\overset{(1)}{p}{}^{e} = \overset{(2)}{p}{}^{e} = p^{e}\right)$, substitution of Eq. (23) into Eq. (18c) yields

$$\rho\left(\overset{(1)}{\psi}_s + \overset{(2)}{\psi}_s\right) = \rho_0 \eta \left(\overset{(2)}{v}_1 - \overset{(1)}{v}_1\right) \tag{24a}$$

For wave propagation in geologic composites, Garg [16] assumed that the diffusive force $\rho_0 \eta$ depends only on the two velocity fields, i.e.,

$$\rho_0 \eta = \rho_0 d \left(\overset{(2)}{v}_1 - \overset{(1)}{v}_1\right) \tag{24b}$$

where d is a constant with the dimension of the reciprocal of time. Combining Eqs. (24a) and (24b), one obtains

$$\rho\left(\overset{(1)}{\psi}_s + \overset{(2)}{\psi}_s\right) = \rho_0 d \left(\overset{(2)}{v}_1 - \overset{(1)}{v}_1\right)^2 \geq 0 \tag{24c}$$

Thus the total diffusive contribution is equal to or greater than zero.

This result is an example of a general requirement one may stipulate for any realistic interaction models, i.e.,

$$\rho\left(\overset{(1)}{\psi}_s + \overset{(2)}{\psi}_s\right) \geq 0 \tag{25}$$

The total internal energy contribution due to diffusive interactions between the constituents must be non-negative. This is equivalent to applying the second law of thermodynamics to a suitable control volume wherein thermal equilibrium may occur within the time scale of interest.

A more restrictive prescription for the internal energy rise due to diffusive interaction in the constituents is

$$\overset{(1)}{\rho}\,\psi_s \geq 0, \quad \overset{(2)}{\rho}\,\psi_s \geq 0 \tag{26}$$

The latter should hold when the thermal time scale is large relative to the time scale associated with particle motion.

Garg and Kirsch [9] assumed that

$$\overset{(\alpha)}{\rho}\,\psi = -\rho\,\overset{(\alpha)}{\beta_1}\,\overset{(\alpha)}{v_1} \tag{27a}$$

This implies that

$$\begin{aligned} \overset{(1)}{\rho}\,\psi_s &= -\rho_0 \eta\,\overset{(1)}{v_1} \\ \overset{(2)}{\rho}\,\psi_s &= +\rho_0 \eta\,\overset{(2)}{v_1} \end{aligned} \tag{27b}$$

Equations (27b) cannot satisfy the thermally isolated constituent relations (26) unless $\overset{(1)}{v_1} = \overset{(2)}{v_2} = 0$ or $\eta = 0$. Moreover, Eq. (25) could only be satisfied if $\overset{(2)}{v_1} \geq \overset{(1)}{v_1}$. Since one cannot, *a priori*, know if the inequality is valid, we conclude that Eq. (27) is appropriate only for the case $\overset{(2)}{v_1} = \overset{(1)}{v_1}$, and the materials are in local thermal equilibrium. Thus the identification of the interaction terms by Garg and Kirsch [9], i.e., Eq. (27a), is not generally valid.

IV. HUGONIOT RELATIONS—BINARY MIXTURE*

Let us apply the integral equations (3)-(5) to the case where a disturbance is propagating through a binary

*General jump relations for a mixture have been derived by Kelly [19]. We, however, prefer to follow the somewhat simpler treatment outlined here.

mixture in the x_1 direction at a steady speed, U_{sh}. The material is assumed to be infinite in extent. The disturbance is of infinite extent transverse to the direction of motion but has a finite thickness which does not vary with time. This reduces the problem to that of uniaxial wave propagation.

Let us consider the coordinate system to be moving with the wave, and utilize the finite control volume shown in Fig. 1. The flow, in both constituents, is assumed to be perfectly one-dimensional on both sides of the shock front.

The control volume is selected so that no fluid leaks across the outer surface in the shock front. In addition, the effective pressures in the two constituents are assumed to be equal, i.e.,

$$\overset{(1)}{p}{}^{e} = \overset{(2)}{p}{}^{e} = p^{e} \tag{28}$$

The particle velocities are assumed to be equal on both sides of the control volume $\left(\overset{(1)}{v}_1 = \overset{(2)}{v}_1 = v\right)$. They may, however, differ within the control volume itself.

The continuity equation for the $\overset{(\alpha)}{s}$ constituent is

$$\int_S \overset{(\alpha)}{\rho}{}^{e} \overset{(\alpha)}{V} \overset{(\alpha)}{v}_j n_j \, dS = 0 \tag{29}$$

where the surface integral is over the entire control volume. Since there is no outflow across sides 3 and 4, we obtain the following mass conservation relations:

$$\overset{(1)}{V} \overset{(1)}{\rho}{}^{e} (U_{sh} - v) - \overset{(1)}{V}_0 \overset{(1)}{\rho}_0{}^{e} U_{sh} \tag{30}$$

$$\left(1 - \overset{(1)}{V}\right) \overset{(2)}{\rho}{}^{e} (U_{sh} - v) = \left(1 - \overset{(1)}{V}_0\right) \overset{(2)}{\rho}_0{}^{e} U_{sh} \tag{31}$$

where

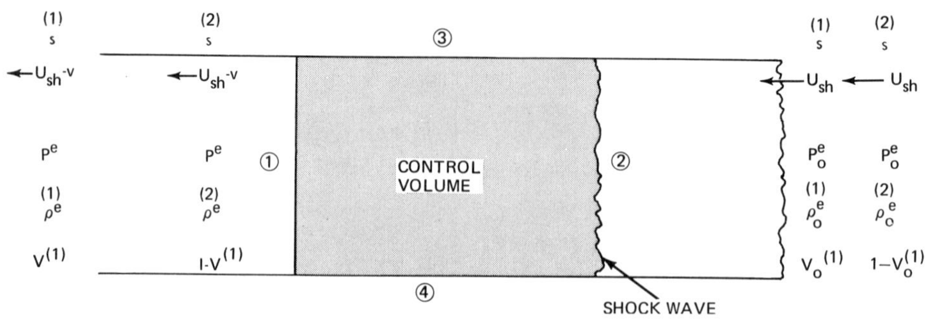

FIG. 1. Schematic of control volume.

$$\overset{(2)}{V} = 1 - \overset{(1)}{V} \tag{32}$$

Here $\overset{(1)}{V}_0$ denotes the initial volume fraction of $\overset{(1)}{s}$ and v is the particle velocity of the $\overset{(\alpha)}{s}$ constituent in the x_1 direction.

The momentum conservation equation (4) may be integrated to give

$$\overset{(1)}{V}_0 \overset{(1)}{\rho}{}^e_0 U_{sh} v = \overset{(1)}{V} p^e - \overset{(1)}{V}_0 p^e_0 - F \tag{33}$$

$$\left(1 - \overset{(1)}{V}_0\right) \overset{(2)}{\rho}{}^e_0 U_{sh} v = \left(1 - \overset{(1)}{V}\right) p^e - \left(1 - \overset{(1)}{V}_0\right) p^e_0 + F \tag{34}$$

where $p^e \delta_{ij} = - \overset{(\alpha)}{\sigma}{}^e_{ij}$ has been substituted into the integrands in Eq. (4). The interaction term F is given by

$$F = - \int_{\Delta x_s} \rho \overset{(1)}{\beta}_1 dx_1 = \int_{\Delta x_s} \rho \overset{(2)}{\beta}_1 \, dx_1 \tag{35}$$

where Δx_s denotes the length of the control volume. Using the definition of $\rho \overset{(\alpha)}{\beta}_1$ from Eq. (16), we obtain

$$F = -\int_{\Delta x_s} p^e(x_1)\left(\partial \overset{(1)}{V}/\partial x_1\right) dx_1 - \rho_0 \int_{\Delta x_s} \eta \; dx_1 \tag{36}$$

We may further redefine these terms as follows:

$$p_m\left(\overset{(1)}{V} - \overset{(1)}{V}_0\right) = -\int_{\Delta x_s} p^e(x_1)\left(\partial \overset{(1)}{V}/\partial x_1\right) dx_1 \tag{37}$$

$$f = -\rho_0 \int_{\Delta x_s} \eta \; dx_1 \tag{38}$$

so that

$$F = p_m\left(\overset{(1)}{V} - \overset{(1)}{V}_0\right) + f \tag{39}$$

Here, p_m is an average pressure acting on the material interfaces in the disturbance region. Note that although the integral to determine p_m is path dependent, it will always be some finite value, even if the shock is very thin. Moreover, for materials in a fine-grain mixture, the dilatancy term $p_m\left(\overset{(1)}{V} - \overset{(1)}{V}_0\right)$ will still be present. This is evident if we note that in the disturbance zone, the fluids are (effectively) moving through a tube with varying cross section, and this additional force must be taken into account.

Substituting Eqs. (39) into the momentum conservation yields

$$\overset{(1)}{V}_0 \overset{(1)}{\rho}{}_0^{e} U_{sh} v = \overset{(1)}{V} p^e - \overset{(1)}{V}_0 p_0^e - p_m\left(\overset{(1)}{V} - \overset{(1)}{V}_0\right) - f \tag{40}$$

$$\left(1 - \overset{(1)}{V}_0\right) \overset{(2)}{\rho}{}_0^{e} U_{sh} v = \left(1 - \overset{(1)}{V}\right) p^e - \left(1 - \overset{(1)}{V}_0\right) p_0^e$$

$$+ p_m\left(\overset{(1)}{V} - \overset{(1)}{V}_0\right) + f \tag{41}$$

The energy integrals in Eq. (5) may be evaluated to obtain the conservation relations for each constituent*:

$$\overset{(1)}{V}_0 \overset{(1)}{\rho}_0 {}^eU_{sh}\left[\overset{(1)}{E} - \overset{(1)}{E}_0 + \tfrac{1}{2}v^2 - vU_{sh}\right] = -\left[\overset{(1)}{V}p^e\left(U_{sh} - v\right)\right.$$

$$\left. - \overset{(1)}{V}_0 p^e_0 U_{sh}\right] + Q_1 + Q_2 \tag{42}$$

$$\left(1 - \overset{(1)}{V}_0\right)\overset{(2)}{\rho}_0 {}^eU_{sh}\left[\overset{(2)}{E} - \overset{(2)}{E}_0 + \tfrac{1}{2}v^2 - vU_{sh}\right] = -\left[\left(1 - \overset{(1)}{V}\right)p^e(U_{sh}\right.$$

$$\left. - v) - \left(1 - \overset{(1)}{V}_0\right)p^e_0 U_{sh}\right] - Q_1 - Q_2 \tag{43}$$

where

$$Q = Q_1 + Q_2$$

$$Q_1 = -\int_{\Delta x_s} \left(\partial \overset{(1)}{q}_1/\partial x_1\right)dx_1$$

$$Q_2 = \int_{\Delta x_s} \left[\rho_0 \eta\left(U_{sh} - \overset{(1)}{v}_1\right) + \rho \overset{(1)}{\psi}_s\right]dx_1 \tag{44}$$

V. GENERAL DISCUSSION

A. Macroscopically Isotropic Composites

The conservation equations (30), (31), (33), (34), (42) and (43) combined with the equations of state

*In deriving Eqs. (42) and (43), the continuity equations (30) and (31) have been utilized to simplify the expressions obtained from integrating Eq. (5).

$$\overset{(\alpha)}{p}{}^{e} = \overset{(\alpha)}{p}{}^{e}\left(\overset{(\alpha)}{\rho}{}^{e}, \overset{(\alpha)}{E}\right), \quad \alpha = 1, 2 \tag{45}$$

provide a system of eight equations for 12 unknowns (i.e., p^e, U_{sh}, v, $\overset{(1)}{\rho}{}^{e}$, $\overset{(2)}{\rho}{}^{e}$, $\overset{(1)}{E}$, $\overset{(2)}{E}$, $\overset{(1)}{V}$, p_m, f, Q_1, Q_2). It is possible to eliminate one equation and the two momentum interaction terms by combining Eqs. (33) and (34) to obtain the overall momentum equation

$$vU_{sh}\left[\overset{(1)}{V}_0 \overset{(1)}{\rho}{}^{e}_0 + \left(1 - \overset{(1)}{V}_0\right)\overset{(2)}{\rho}{}^{e}_0\right] = p^e - p^e_0 \tag{46}$$

The problem is made determinate if it is further assumed that

$$Q = Q_1 + Q_2 = 0 \tag{47}$$

This reduces the algebraic set to seven equations and eight unknowns. The equations obtained by substituting Eq. (47) into Eqs. (30), (31), (42), (43), and (46) along with the equations of state (45) are equivalent to those derived by Tsou and Chou [1,2]. It should be noted that the interaction terms represented by p_m, f, Q_1, and Q_2 are still present but cannot be calculated unless two additional (constitutive) relations are specified. The overall interaction momentum term F can be readily obtained once the Hugoniot states are known, by solving for F in Eqs. (33) and (34) (see Section V.C).

This model was first introduced by Tsou and Chou [1,2]. Garg and Kirsch [9] denoted this model by ADEQ (adiabatic equilibrium). This is somewhat of a misnomer since no heat transfer condition ($Q_1 = 0$) together with Eq. (47) implies that Q_2 is zero. However, $Q_2 = 0$, would violate the inequality (25).

It is appropriate to discuss another pressure equilibrium Hugoniot model in which different implicit assumptions are made as to the nature of the interaction terms. This is the formulation whereby the Hugoniot curve for the mixture is obtained by assuming each constituent (shock) compresses as if it is in the pure state [4-6, 11-15]. Such a pressure equilibrium (PEQ) Hugoniot is readily computed from solutions to

$$p_H(\rho_H) = \overset{(\alpha)}{p}{}_H^{e}\left(\overset{(\alpha)}{\rho}{}_H^{e}\right) \tag{48}$$

$$1/\rho_H = \sum_{\alpha=1}^{r} \overset{(\alpha)}{m} / \overset{(\alpha)}{\rho}{}_H^{e}, \quad \alpha = 1, 2, \ldots r \tag{49}$$

where $\overset{(\alpha)}{m}$ is the mass fraction of $\overset{(\alpha)}{s}$,

$$\sum_{\alpha=1}^{r} \overset{(\alpha)}{m} = 1 \tag{50}$$

and the subscript H denotes Hugoniot states.

It is readily apparent that for equations of state of the form

$$\overset{(\alpha)}{p}{}^{e} = \overset{(\alpha)}{p}{}^{e}\left(\overset{(\alpha)}{\rho}{}^{e}\right)$$

[like Eq. (48)], the overall continuity* and momentum equations are automatically satisfied. The shock and particle velocities in the mixture may be readily computed, i.e.,

$$U_{sh} = \left(\frac{p_H - p_0}{\rho_0}\right) \Bigg/ \left(\frac{1}{\rho_0} - \frac{1}{\rho_H}\right)^{1/2}, \quad v = \left(\frac{\rho_H - \rho_0}{\rho_H}\right) U_{sh}$$

$$\frac{1}{\rho_0} = \sum_{\alpha=1}^{r} \frac{\overset{(\alpha)}{m}}{\overset{(\alpha)}{\rho}{}_0^{e}} \tag{51}$$

*The overall continuity equation is $\rho_H(U_{sh} - v) = \rho_0 U_{sh}$.

In general, these differ from those associated with the pure material at each shock pressure. Consequently, the F term in the constituent momentum equation makes up for the difference in velocities between the mixture and that of the pure constituent. It is shown in Section V.C that this deviation in the velocity state requires a compensating energy exchange term Q_{PEQ} to ensure thermodynamic consistency of the theory.

A third pressure equilibrium model has been independently studied by Tsou and Chou [1,2] and Riney et al. [14, 15]. In these analyses, the two constituents are presumed to thermally equilibrate behind the shock front. Given the thermal equations of state for the materials,

$$\overset{(\alpha)}{E} = \overset{(\alpha)}{E}\left(\overset{(\alpha)}{\rho}{}^{e}, \overset{(\alpha)}{T}\right) \tag{52}$$

where $\overset{(\alpha)}{T}$ is the temperature of $\overset{(\alpha)}{s}$, the thermal equilibrium condition,

$$T = \overset{(\alpha)}{T}, \quad \alpha = 1, 2, \ldots, r \tag{53}$$

may be utilized to determine the pressure-thermal equilibrium (PTEQ) Hugoniot. An overall energy equation is obtained by summing the constituent energy relations, which for the binary mixture, is

$$\begin{aligned} &\overset{(1)}{V}_0 \overset{(1)}{\rho}{}_0^{e}\left(\overset{(1)}{E} - \overset{(1)}{E}_0\right) + \left(1 - \overset{(1)}{V}_0\right)\overset{(2)}{\rho}{}_0^{e}\left(\overset{(2)}{E} - \overset{(2)}{E}_0\right) \\ &\quad + \left[\overset{(1)}{V}_0 \overset{(1)}{\rho}{}_0^{e} + \left(1 - \overset{(1)}{V}_0\right)\overset{(2)}{\rho}{}_0^{e}\right]\left(\frac{v^2}{2} - vU_{sh}\right) \\ &\quad = -p^{e}\left(1 - \frac{v}{U_{sh}}\right) + p_0^{e} \end{aligned} \tag{54}$$

This equation, along with that for the overall momentum (46), the two continuity relations (30) and (31), and the state equations (45) and (52) provide a set of eight algebraic equations for nine unknowns, $\left(\overset{(1)}{V}, \overset{(1)}{\rho}^e, \overset{(2)}{\rho}^e, U_{sh}, v, p^e, \overset{(1)}{E}, \overset{(2)}{E}, T\right)$. It can be demonstrated that this problem may be further reduced to five equations for six unknowns if U_{sh}, v, and $\overset{(1)}{V}$ are eliminated from the algebraic system.

An alternative approach to the PTEQ problem is provided by Duvall and Taylor [10]. They assume the composite Gibbs function to be a mass-weighted average of the constituent Gibbs functions. Thus, given the constituent mass fractions and Gibbs functions, it is shown that all the composite thermodynamic variables may be computed by invoking suitable thermodynamic identities. It is important to note that the basic assumptions involved in the Duvall-Taylor Hugoniot analysis are the same as those utilized in the PTEQ formulation, i.e., pressure and thermal equilibrium. The Hugoniot states they calculate for a polyethylene-quartz composite are the PTEQ Hugoniot states.

Analogous to the preceding formulations, the momentum interaction and energy exchange terms F and Q may be calculated from the final Hugoniot states. The calculations in Refs. [1, 15] indicate that for the structural and geologic mixtures considered, the order of magnitudes of the F terms are the same as in the ADEQ and PEQ models.* Consequently, in the U_{sh} versus v plane, there are only small differences in the ADEQ, PEQ, and PTEQ Hugoniots. However, the shock temperatures of the constituents may vary significantly from model to model (see e.g., Refs. [14,15] for wet geologic materials). These effects are important when chemical reactions, phase changes, etc. must be taken into account.

Before discussing the laminate models, it should be mentioned that an alternative to the three Hugoniot for-

*The Q terms are of the same order of magnitude in the PEQ and PTEQ formulations.

mulations described above is that proposed by McQueen et al. [8]. In their approach, a new material equation of state is derived by taking the pressure equilibrium mixture of the 0°K isotherms as the 0°K isotherm of the mixture material [instead of employing the constituent Hugoniots, from the 0°K isotherm are treated by taking a mass-weighted sum of the partial derivates $\left(\partial \overset{(\alpha)}{E}/\partial p^e\right)_{\rho^e}$, assumed to be constant, to form a $(\partial E/\partial p)_{\rho_{MIX}}$ for the "new" material. The mixture pressure is given by

$$p(E_{MIX}, \rho_{MIX}) = p_{0°K}(\rho_{MIX}) + \frac{1}{(\partial E/\partial p)_{\rho_{MIX}}}\left[E_{MIX} - E_{0°K}(\rho_{MIX})\right] \tag{55}$$

Such a model, although limited to specific forms of the energy contribution, is comparable in simplicity and concept to the PEQ case. It is not quite the same, however, since $\overset{(\alpha)}{V}$ is determined by the 0°K conditions and not the Hugoniots of the pure materials (as in the PEQ model). If one were to assume a pressure equilibrium between the constituents, the F and Q terms could be readily calculated (see Section V.C).

B. Laminated Composites

Hugoniot relations for laminar and unidirectionally fiber-reinforced composites (wave propagation in the direction of layering) have been developed by Tsou and Chou [1] and Munson and Schuler [6]. The formulations by Tsou and Chou are identical with the ADEQ and PTEQ models discussed in the preceding section. It should be emphasized here that it is not possible to determine p_m and f separately without introducing an additional constitutive assumption. In their analysis, Tsou and Chou assume p_m to be zero.

Naturally, any other assumption for p_m would lead to a different partitioning of the total interaction force F. Munson and Schuler [6] formulate Hugoniot models for longitudinal and transverse laminates within the PEQ assumptions. Since the Hugoniot states are uniquely specified in the PEQ model, the Munson and Schuler conclusion, that the Hugoniot states are independent of laminate orientation, is implicit in their formulation of the problem.

C. Interaction Terms Implicit in Mixture Models

The momentum conservation equation (33) for the first material may be written as

$$\overset{(1)}{\rho}{}^{e} v_i (U_i - v_i) + \left(F_i / \overset{(1)}{V_i}\right) = p_i^e - \left(\overset{(1)}{V_0} / \overset{(1)}{V_i}\right) p_0^e \tag{56}$$

where the subscript i denotes a particular mixture model (PEQ, McQueen et al., etc.). In the pure material (F = 0, $\overset{(1)}{V} = 1$), we have

$$\overset{(1)}{\rho_H} v_H (U_H - v_H) = p_H - p_0 \tag{57}$$

where the subscript H refers to the pure material Hugoniot properties. Thus, if we take the same stress level in Eqs. (56) and (57), we may solve for F_i, i.e.,

$$F_i = \overset{(1)}{V_i}\left[\overset{(1)}{\rho_H} v_H (U_H - v_H) - \overset{(1)}{\rho_i}{}^{e} v_i (U_i - v_i)\right] \tag{58}$$

where the p_0 term has been neglected. In general, the right-hand side is nonzero, so that such formulations implicitly specify finite values of the momentum interaction term. (Note that for the PEQ model, $\overset{(1)}{\rho_H} = \overset{(1)}{\rho_i}{}^{e}$.)

It can also be shown that the energy exchange term is nonzero in the PEQ and McQueen et al. (MIX) models. The energy conservation equation (42) for the PEQ model may be written in a more convenient form if one divides Eq. (42) by $\overset{(1)}{V}_0 \overset{(1)}{\rho}{}_0^e$, and utilizes the continuity Eq. (30) to obtain

$$\Delta \overset{(1)}{E}_i = -\frac{1}{2} v_i^2 + v_i U_i - \left(p_i^e / \overset{(1)}{\rho}{}_i^e\right) + \left(p_0^e / \overset{(1)}{\rho}{}_0^e\right) + \left(Q_i / \overset{(1)}{V}_0 \overset{(1)}{\rho}{}_0^e U_i\right) \tag{59}$$

Then, substituting the expression for $p_i^e / \overset{(1)}{\rho}{}_i^e$ from Eq. (56) and neglecting p_0 terms yields

$$\Delta \overset{(1)}{E}_i = \frac{1}{2} v_i^2 - \left(F_i / \overset{(1)}{V}_i \overset{(1)}{\rho}{}_i^e\right) + \left(Q_i / \overset{(1)}{V}_0 \overset{(1)}{\rho}{}_0^e U_i\right) \tag{60}$$

Employing the expression for F_{PEQ} [Eq. (58)], and noting that in the PEQ model $\overset{(1)}{\rho}_H = \overset{(1)}{\rho}{}_{PEQ}^e$ at a given stress level, Eq. (60) reduces to

$$\Delta \overset{(1)}{E}_{PEQ} = v_{PEQ}\left(U_{PEQ} - \frac{1}{2} v_{PEQ}\right) - v_H(U_H - v_H) + \left(Q_{PEQ} / \overset{(1)}{V}_0 \overset{(1)}{\rho}{}_0^e U_{PEQ}\right) \tag{61}$$

In the PEQ model, a given p versus ρ (Hugoniot) state in the constitutent corresponds to an energy increase given by the pure material Hugoniot,

$$\Delta \overset{(1)}{E} = \frac{1}{2} v_H^2 \tag{62}$$

Thermodynamic consistency of the PEQ model requires that the above changes in energy be the same. Equating the right-hand sides of Eqs. (61) and (62) results in an expression for Q_{PEQ},

$$Q_{PEQ}/\overset{(1)}{V}_0\,\overset{(1)}{\rho}_0\,{}^{e}U_{PEQ} = v_H\left(U_H - \frac{1}{2}\,v_H\right) - v_{PEQ}\left(U_{PEQ} - \frac{1}{2}\,v_{PEQ}\right) \tag{63}$$

Once again, this term is generally nonzero for an arbitrary mixture.

The McQueen et al. mixture formulation (MIX) specifies a different form for Q. That is, the energy change for a constituent is the product of its mass fraction and the total energy rise in the mixture at a given stress level,

$$\Delta\,\overset{(1)}{E}_{MIX} = \overset{(1)}{m}\;\Delta E = \overset{(1)}{m}\;v_{MIX}^2/2 \tag{64}$$

Analogous to the derivation of Eq. (61) the energy gain in $\overset{(1)}{s}$ for this formulation is given by

$$\Delta\,\overset{(1)}{E}_{MIX} = v_{MIX}\left(U_{MIX} - \frac{1}{2}\,v_{MIX}\right) - \left(\overset{(1)}{\rho}_H/\,{}^{e}\overset{(1)}{\rho}_{MIX}\right)v_H(U_H - v_H)$$

$$+ \left(Q_{MIX}/\,\overset{(1)}{V}_0\,\overset{(1)}{\rho}_0\,{}^{e}U_{MIX}\right) \tag{65}$$

Then, equating the right-hand side of Eqs. (64) and (65), one obtains

$$Q_{MIX}/\,\overset{(1)}{V}_0\,\overset{(1)}{\rho}_0\,{}^{e}U_{MIX} = \left(\overset{(1)}{\rho}_H/\,{}^{e}\overset{(1)}{\rho}_{MIX}\right)v_H(U_H - v_H)$$

$$- v_{MIX}\left[U_{MIX} - \frac{1}{2}\left(\overset{(1)}{m} + 1\right)v_{MIX}\right] \tag{66}$$

VI. CONCLUDING REMARKS

A formulation of the conservation relations across a steadily propagating shock front in composite materials has been presented. Interaction terms, either implicitly or explicitly assumed, have been shown to differ for the various models considered. Table 1 provides a summary, where the mixture Hugoniot hypotheses are classified by the assumptions on particle velocity, F, and Q. It should be remarked that only steady shock fronts have been considered and that shock wave studies in the laboratory may only approach this

TABLE 1

Pressure Equilibrium, Binary Mixture Hugoniot Analyses

Model	Implicit specification of momentum interaction term F	Implicit specification of Q
ADEQ	Yes; but f and p_m cannot be determined without additional constitutive relation	No; Q = 0
PTEQ	Yes; but f and p_m cannot be determined without additional constitutive relation	Yes; but Q fixed by the condition behind shock front, $T = T^{(1)} = T^{(2)}$
PEQ	Yes; but f and p_m cannot be determined without additional constitutive relation	Yes
McQueen et al. [8]	Yes; but f and p_m cannot be determined without additional constitutive relation	Yes (differs from PEQ)

condition. Moreover, only the case of pressure equilibrium between constituents has been discussed, and strength effects, which may be significant at low shock intensities, have been ignored.

REFERENCES

1. F. K. Tsou and P. C. Chou, "Analytical Study of Hugoniot in Unidirectional Fiber-Reinforced Composites," J. Composite Materials, 3, 500-514 (1969).

2. F. K. Tsou and P. C. Chou, "The Control-Volume Approach to Hugoniot of Macroscopically Homogeneous Composites," J. Composite Materials, 4, 526-537 (1970).

3. P. J. Torvik, "Shock Propagation in a Composite Material," J. Composite Materials, 4, 296-309 (1970).

4. P. J. Torvik, "A Simple Theory for Shock Propagation in Homogeneous Mixtures," Air Force Institute of Technology Report AFIT TR 70-3, May 1970.

5. A. K. Hopkins, "On the Response of Mixed Materials to One-Dimensional Shock Waves," Air Force Materials Laboratory Report AFML-TR-70-158, July 1970.

6. D. E. Munson and K. W. Schuler, "Hugoniot Predictions in Laminated Composites and Mechanical Mixtures," Sandia Laboratories Report SC-DR-70-34, 1970.

7. G. A. Gurtman, A. H. Nayfeh, G. A. Hegemier, E. W. Sims, and D. H. Brownell, "A Continuum Theory Applied to Wave Propagation in Three-Dimensionally Reinforced Composites," Systems, Science and Software Report SSS-R-72-1462, February 1973.

8. R. G. McQueen, S. P. Marsh, J. W. Taylor, J. N. Fritz, and W. J. Carter, "The Equation of State of Solids from Shock Waves Studies," in High-Velocity Impact Phenomena, R. Kinslow, ed., Academic Press, New York, 1970, pp. 293-417.

9. S. K. Garg and J. W. Kirsch, "Hugoniot Analysis of Composite Materials," J. Composite Materials, 5, 428-445 (1971).

10. G. E. Duvall and S. M. Taylor, Jr., "Shock Parameters in a Two-Component Mixture," J. Composite Materials, 5, 130-139 (1971).

11. J. T. Rosenberg, T. J. Ahrens, and C. F. Petersen, "Dynamic Properties of Rocks," Stanford Research Institute, Defense Atomic Support Agency Report DASA 2112, July 1968.

12. P. C. Lysne, "A Comparison of Calculated and Measured Low-Stress Hugoniots and Release Adiabats for Dry and Water-Saturated Tuff," J. Geophys. Res., 75 (23), 4375-4386 (1970).

13. T. R. Butkovich, "The Influence of Water in Rocks on Underground Nuclear Explosion Effects," J. Geophys. Res., 76 (8), 1993-2011 (1971).

14. T. D. Riney, S. K. Garg, J. W. Kirsch, L. W. Morland, and C. R. Hastings, "Stress Wave Effects in Inhomogeneous and Porous Earth Materials," Systems, Science and Software Report 3SR-267, Defense Atomic Support Agency Report DASA 2498, March 1970.

15. T. D. Riney, S. K. Garg, J. W. Kirsch, C. R. Hastings, and K. G. Hamilton, "Wave Propagation in Porous Geologic Composites," Systems, Science and Software Report 3SR-648, Defense Nuclear Agency Report DNA 27251, July 1971.

16. S. K. Garg, "Wave Propagation Effects in a Fluid Saturated Porous Solid," J. Geophys. Res., 76 (32), 7947-7962 (1971).

17. L. W. Morland, "A Simple Constitutive Theory for a Fluid Saturated Porous Solid," J. Geophys. Res., 77 (5), 890-900 (1972).

18. A. E. Green and P. M. Naghdi, "A Theory of Mixtures," Arch. Rat. Mech. Anal., 24, 243-263 (1967).

19. P. D. Kelly, "A Reacting Continuum," Int. J. Engng. Sci., 2, 129-153 (1964).

Appendix

INTRODUCTION TO PROBABILITY AND STATISTICS

I. STATISTICS

The term "statistics" refers to the collection, organization, and analysis of "samples" of data on the basis of the laws of probability for the purpose of arriving at valid conclusions regarding the characteristics or the behavior of the entire group called the "population." The characteristics of the population are referred to as "population parameters." The population may be "discrete" or "continuous" depending upon whether it contains enumerated or measured data. In the statistical analysis, a small part of the population called the "sample" is examined. The sample must be representative of the population in order that the conclusions arrived at by examining the sample may be extended to the entire population. This can be achieved by selecting the sample at random, i.e., each element of the population has an equal chance of being selected for inclusion in the sample. A sample characteristic is referred to as a "statistic." It may be utilized as a measure of the corresponding population parameter. The problem of selecting a sample and obtaining thereby an estimate of the population parameters forms a well-developed branch of statistics. However, it is outside the scope of this brief appendix. In the following, no distinction is made between the sample and the population characteristics.

II. CUMULATIVE PROBABILITY DISTRIBUTION FUNCTION

One of the simplest ways to reveal the general nature of a population is to construct a histogram or a bar diagram. For example, consider a large number of material strength measurements x_i $(i = 1, 2, \ldots, N)$. The range of the strength data is first divided into n equal-sized intervals, $d = (x_N - x_1)/n$. The number of times that the strength measurements fall into each interval (d), referred to as the "frequency" $\bar{f}_i(x)$ $(i = 1, 2, \ldots, n)$, is then tabulated. A bar diagram is now drawn such that the height of a rectangle equals the frequency $\bar{f}_i(x)$ for the interval in question (see Fig. 1). The frequency $\bar{f}_i(x)$ is usually

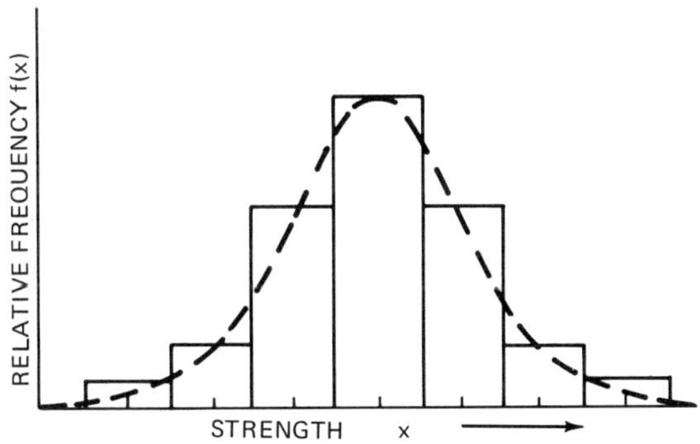

FIG. 1. Typical histogram.

divided by the total population size, giving a relative frequency $f_i(x) = \bar{f}_i(x)/N$. As the population size is increased and the interval width d decreased, the histogram will approach a continuous curve (dashed curve in Fig. 1). The total area under the curve is unity. Such a curve is referred to as the "probability density function" f(x). The area of the curve between any two abscissas (x_i, x_{i+1}) represents the probability that the random variable x (see Section V) falls in the interval (x_i, x_{i+1}). Note that the probability density function is not a direct measure of probability. The "cumulative probability distribution" F(x) is defined as

$$F(x) = \int_{-\infty}^{x} f(x)\ dx \tag{1}$$

It equals the area under the probability density curve between the abscissas $-\infty$ and x, and represents the probability that the random variable lies within the interval $(-\infty, x)$. Several of the more important probability density functions will be discussed later.

III. BASIC PROBABILITY THEORY

A. Permutations and Combinations

For convenient reference, we start by stating a few of the principles of combinatorial analysis.

A permutation is an arrangement of items where the order of selection is noted. The number of permutations of n different things taken r at a time, with replacement, is n^r, where $1 \leq r \leq n$. The number of permutations of n different things taken r at a time, without replacement, is

$$P_r^n = n!/(n - r)! \tag{2}$$

where

$$n! = n(n - 1)(n - 2) \ldots 1, \quad 1! = 0! = 1$$

The number of arrangements of n things taken all at a time, where there are n_i $(i = 1,\ldots, k)$ of each kind, is given by

$$n!/n_1!\, n_2! \ldots n_k! \tag{3}$$

where

$$\sum_{i=1}^{k} n_i = n$$

A combination is defined as a group of items where the order of selection is not considered. Thus, the number of combinations of n things taken r at a time is

$$C_r^n = n!/r!(n - r)! \tag{4}$$

B. Probability

Modern probability theory is a deductive science and can be developed axiomatically. However, to do this it is essential to introduce measure (or set) theory. For a development in terms of set theory, the reader is referred to the book by Papoulis [1]. In the following, we shall only discuss the classical *a priori* definition and frequency interpretation of probability.

If the number of ways an event may result can be resolved into a successes and b failures, each equally likely to occur, then the *a priori* probability of success in a single trial is $\overline{p} = a/(a + b)$ and the corresponding probability of failure is $\overline{q} = b/(a + b)$.

Let us consider a series of n trials in which n_a successes are observed. Then the probability of successes (frequency interpretation) is approximately n_a/n. It is assumed that n_a/n approaches a limit as n approaches infinity. This limiting value of n_a/n is called the *a posteriori* probability of success in a single trial.

C. Probability Laws

(i) MUTUALLY EXCLUSIVE EVENTS

Two events A and B are called mutually exclusive if the happening of one precludes the happening of the other.

(ii) ADDITION OF PROBABILITIES

If two events A and B are mutually exclusive, then the probability of A or B occurring is given by

$$P(A + B) = P(A) + P(B) \tag{5}$$

where $P(X)$ denotes the probability of the occurrence of

event X. The above definition can be generalized to n mutually exclusive events $A_1, A_2, \ldots, A_n$:

$$P(A_1 + A_2 + \ldots + A_n) = P(A_1) + P(A_2) + \ldots + P(A_n) \tag{6}$$

Let us now consider n events $A_1, A_2, \ldots, A_n$ which are not necessarily mutually exclusive. For this case, Eq. (6) becomes

$$P(A_1 + A_2 + \ldots + A_n) = P(A_1) + \ldots + P(A_n) - \left[P(A_1A_2) + \ldots + P(A_{n-1}A_n)\right] + \ldots + (-1)^{n-1}P(A_1A_2 \ldots A_n) \tag{7}$$

where $P(A_1A_2 \ldots)$ denotes the probability of the joint occurrence of the events $A_1, A_2, \ldots$.

(iii) INDEPENDENT EVENTS AND MULTIPLICATION

Two events A and B are called independent if the probability of the joint occurrence of A and B is given by the multiplication of their individual probabilities:

$$P(AB) = P(A)P(B) \tag{8}$$

If A and B are mutually exclusive, then it is easily seen from Eq. (8) that the probability of their joint occurrence must be zero.

The definition of Eq. (8) may be extended to n independent events $A_1, A_2, \ldots, A_n$:

$$P(A_1A_2 \ldots A_n) = P(A_1)P(A_2) \ldots P(A_n) \tag{9}$$

(iv) CONDITIONAL PROBABILITY

Given an event B with nonzero probability,

$$P(B) > 0$$

the conditional probability of A assuming B is defined as

$$P(A/B) = P_{A/B} = P(AB)/P(B) \tag{10}$$

Clearly, if A and B are mutually exclusive, then

$$P(A/B) = 0$$

On the other hand, if A and B are independent, then

$$P(A/B) = P(A)$$

It is also easily deduced from Eq. (10) that

$$P(AB) = P(A/B)P(B) = P(B/A)P(A)$$

IV. MEASURES OF CENTRAL TENDENCY AND SCATTER

The basic statistical measures deal with central values and scatter of data.

The "mean" or the "expectation" for a probability density distribution is defined as

$$\langle x \rangle = \int_{-\infty}^{\infty} xf(x)\, dx \tag{11}$$

It corresponds to the abscissa of the center of gravity of the area under the probability density curve.

The "median" is defined as the value of x which

divides the area under the probability density curve into two equal parts.

The "mode" is the most frequent value in a set of observations. For a continuous distribution, it is found by solving the equation

$$df(x)/dx = 0 \tag{12}$$

The mean, median, and mode coincide for symmetric distributions (e.g., normal distribution).

The most useful measure of the data spread is the "variance." It is defined as the average of the squared deviations from the mean value. Mathematically, the variance Ξ^2 is defined as

$$\Xi^2 = \int_{-\infty}^{\infty} (x - \langle x \rangle)^2 f(x)\, dx \tag{13}$$

Physically, the quantity equals the moment of inertia of the probability masses and gives some notion of their concentration near $\langle x \rangle$. The "standard deviation Ξ" is defined as the square root of the variance. Sometimes, "coefficient of variation (c.v.)" is utilized as a measure of data scatter. It is defined as

$$\text{c.v.} = \Xi/\langle x \rangle \tag{14}$$

The third moment, μ, about the mean serves as a measure of skewness:

$$\mu^3 = \int_{-\infty}^{\infty} (x - \langle x \rangle)^3 f(x)\, dx \tag{15}$$

The "coefficient of skewness (c.s.)" is defined as

$$\text{c.s.} = \mu^3/\Xi^3 \tag{16}$$

For symmetric distributions (e.g. normal), the third moment about the mean and hence the coefficient of skewness are zero.

V. FUNCTIONS OF A RANDOM VARIABLE

We are given an experiment specified by its possible outcomes (ξ_i, $i = 1, 2, \ldots, n$). To every ξ_i, we assign a number $x(\xi_i)$. This defines a function x whose domain is the set of all outcomes and range a set of numbers. This function provided it satisfies certain conditions of a general nature [1] is called a random variable.

Let us now consider a function $y = g(x)$ of the random variable x. It is a random variable with a probability density function

$$f(y) = f(x)/|dg/dx| \tag{17}$$

The mth moment of y around zero is given by

$$\langle y^m \rangle = \int_{-\infty}^{\infty} y^m f(y)\, dy = \int_{-\infty}^{\infty} [g(x)]^m f(x)\, dx \tag{18}$$

VI. JOINT DISTRIBUTION AND DENSITY FUNCTIONS

Consider two random variables x and y with cumulative distribution functions $F(x)$ and $F(y)$, respectively. In general, the joint cumulative distribution function $F(x,y)$ cannot be found from $F(x)$ and $F(y)$. The distribution function $F(x,y)$ is defined as the joint probability of random variable x lying between $-\infty$ and x and random variable y lying between $-\infty$ and y. If it is now assumed that $F(x,y)$ has partial derivatives of order up to two, then the joint density function is given by

$$f(x,y) = \partial^2 F(x,y)/\partial x\, \partial y \tag{19}$$

The cumulative distribution functions $F(x)$ and $F(y)$ may be easily calculated from the joint statistics. Thus

$$F(x) = F(x,\infty) = \int_{-\infty}^{\infty}\int_{-\infty}^{x} f(\xi, y)\, d\xi\, dy$$

and

$$F(y) = F(\infty,y) = \int_{-\infty}^{\infty}\int_{-\infty}^{y} f(x, \eta)\, d\eta\, dx \qquad (20)$$

Finally, if x and y are independent random variables, then one has

$$F(x,y) = F(x)F(y)$$

and

$$f(x,y) = f(x)f(y) \qquad (21)$$

VII. PROBABILITY DISTRIBUTIONS

A. Gaussian (Normal) Distribution

The Gaussian (normal) probability distribution is probably the most common density function in applied statistics. It provides a good approximation to many physical phenomena (e.g., Maxwellian gas flow), and is the basis of the tests of significance. The Gaussian probability density function $f(x)$ is given by

$$f(x) = \frac{1}{(2\pi)^{1/2}\Xi} \exp\left[-\frac{1}{2}\left(\frac{x - \langle x\rangle}{\Xi}\right)^2\right] \qquad (22)$$

where $\langle x \rangle$ and Ξ denote the mean and the standard deviation of the random variable x, respectively. This is the familiar bell-shaped curve, symmetric about its mean value (see Figs. 1 and 2).

The area under the curve represents the probability of the occurrence of the random variable x. The total area under f(x) is unity, by definition. Note that 68.2% of this area lies within $(\langle x \rangle + \Xi, \langle x \rangle - \Xi)$, and 99.7% within $(\langle x \rangle + 3\Xi, \langle x \rangle - 3\Xi)$. The cumulative probability distribution function F(x) is given by

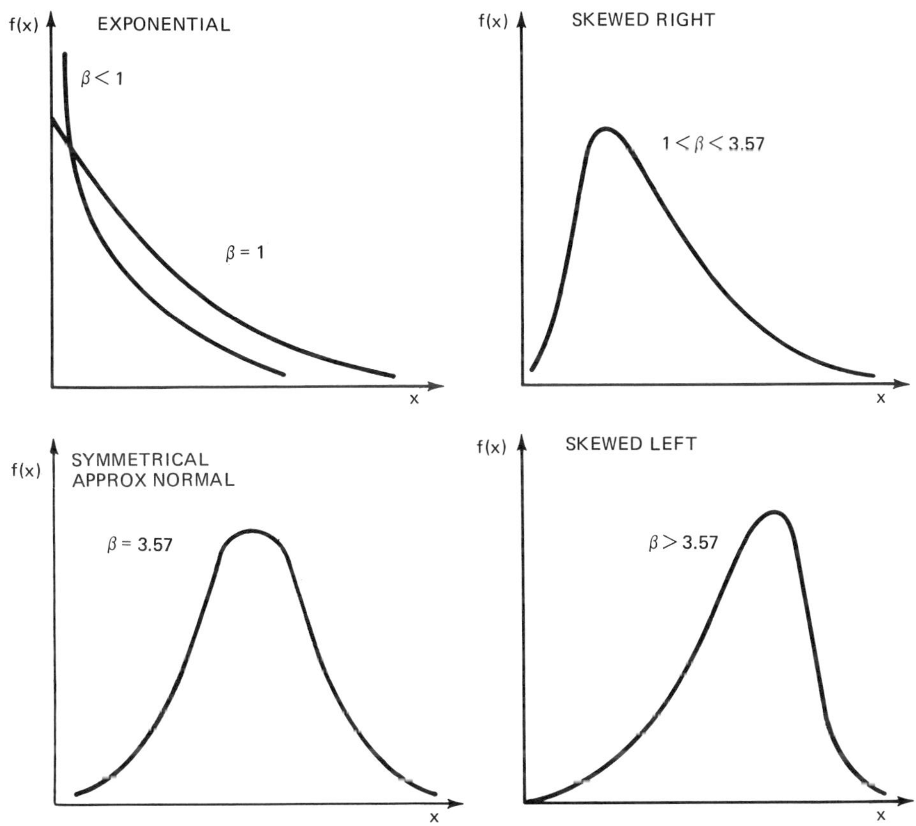

FIG. 2. Weibull probability density curves.

$$F(x) = \int_{-\infty}^{x} f(x)\, dx \tag{23}$$

Both the probability density and the cumulative distribution functions are frequently expressed in terms of the standard normal deviate Z:

$$Z = (x - \langle x \rangle)/\Xi \tag{24}$$

The probability density and cumulative distribution functions can thus be written as

$$f(Z) = [1/(2\pi)^{1/2}] \exp(-Z^2/2), \; F(Z) = \int_{-\infty}^{Z} f(Z)\, dZ \tag{25}$$

Random variable Z has zero mean and unity standard deviation. Tables for the function $F(Z)$ are readily available [1].

B. Weibull Distribution

The Weibull probability density function $f(x)$ changes its shape with the distribution parameters. It can adopt anywhere from exponential to highly skewed shapes. The probability density function $f(x)$ is given by

$$f(x) = \alpha\beta x^{\beta-1} \exp(-\alpha x^{\beta}) \tag{26}$$

where $x \geq 0$, $\beta > 0 \sim$ shape parameter, and $\alpha > 0 \sim$ characteristic parameter.

The cumulative distribution function $F(x)$, corresponding to $f(x)$ of Eq. (26), is

$$F(x) = 1 - \exp(-\alpha x^{\beta}) \tag{27}$$

In the literature, sometimes a generalized version of $F(x)$ is employed

$$F(x) = 1 - \exp\left[-\left(\frac{x - x_0}{\theta - x_0}\right)^{\beta}\right] \tag{28}$$

where $x_0 \geq 0$ is the origin of the distribution and θ is the characteristic parameter. The distribution (28) can be reduced to distribution (27) by setting

$$x_0 = 0, \quad \theta = \alpha^{-1/\beta}$$

The wide range of possible shapes for the Weibull probability density function is shown in Fig. 2.

C. Binomial Distribution

This is one of the more common discrete distributions in applied statistics and forms the basis for the Gaussian and the Poisson distributions. It applies where each outcome of an experiment is judged on some go/no go basis, such as heads or tails, etc. The probability of success for a single outcome is denoted by $\bar{p}$ and the corresponding probability of failure by $\bar{q}$ (= $1 - \bar{p}$). Thus, if $\bar{p}$ is the probability of a single event occurring and $\bar{q}$ the probability that the event does not occur, then the probability that the event will occur exactly r times in n independent trials is given by $P_n(x=r)$

$$P_n(x=r) = C_r^n \, \bar{p}^r \, \bar{q}^{n-r} \tag{29}$$

where $0 \leq r \leq n$. The binomial distribution probability density function $f(x)$ is a sequence of impulses

$$f(x) = \sum_{r=0}^{n} C_r^n \, \bar{p}^r \, \bar{q}^{n-r} \, \delta(x - r) \tag{30}$$

where $\delta(x - r)$ denotes the Dirac delta function. The

function $\delta(x - r)$ is specified by its integral property

$$\int_{-\infty}^{\infty} \phi(x)\, \delta(x - r)\, dx = \phi(r)$$

The cumulative distribution function $F(x)$ is given by

$$F(x) = \sum_{r=0}^{r} C_r^n\, \bar{p}^r\, \bar{q}^{n-r} \tag{31}$$

It denotes the probability that the event will occur r times or less in n trials. The mean $\langle x \rangle$ and the variance Ξ^2 for the binomial distribution are $n\bar{p}$ and $n\bar{p}\,\bar{q}$, respectively.

There exist two important limiting cases of the binomial distribution. When $n\bar{p}\,\bar{q} \gg 1$ (large n) and $(n\bar{p}\,\bar{q})^{1/2} \sim O(r - n\bar{p})$, the binomial distribution approaches the normal distribution with r as the random variable. On the other hand, when n is large but $\bar{p}$ is small, such that $n\bar{p}$ is finite, the binomial distribution approaches the Poisson distribution.

D. Poisson Distribution

This discrete distribution is utilized when the probability of occurrence of an event $\bar{p}$ is small and the number of trials n is very large, such that $n\bar{p}$ is finite. As an example, one may consider the number of meteorites striking a specific area of the surface of the moon in a month. The probability of exactly r occurences is given by

$$P(x = r) = (m^r e^{-m})/r! \tag{32}$$

where $m = n\bar{p}$ is the expected number of occurrences. The

Poisson distribution density function is a sequence of impulses,

$$f(x) = e^{-m} \sum_{r=0}^{\infty} (m^r/r!)\delta(x - r) \tag{33}$$

where $\delta(x - r)$ denotes the Dirac delta function defined previously.

The cumulative distribution function $F(x)$ is

$$F(x) = e^{-m} \sum_{r=0}^{r} (m^r/r!) \tag{34}$$

It denotes the probability of r or fewer occurrences. The mean $\langle x \rangle$ and the variance Ξ^2 for the Poisson distribution are both equal to m.

E. Other Distributions

There are several other important probability density functions encountered in the engineering literature. These are briefly introduced below.

(i) Rectangular distribution: The probability density function is

$$f(x) = 1/(b - a), \quad a \le x \le b$$

$$f(x) = 0 \qquad \text{elsewhere} \tag{35}$$

(ii) Cauchy distribution: The probability density function is given by

$$f(x) = (1/\pi)\ \lambda/[\lambda^2 + (x - \langle x \rangle)^2] \tag{36}$$

where λ and $\langle x \rangle$ are the distribution parameters. It is straightforward to show that the mean of the distribution is given by $\langle x \rangle$.

(iii) Laplace distribution: The Laplace probability density function is

$$f(x) = (1/2\lambda)[\exp(-|x - \langle x \rangle|/\lambda)] \tag{37}$$

where $|\cdots|$ denotes the absolute value and λ and $\langle x \rangle$ are the distribution parameters. The mean of the distribution is $\langle x \rangle$. Both the Cauchy and the Laplace distributions are symmetric about the mean value $\langle x \rangle$.

VIII. STATISTICAL THEORY OF THE DISTRIBUTION OF EXTREME VALUES

It was remarked in the text that the weakest-link hypothesis corresponds to an important problem in mathematical statistics, namely the problem of the distribution of the smallest value, in samples of size n, drawn from a population having the probability density function $f(x)$. In the following, we will discuss the mathematical theory of the distribution of the smallest value.

Let the underlying distribution be given by the continuous probability density function $f(x)$, with the associated cumulative distribution function $F(x)$. The distribution of the smallest value in samples of size n, drawn from this population, is given by the probability density function $g_n(x)$:

$$g_n(x) = nf(x)[1 - F(x)]^{n-1} \tag{38}$$

or the cumulative distribution function $G_n(x)$:

$$G_n(x) = \int_{-\infty}^{x} g_n(x)\, dx = 1 - [1 - F(x)]^n \tag{39}$$

The above equations can be obtained in the following manner. Consider a fiber made of n links. The strength of the links is characterized by the probability density function f(x). According to the weakest-link hypothesis, the fiber fractures when any one of the n links gives way. The probability of the survival of a link under stress x is [1 - F(x)], and therefore the probability of survival of the entire fiber is $[1 - F(x)]^n$. The probability density function $g_n(x)$ is obtained by differentiating $G_n(x)$.

The most probable value (mode of the smallest value in samples of size n) can be found by maximizing $g_n(x)$. If the solution to the equation $dg_n/dx = 0$ exists, it may be written as

$$(n - 1)f^2(x_n^*) = [df(x)/dx]_{x=x_n^*}[1 - F(x_n^*)] \tag{40}$$

where $x_n{}^*$ is the most probable value.

In order to find the distribution of the smallest value in samples of size n, it is convenient to introduce a new random variable.

$$\xi = nF(x) \tag{41}$$

The equation for $g_n(x)$ along with Eqs. (17) and (41) gives

$$h_n(\xi) = (1 - \xi/n)^{n-1} \tag{42}$$

where $h_n(\xi)$ denotes the probability density function of ξ.

In the limit as n approaches infinity, the probability density function $h_n(\xi)$ approaches $h(\xi)$:

TABLE 1. Distribution of Extreme Values for

Probability Density Function	x_n^*, Mode of smallest value in samples of size n
1. Rectangular distribution $f(x) = 1/(b - a), a \leq x \leq b$ $f(x) = 0,$ elsewhere	a
2. Cauchy distribution $f(x) = (1/\pi)\ \lambda/[\lambda^2 + (x - \langle x\rangle)^2]$	$\langle x\rangle - (\lambda n/2\pi)$
3. Laplace distribution $f(x) = (1/2\lambda)\ \exp(-\lvert x - \langle x\rangle\rvert/\lambda)$	$\langle x\rangle - \lambda \log(n/2)$
4. Gaussian (or normal) distribution $f(x) = [1/(2\pi)^{1/2}\Xi]\ \exp[-(x - \langle x\rangle)^2/2\Xi^2]$	$\langle x\rangle - \Xi(2\log n)^{1/2} + \Xi\ [\log(\log n) + \log 4\pi]/2(2\log n)^{1/2}$
5. Weibull distribution $f(x) = \alpha\beta x^{\beta-1} \exp(-\alpha x^{\beta})$ $\alpha > 0, \beta > 1, x \geq 0$ $F(x) = 1 - \exp(-\alpha x^{\beta})$	$[1/(\alpha n)^{1/\beta}](1 - 1/\beta)^{1/\beta}$

Various Probability Density Functions

Distribution of smallest values in samples of size n (for n large)	Remarks on general characteristics of the distribution of x, the smallest value in samples of size n (n large)
$x = a + (b - a)\xi/n$	x is nearly independent of n. Variance $= (b - a)^2/n^2$.
$x = \langle x\rangle - (\lambda n/\pi\xi)$	Most probable value of x decreases linearly with n. Both mean value and variance are infinite.
$x = \langle x\rangle - \lambda \log(n/2) + \lambda \log \xi$	Most probable value decreases as a multiple of $(\log n)$. Variance becomes $\lambda^2\pi^2/6$ and is independent of n.
$x = \langle x\rangle - \Xi(2 \log n)^{1/2} + \Xi[\log(\log n) + \log 4\pi]/2(2 \log n)^{1/2} + \Xi[\log\xi/(2 \log n)^{1/2}]$	Most probable value decreases as a multiple of $(\log n)^{1/2}$. Variance decreases as n increases and is given by $\pi^2\Xi^2/(12 \log n)$.
$x = (\xi/n\alpha)^{1/\beta}$	Most probable value decreases as $n^{-1/\beta}$. Mean value decreases as $n^{-1/\beta}$ and variance as $n^{-2/\beta}$.

$$h(\xi) = \lim_{n\to\infty} h_n(\xi) = \lim_{n\to\infty} [1 - (\xi/n)]^{n-1} = e^{-\xi} \tag{43}$$

where $\xi \geq 0$.

The random variable ξ has thus a very simple probability density function. The various moments of ξ, $\log \xi$, $1/\xi$, and $\xi^{1/\beta}$ can be obtained in a straightforward manner.

It is now a simple step to the calculation of the moments of the smallest value. In Table 1, compiled from Ref. [2], pertinent information is available regarding the distribution of the smallest value for several important probability density functions.

REFERENCES

1. A. Papoulis, Probability, Random Variables, and Stochastic Processes, McGraw-Hill, New York, 1965.
2. B. Epstein, "Statistical Aspects of Fracture Problems," J. Appl. Phys., 19, 140-147 (1948).

AUTHOR INDEX

Numbers in brackets are reference numbers and indicate that an author's work is referred to although his name is not cited in the text. Numbers in italics give the page on which the complete reference is listed.

SUBJECT INDEX

Other books of interest to you...

Because of your interest in our books, we have included the following catalog of books for your convenience.

Any of these books are available on an approval basis. This section has been reprinted in full from our ***material science*** catalog.

If you wish to receive a complete catalog of MDI books, journals and encyclopedias, please write to us and we will be happy to send you one.

material science

including

Polymers, Plastics, Fibers, and Coatings

Metals and Metallurgy

Ceramics and Glass

Vacuum Science

ALTGELT and SEGAL

Gel Permeation Chromatography

edited by KLAUS H. ALTGELT, *Chevron Research Company, Richmond, California,* and LEON SEGAL, *South Regional Research Laboratory, U.S.D.A., New Orleans, Louisiana*

672 pages, illustrated. 1971

Demonstrates the manifold applications of gel permeation chromatography in the field of polymer chemistry. Directed to all research, quality-control, and analytical chemists working with conventional and unconventional polymers and other large molecules in the fields of polymer, cellulose, and petroleum chemistry.

CONTENTS: The sizes of polymer molecules and the GPC separation, *F. W. Billmeyer, Jr. and K. H. Altgelt.* Gel permeation chromatography column packings – types and uses, *D. J. Harmon.* Chromatographic instrumentation and detection of gel permeation effluents, *E. M. Barrall, II and J. F. Johnson.* Peak resolution and separation power in gel permeation chromatography, *D. J. Harmon.* A review of peak broadening in gel chromatography, *R. N. Kelley and F. W. Billmeyer, Jr.* Mathematical methods of correcting instrumental spreading in GPC, *L. H. Tung.* Comparison of different techniques of correcting for band broadening in GPC, *J. H. Duerksen.* Separation mechanisms in gel permeation chromatography, *W. W. Yau, C. P. Malone, and H. L. Suchan.* Gel permeation chromatography and thermodynamic equilibrium. *E. F. Casassa.* Calibration of GPC columns, *H. Coll.* Data treatment in GPC, *L. H. Tung.* The overload effect in gel permeation chromatography, *J. C. Moore.* Gel permeation chromatography using a bio-glas substrate having a broad pore size distribution, *A. R. Cooper, J. H. Cain, E. M. Barrall, II, and J. F. Johnson.* High resolution gel permeation chromatography – using recycle, *K. J. Bombaugh and R. F. Levangie.* Gel permeation chromatography with high loads, *K. H. Altgelt.* Fast gel permeation chromatography, *J. N. Little, J. L. Waters, K. J. Bombaugh, and W. J. Pauplis.* Extension of GPC techniques, *G. Meyerhoff.* Phase distribution chromatography (PDC) of polystyrene, *R. H. Casper and G. V. Schulz.* Apparent and real distribution in GPC (experiments with PMMA samples), *K. C. Berger and G. V. Schulz.* The instrument spreading correction in GPC. I: The general shape function using a linear calibration curve, *T. Provder and E. M. Rosen.* The instrument spreading correction in GPC. II: The general shape function using the Fourier transform method with a nonlinear calibration curve, *E. M. Rosen and T. Provder.* Behavior of micellar solutions in gel permeation chromatography: A theory based on a simple model, *H. Coll.* Gel permeation analysis of macromolecular association by an equilibrium method, *B. F. Cameron, L. Sklar, V. Greenfield, and A. D. Adler.* Gel filtration chromatography, *B. F. Cameron.* Determination of polymer branching with gel permeation chromatography. Abstract of a review, *E. E. Drott and R. A. Mendelson.* Fractionation of linear polyethylene with gel permeation chromatography. Part III, *N. Nakajima.* Application of GPC in the study of stereospecific block copolymers, *R. D. Mate and M. R. Ambler.* Composition of butadiene-styrene copolymers by gel permeation chromatography, *H. E. Adams.* A direct GPC calibration for low molecular weight polybutadiene, employing dual detectors, *J. R. Runyon.* Quantitative determination of plasticizers in polymeric mixtures by GPC, *D. F. Alliet and J. M. Pacco.* Evaluation of pulps, rayon fibers, and cellulose acetate by GPC and other fractionation methods, *W. J. Alexander and T. E. Muller.* Characterization of the internal pore structures of cotton and chemically modified cottons by gel permeation, *L. F. Martin, F. A. Blouin, and S. P. Rowland.* Application of GPC to studies of the viscose process. I: Evaluation of the method, *L. H. Phifer and J. Dyer.* Application of GPC to studies of the viscose process. II: The effects to steeping and alkali-crumb aging, *J. Dyer and L. H. Phifer.* Gel permeation chromatography calibration. I. Use of calibration curves based on polystyrene in THF and integral distribution curves of elution volume to generate calibration curves for polymers in 2,2,2-trifluoroethanol, *T. Provder, J. C. Woodbrey, and J. H. Clark.* Modification of a gel permeation chromatograph for automatic sam-

(continued)

ALTGELT and SEGAL *(continued)*

ple injection and on-line computer data recording, *A. R. Gregges, B. F. Dowden, E. M. Barral, II, and T. T. Horikawa.* Characterization of crude oils by gel permeation chromatography, *H. H. Oelert, D. R. Latham, and W. E. Haines.* Separation and characterization of high-molecular-weight saturate fractions by gel permeation chromatography, *J. H. Weber and H. H. Oelert.* Fractionation of residuals by gel permeation chromatography, *E. W. Albaugh, P. C. Talarico, B. E. Davis, and R. A. Wirkkala.* Combined gel permeation chromatography–NMR techniques in the characterization of petroleum residuals, *F. E. Dickson, R. A. Wirkkala, and B. E. Davis.* A rapid method of identification and assessment of total crude oils and crude oil fractions by gel permeation chromatography, *J. N. Done and W. K. Reid.* Gel permeation analysis of asphaltenes from steam stimulated oil wells, *C. A. Stout and S. W. Nicksic.* GPC separation and integrated structural analysis of petroleum heavy ends, *K. H. Altgelt and E. Hirsch.*

AMERICAN VACUUM SOCIETY *Experimental Vacuum Science and Technology*

edited by THE AMERICAN VACUUM SOCIETY EDUCATION COMMITTEE

288 pages, illustrated. 1973

A collection of experiments, which are graded from simple procedures to sophisticated vacuum processes, and designed to aid instructors and introduce students to the basic concepts and techniques of the field of vacuum science. Includes an extensive bibliography to stimulate further investigation. Especially useful for all students and teachers in the many fields of the basic sciences and engineering where vacuum methods and techniques are important.

CONTENTS: **Section 1: Procedures in Vacuum Production and Measurement,** *W. Brunner and H. Patton.* **Section 2: Experiments which Illustrate the Characteristics of the Vacuum Environment:** Demonstration of the outgassing of different vacuum materials, *F. Rosebury.* Comparison of gas evolution phenomenon from glass and metal system envelopes during baking, *R. Lawson.* Determination of the net quantity of gas flowing through a cylindrical tube, *K. Busen.* **Section 3: Experiments which Illustrate the Dependence of the Physical Properties of Gases on Gas Density:** Measurement of the pumping action of an ionization gauge, *H. Farber.* Study of the linearity of an ionization gauge, *J. Miller, III.* Calibration of gauges, *C. Morrison.* **Section 4: Experiments which Examine Physical and Chemical Interactions at Surfaces:** Study of the sorption of gases for different gas–sorbent combinations, *K. Wear.* The use of sorbents as traps and pumps, *H. Farber.* Sorption of gases by titanium, *H. Farber.* Investigation of the passage of oxygen across a silver barrier, *K. Busen.* **Section 5: Processes Requiring a Vacuum Environment:** Thin film evaporation, *M. Thomas.* Fabrication of a nichrome resistor, *R. Riegert and G. Breitweiser.* Sputtering, *P. Grosewald.* Ejection patterns in single crystal sputtering, *G. Wehner.* **Section 6: Special Projects:** Study of the sublimation of ice at various pressures, *W. Parker.* Study of friction, *P. McElligott.* Measurement of the mean free path of conduction electrons in silver, *R. Olson and J. Wilson.* Construction and use of a cathode ray tube, *B. Kendall and H. Luther.* Construction of a vacuum triode using solder glass techniques, *J. King and J. Orsula.* Experiments using solder glass techniques, *D. Whitcomb.* **Section 7: Speculations:** Original thought experiments, *M. Carbone.* Provocative ideas and questions, *N. Milleron.*

BEER *Liquid Metals: Chemistry and Physics*

(Monographs and Textbooks in Material Science Series, Volume 4)

edited by SYLVAN Z. BEER, *Converta Enterprises, Inc., Syracuse, New York*

742 pages, illustrated. 1972

Presents a comprehensive review of the research done on the liquid state of metals, bringing together the latest advances, as well as data previously scattered among a wide variety of publications. Of prime importance to chemists, physicists, research metallurgists, metallurgical engineers, and materials scientists working in the areas of liquid-state theory, the theory of metals, process metallurgy involving liquid metals, and high-temperature chemistry.

CONTENTS: On the thermodynamic formalism of metallic solutions, *C. H. P. Lupis.* Kinetics of evaporation of various elements from liquid iron alloys under vacuum, *R. Ohno.* Relation between thermodynamic and electrical properties of liquid alloys, *D. N. Lee and B. D. Lichter.* The surface tension of liquid metals, *B. C. Allen.* Significant structure theory applied to liquid metals, *S. M. Breitling and H. Eyring.* Diffraction analysis of liquid metals and alloys, *C. N. J. Wagner.* The optical properties of liquid metals, *J. N. Hodgson.* Effect of pressure on the properties of liquid metals, *A. Rapoport.* Sound propagation in liquid metals, *R. T. Beyer and E. M. Ring.* The viscosity of liquid metals, *R. T. Beyer and E. M. Ring.* Magnetic properties of liquid metals, *R. Dupree and E. F. W. Seymour.* Diffusion in liquid metals, *N. H. Nachtrieb.* Electromigration in liquid alloys, *S. G. Epstein.* Electronic nature of liquid metals and liquid metal theory, *J. E. Enderby.* Structure and properties of noncrystalline metallic alloys produced by rapid quenching of liquid alloys, *B. C. Giessen and C. N. J. Wagner.*

BLACK and PRESTON *High-Modulus Wholly Aromatic Fibers*

(Fiber Science Series, Volume 5)

edited by W. BRUCE BLACK, *Monsanto Textiles Company, Pensacola, Florida* and JACK PRESTON, *Monsanto Textiles Company, Chemstrand Research Center, Durham, North Carolina*

304 pages, illustrated. 1973

Based on a symposium on high-modulus aromatic fibers held by the American Chemical Society in Boston on April 13, 1972. The first formal publication of research which shows the relationship of fiber properties to polymer structure. Extremely significant reading for all fiber scientists and material scientists; plastics scientists and engineers interested in fiber-reinforced plastics; spacecraft and aircraft oriented engineers; and scientists in the industrial fiber, sports equipment, and airframe fields.

CONTENTS: High-modulus wholly aromatic fibers: Introduction to the Symposium and historical perspective, *W. Black.* High-modulus wholly aromatic fibers. I. Wholly ordered polyamide-hydrazines and poly-1,3,4,-oxadizole-amides, *J. Preston, W. Black and W. Hofferbert, Jr.* High-modulus wholly aromatic fibers. II. Partially ordered polyamide-hydrazides, *J. Preston, W. Black, and W. Hofferbert, Jr.* Self-regulating polycondensations. II. A study of the order present in polyamide-hydrazides derived from terephthaloyl chloride and p-aminobenzhydrazide, *R. Morrison, J. Preston, J. Randall, and W. Black.* Self-regulating polycondensations. III. NMR analysis of oligomers derived from terephthaloyl chloride and p-aminobenzhydrazide, *J. Randall, R. Morrison, and J. Preston.* Some physical and mechanical properties of some high-modulus fibers prepared from all-para aromatic polyamide-hydrazides, *W. Black, J. Preston, H. Morgan, G. Raumann, and M. Lilyquist.* Morphology and crystal structure of wholly aromatic all-para polyamide-hydrazide polymers, *V. Holland.* X-ray study of an all-para wholly aromatic polyamide-hydrazide[a,b], *R. Miller.* Molecular weight characterization of wholly para-oriented, aromatic polyamide-hydrazides and wholly aromatic polyamides, *J. Burke.* Construction and properties of fabrics of high-modulus organic fibers useful for composite reinforcing, *M. Lilyquist, R. DeBrunner, and J. Fincke.* Mechanical properties of a high-modulus polyamide-hydrazide fiber in composites and of the polyamide-hydrazide fiber and fabric composites, *D. Zaukelies and B. Daniels.* Tire cord application of high-modulus fibers derived from polyamide-hydrazides, *G. Raumann and J. Brownlee.* The application of high-modulus fibers to ballistic protection, *R. Laible, F. Figucia, and W. Ferguson.* High-modulus wholly aromatic fibers. III. Random copolymers containing hydrazide and/or carbonamide linkages, *J. Preston, H. Morgan, and W. Black.*

BOLKER *Natural and Synthetic Polymers: An Introduction*

by HENRY I. BOLKER, *Department of Chemistry, McGill University, Montreal, Quebec*

in preparation. 1973

Presents a unified approach to polymer chemistry, with equal emphasis on natural and synthetic polymers, and is arranged in a logical sequence of topics based on increasing complexity of molecular architecture. Useful as a textbook for a first course in polymer chemistry and as a reference book for workers in the field.

CONTENTS: Introduction • Natural condensation polymers: The linear polysaccharides • Synthetic condensation (step-growth) polymers • Addition (chain-growth) polymers • Stereoregularity in addition polymers • Branched homopolymers: Synthetic and natural • Natural heteropolymers: I. Heteropolysaccharides • Natural heteropolymers: II. Nucleic acids • Copolymers and copolymerization • Cross-linking in synthetic polymers • Natural heteropolymers: III. Polypeptides and proteins • Lignins.

BROWNING *Analysis of Paper*

by B. L. BROWNING, *The Institute of Paper Chemistry, Appleton, Wisconsin*

352 pages, illustrated. 1969

Provides comprehensive coverage of methods for chemical analysis of paper. Is of value to manufacturers of paper and paper board, suppliers of components or of additives introduced into paper, converters and printers, purchasers and users, librarians, and others concerned with the properties, behavior, and applications of paper that are related to composition.

CONTENTS: Paper as a commodity • Sampling and preparation of sample • Determination of moisture • Fiber analysis • Fiber quality methods • Lignin • Rosin size • Starch • Proteins • Coatings • Waxes and oils • Fillers and white coating pigments • Dyes and colored pigments • Acidity and alkalinity • Residues and impurities • Biological control agents • General identification of additives in paper • Synthetic resins • Wet-strength agents • Polysaccharides and gums • Miscellaneous additives • Noncellulose fibers • Specks and spots • Permanence of paper • Paper in forensic science.

BUTLER, O'DRISCOLL, and SHEN *Reviews in Macromolecular Chemistry*

(Book Edition)

edited by GEORGE B. BUTLER, *Department of Chemistry, University of Florida,*

(continued)

BUTLER, O'DRISCOLL, and SHEN *(continued)*

Gainesville, and KENNETH F. O'DRISCOLL, *Department of Chemical Engineering, University of Waterloo, Ontario, Canada* and MITCHEL SHEN, *Department of Chemical Engineering, University of California, Berkeley*

Vol. 1 *out of print*

Vol. 2 388 pages, illustrated. 1968

Vol. 3 430 pages, illustrated. 1969

Vol. 4 428 pages, illustrated. 1970

Vol. 5, Part I see NEUSE and ROSENBERG

Vol. 5, Part II 250 pages, illustrated. 1970

Vol. 6 498 pages, illustrated. 1971

Vol. 7 314 pages, illustrated. 1972

Vol. 8 346 pages, illustrated. 1972

Vol. 9 380 pages, illustrated. 1973

Reviews of the currently published literature for those who wish to keep abreast of the new and rapidly advancing developments in macromolecular chemistry. Of interest to organic and physical chemists, biochemists, engineers, and all students and research workers in polymer chemistry and related fields.

CONTENTS:

Volume 1: Application of molecular orbital theory to vinyl polymerization, *K. F. O'Driscoll and T. Yonezawa.* Poly(alkylene oxides), *A. E. Gurgiolo.* Polyurethanes, *D. J. Lyman.* Uncatalyzed, uninhibited thermal oxidation of saturated polyolefins, *L. Reich and S. S. Stivala.* Double-strand polymers, *W. De Winter.* Biomedical polymers, *D. J. Lyman.* Gel permeation chromatography with organic solvents, *J. F. Johnson, R. S. Porter, and M. J. R. Cantow.*

Volume 2: Phosphorus-containing polymers: Introduction, *M. Sander and E. Steininger.* Linear polymers with phosphorus in side chains, *M. Sander and E. Steininger.* Linear polymers with phosphorus and carbon in the main chain, *M. Sander and E. Steininger.* Reassessment of the theory of polyesterification with particular reference to alkyd resins, *D. H. Solomon.* Symmetry considerations for stereoregular polymers, *A. M. Liquori.* Copolymerization of vinyl monomers with ring compounds, *R. A. Patsiga.* Application of high-resolution nuclear magnetic resonance to polymer structure determination, I., *K. C. Ramey and W. S. Brey, Jr.* Ten years of polymer single crystals, *D. A. Blackadder.* Thermal degradation of polystyrene, *G. G. Cameron and J. R. MacCallum.*

Volume 3: Phosphorous-containing resins, *M. Sander and E. Steininger.* Inorganic phosphorous polymers, *M. Sander and E. Steininger.* Phosphorylation of polymers, *M. Sander and E. Steininger.* Sulphur-containing polymers, *E. J. Goethals.* Polymer molecular weight distributions, *N. Amundson and D. Luss.* Heteroatom ring-containing polymers, *A. D. Delman.* Molecular theories of rubber-like elasticity and polymer viscoelasticity, *M. Shen, W. F. Hall, and R. E. DeWames.* End-group studies using dye techniques, *S. R. Palit and B. M. Mandal.* Free-radical spin labels for macromolecules, *J. D. Ingham.* The synthesis of thermally stable polymers: A progress report, *J. I. Jones.*

Volume 4: Polymer enzymes and enzyme analogs, *A. S. Lindsey.* Stability of polycarbonate, *A. Davis and J. Golden.* Cross-linking – effect on physical properties of polymers, *L. E. Nielsen.* The synthesis of thermally stable polymeric azomethines by polycondensation reactions, *G. F. D'Alelio and R. K. Schoenig.* On the dehydrochlorination and the stabilization of polyvinyl chloride, *M. Onozuka and M. Asahina.* Recent advances in the development of flame-retardant polymers, *A. D. Delman.* Thermodynamics of polymerization. I, *H. Sawada.*

Vol. 5, Part II: Ring-chain equilibria, *H. Allcock.* Occupied volume of liquids and polymers, *R. Haward.* The application of ESR techniques to high polymer fracture, *H. Kausch-Blecken von Schmeling.* The science of determining copolymerization reactivity ratios, *P. Tidwell and G. Mortimer.* Block polymers and related heterophase elastomers, *G. Estes, S. Cooper, and A. Tobolsky.*

Volume 6: Proton magnetic resonance of molecular interactions in polymer solutions, *K.-J. Liu and J. E. Anderson.* Preparation and polymerization of vinyl heterocyclic compounds, *K. Takemoto.* Catalysis in isocyanate reactions, *K. C. Frisch and L. P. Rumao.* Thermodynamics of polymerization. II. Thermodynamics of ring—opening polymerization, *H. Sawada.* Copolymers of naturally occurring macromolecules, I. *C. Watt.* Molecular configuration and pyrolysis of phenolic-novolaks, *E. L. Winkler and J. A. Parker.* Physical properties of ionic polymers, *E. P. Otocka.* Synthesis and properties of polyphenyls and polyphenylenes, *J. G. Speight, P. Kovacic, and F. W. Koch.* Dependence of flow properties on molecular weight, temperature, and shear, *A. Casale, R. S. Porter, and J. F. Johnson.* Synthesis methods and properties of polyazoles, *V. V. Korshak and M. M. Teplyakov.*

Volume 7: Linear polyquinoxalines, *P. M. Hergenrother.* Nylons—known and unknown. A comprehensive index of linear aliphatic polyamides of regular structure, *H. K. Livingston, M. S. Sioshansi, and M. D. Glick.* Recent advances in polymer fractionation, *L. H. Tung.* Rheology of adhesion, *D. H. Kaelble.* Solvation of synthetic and natural polyelectrolytes, *B. E. Conway.* Hydrogen transfer polymerization with anionic catalysts and the problem of anionic isomerization polymerization, *J. P. Kennedy and T. Otsu.*

Volume 8: Polymerization by carbenoids, carbenes, and nitrenes, *M. Imoto and T. Nakaya.* Collagen and gelatin in the solid state, *I. V. Yannas.* Ring-opening polymerization of cycloolefins, *N. Calderon.* Thermodynamics of polymerization. III, *H. Sawada.* Polymerization of N-vinylcarbazole initiated by metal salts, *M. Biswas and D. Chakravarty.* Vibrational spectroscopy of polymers, *F. J. Boerio and J. L. Koenig.* Polymer compatibility, *S. Krause.*

Volume 9: Mechanical properties of polymers: The influence of molecular weight and molecular weight distribution, *J. Martin, J. Johnson, and A. Cooper*. On the mathematical modeling of polymerization reactors, *W. Ray*. Anionic cyclopolymerization, *C. McCormick and G. Butler*. Carbon-13 NMR of polymers, *V. Mochel*. Thermodynamics of polymerization. IV. Thermodynamics of equilibrium polymerization, *H. Sawada*.

CARROLL *Physical Methods in Macromolecular Chemistry*

a series edited by BENJAMIN CARROLL, *Rutgers—The State University, Newark, New Jersey*

Vol. 1 400 pages, illustrated. 1969

Vol. 2 384 pages, illustrated. 1972

A series which reviews why and how analytical methods are used in the study of macromolecules. Each method is critically discussed by experts in the field. Directed to researchers in polymer chemistry, biopolymers, and organic and inorganic chemistry.

CONTENTS:

Volume 1: Surface chemistry and polymers, *M. Rosoff*. Internal reflection spectroscopy, *J. K. Barr and P. A. Flournoy*. Electric properties of synthetic polymers, *E. O. Forster*. Assessing radiation effects in polymers, *P. Y. Feng and E. S. Freeman*. Fluorescence techniques for polymer solutions, *D. J. R. Laurence*. Insoluble polymers: Molecular weights and their distributions, *H. C. Cheung*.

Volume 2: Gel permeation chromatography in polymer chemistry, *D. D. Bly*. Interactions of polymers with small ions and molecules, *D. J. R. Laurence*. Electric properties of biopolymers: Proteins, *E. O. Forster and A. P. Minton*. Thermal methods, *E. P. Manche and B. Carroll*.

CARTER *Essential Fiber Chemistry*

(Fiber Science Series, Volume 2)

by MARY E. CARTER, *FMC Corporation, American Viscose Division, Marcus Hook, Pennsylvania*

232 pages, illustrated. 1971

Discusses the chemical and physical structure and properties of ten commercially important fibers. Useful to all chemists interested in the research and development of natural and man-made fibers.

CONTENTS: Cotton • Rayon • Cellulose acetate • Wool • Polyamide • Acrylic fibers • Polyethylene terephthalate • Polyolefins • Spandex • Glass.

CONLEY *Thermal Stability of Polymers*

In 2 Volumes

(Monographs in Macromolecular Chemistry Series)

edited by R. T. CONLEY, *Wright State University, Dayton, Ohio*

Vol. 1 656 pages, illustrated. 1970

Vol. 2 in preparation. 1974

CONTENTS: Introduction, *R. T. Conley*. Molecular structure and stability criteria, *R. T. Conley*. The relationship between the kinetics and mechanism of thermal depolymerization, *R. H. Boyd*. Random scission processes, *A. V. Tobolsky, A. M. Kotliar, and T. C. P. Lee*. Fundamental reactions in oxidation chemistry, *P. M. Norling and A. V. Tobolsky*. Thermal and oxidative degradation of polyethylene, polypropylene, and related olefin polymers, *R. H. Hansen*. Thermal and oxidative degradation of natural rubber and allied substances, *E. M. Bevilacqua*. Vinyl and vinylidene polymers, *R. T. Conley and R. Malloy*. Fluorocarbon polymers, *W. W. Wright*. Thermal and thermo-oxidative degradation of polyamides, polyesters, polyethers, and related polymers, *R. T. Conley and R. A. Gaudiana*. Thermosetting resins, *R. T. Conley*. Thermal and thermo-oxidative degradation of cellulosic polymers, *R. T. Conley*. Heterocyclic polymers, *G. P. Shulman*. Degradation of inorganic polymers, *J. Economy and J. H. Mason*.

D'ALELIO and PARKER *Ablative Plastics*

edited by GAETANO F. D'ALELIO, *Department of Chemistry, University of Notre Dame, Indiana,* and JOHN A. PARKER, *NASA, Ames Research Center, Moffet Field, California*

504 pages, illustrated. 1971

Provides the comprehensive and rational approach required for the design and production of reliable head shields for future space missions. Includes discussions on the various aspects of heat-rejection mode as a function of heating rate; the nature of the heat transfer, both radiative and conductive; and the nature of degrading polymers. A valuable reference for all aerospace scientists, polymer chemists, physicists, and aerodynamic engineers.

CONTENTS: Ablative polymers in aerospace technology, *D. L. Schmidt*. Hypervelocity heat protection—a review of laboratory experiments, *N. S. Vojvodich*. A review of ablative studies of interest to naval applications, *F. J. Koubek*. Structural design and thermal properties of polymers, *G. F. D'Alelio*. Characterization of an epoxy-anhydride ablative system using com-

(continued)

D'ALELIO and PARKER *(continued)*

puter treatment of analytical results, *C. G. Taylor and E. L. Pendleton.* The synthesis and characterization of some potential ablative polymers, *R. Y. Wen, L. F. Sonnabend, and R. Eddy.* Thermal degradation and curing of polyphenylene, *D. N. Vincent.* Thermosetting polyphenylene resin—its synthesis and use in ablative composites, *N. Bilow and L. J. Miller.* Structural ablative plastics, *R. M. Lurie, S. F. D'Urso, and C. K. Mullen.* Prediction of heat shield performance in terms of epoxy resin structure, *G. J. Fleming.* Ablative resins for hyperthermal environments, *B. S. Marks and L. Rubin.* The development of polybenzimidazole composites as ablative heat shields, *R. R. Dickey, J. H. Lundell, and J. A. Parker.* Ablative degradation of a silicon foam, *T. McKeon.* Thermophysical characteristics of high-performance ablative composites, *M. L. Minges.* The design and development of a high-heating rate thermogravimetric analyzer suitable for use with ablative plastics, *A. M. Melnick and E. J. Nolan.* Pyrolysis kinetics of nylon 6-6, phenolic resin and their composites, *H. E. Goldstein.* Pyrolysis-gas chromatography as a tool for studying the degradation of ablative plastics, *R. M. Ross.* Nonequilibrium flow and the kinetics of chemical reactions in the char zone, *G. C. April, R. W. Pike, and E. G. del Valle.* Arc-image testing of ablation materials, *E. M. Liston.* Development and characterization of a radio frequency-transparent ablator, *E. L. Strauss.* Tailoring polymers for entry into the atmosphere of Mars and Venus, *R. G. Nagler.*

DIGGLE *Oxides and Oxide Films*

in multi-volumes

(The Anodic Behavior of Metals and Semiconductors Series)

edited by JOHN W. DIGGLE, *Research School of Chemistry, The Australian National University, Canberra*

Vol. 1 552 pages, illustrated. 1972

Vol. 2 424 pages, illustrated. 1973

Treats the anodic behavior of metals and semiconductors and peripheral areas in an authoritative and interdisciplinary manner. The initial volumes deal with the physics and chemistry of oxides and oxide films. Of great value for all those involved in electrochemistry, materials science, solid state physics, electrical engineering, metallurgy, corrosion science, semiconductors, and electrochemical technology.

CONTENTS:

Volume 1: Passivation and passivity, *V. Brusić.* Mechanisms of ionic transport through oxide films, *M. J. Dignam.* Electronic current flow through ideal dielectric films, *C. A. Mead.* Electrical double layer at metal oxide-solution interfaces, *S. Ahmed.*

Volume 2: Anodic oxide films: Influence of solid-state properties on electrochemical behavior, *A. Vijh.* Dielectric loss mechanism in amorphous oxide films, *D. M. Smyth.* Porous anodic films in aluminum, *G. C. Wood.* Dissolution of oxide phases, *J. W. Diggle.*

FOURT and HOLLIES

Clothing: Comfort and Function

(Fiber Science Series, Volume 1)

by LYMAN FOURT and NORMAN HOLLIES, *Gillette Research Institute, Rockville, Maryland*

272 pages, illustrated. 1970

A unified review of the present state of knowledge in the science of clothing. Of interest to textile scientists, fiber producers and marketers, textile converters, and garment makers.

CONTENTS: The factors involved in the study of clothing • Clothing considered as a system interacting with the body • Clothing considered as a structural assemblage of materials • Heat and moisture relations in clothing • Physiological and field testing of clothing by wearing it • Physical properties of clothing and clothing materials in relation to comfort • Differences between fibers with respect to comfort • Current trends and new developments in the study of clothing.

FRISCH and REEGEN *Ring-Opening Polymerization*

(Kinetics and Mechanisms of Polymerization Series, Volume 2)

edited by KURT C. FRISCH, and SIDNEY L. REEGEN, *Polymer Institute, University of Detroit, Michigan*

544 pages, illustrated. 1969

Covers the polymerization of important classes of cyclic monomers such as ethylene and propylene oxide, alkylenimines, and sulfides, lactones, lactams, cyclic silicone compounds, and cyclic nitrogen containing heterocycles. Of great interest to the industrial, commercial, and academic worlds as it has application to elastomers, coatings, fibers, films, and foams.

CONTENTS: 1,2 Epoxides, *Y. Ishii and S. Sakai.* 1,3 Epoxides and higher epoxides, *P. Dreyfuss and M. P. Dreyfuss.* Cyclic formals, *J. Furukawa and K. Tada.* Cyclic sulfides, *P. Sigwalt.* Alkylenimines, *M. Hauser.* Lactones, *R. D. Lundberg and E. F. Cox.* Lactams, *H. K. Reimschuessel.* Cyclic siloxanes and silazanes, *E. E. Bostick.* Nitrogen-containing heterocyclic compounds, *V. Kargin and V. Kabanov.* N-carboxy-α-amino acid anhydrides, *Y. Shalitin.*

FRISCH and SAUNDERS
Plastic Foams

(Monographs on Plastic Series, Volume 1)

edited by KURT C. FRISCH, *University of Detroit, Michigan,* and JAMES H. SAUNDERS, *Monsanto Company, Pensacola, Florida*

Part I 464 pages, illustrated. 1972

Part II 704 pages, illustrated. 1973

Gives an integrated picture of the fundamental principles, technology, and applications of foams, and offers a thorough treatment of specific types of plastic foams. Emphasis is placed on the newer trends in this science.

Of particular value to chemists and engineers engaged in research and development, and marketing and production personnel in the polymer and plastics industry.

CONTENTS:

Part I: Introduction, *K. C. Frisch.* The mechanism of foam formation, *J. H. Saunders and R. H. Hansen.* Flexible polyurethane foams, *G. T. Gmitter, H. J. Fabris, and E. M. Maxey.* Sponge rubber and latex foam, *R. L. Zimmerman and H. R. Bailey.* Polyolefin foams, *D. J. Sundquist.* Polyvinyl chloride foams, *A. C. Werner.* Silicone foams, *H. L. Vincent.* Testing of cellular materials, *R. A. Stengard.*

Part II: Rigid urethane foams, *J. K. Backus and P. G. Gemeinhardt.* Polystyrene and related thermoplastic foams, *A. R. Ingram and J. Fogel.* Phenolic foams, *A. J. Papa and W. R. Proops.* Urea-formaldehyde foams, *K. C. Frisch.* Epoxy-resin foams, *H. Lee and K. Neville.* New high-temperature-resistant plastic foams, *E. E. Hardy and J. H. Saunders.* Miscellaneous foams, *K. C. Frisch.* Inorganic foams, *M. Wismer.* Effects of cell geometry on foam performance, *R. H. Harding.* Thermal decomposition and flammability of foams, *P. E. Burgess, Jr. and C. J. Hilado.* Foams in transportation, *M. Kaplan and L. M. Zwolinski.* Architectural uses of foam plastics, *S. C. A. Paraskevopoulos.* Military and space applications of cellular materials, *R. J. F. Palchak.*

GARG, SVALBONAS, and GURTMAN
Analysis of Structural Composite Materials

(Monographs and Textbooks in Material Science Series, Volume 6)

by SABODH GARG, *Systems, Science and Software Company, La Jolla, California,* VYTAS SVALBONAS, *Grumman Aerospace Corporation, Bethpage, New York,* and GERRY GURTMAN, *Systems, Science and Software Company, La Jolla, California*

552 pages, illustrated. 1973

Compares various theories for the static and dynamic analysis of structural composite materials. Deals with the elastic properties of laminated composites and particulate and unidirectional fiber reinforced composites, composite strength, and stress wave propagation. May be used as a textbook for a graduate course in composites and is of interest to researchers and analysts in any industry that uses composites.

CONTENTS: Why composites? • Simple analytic models • Elasticity analyses • Bounds on elastic properties by energy methods • Multilayer laminates • Non-statistical models of composite strength • Statistical tensile strength of fiber and fiber bundles • Composite tensile-strength models • Cumulative weakening model including stress concentrations • Compressive strength of composites • Theory of breaking kinetics • Introduction to elastic wave propagation in composites • Approximate analysis techniques for stress wave propagation in composites • Application of continuum mixture theories to the study of elastic wave propagation in composite materials • Shock waves in composite materials.

HAM *Vinyl Polymerization*
In 2 Parts

(Kinetics and Mechanisms of Polymerization Series, Volume 1)

edited by GEORGE E. HAM, *Geigy Chemical Corporation, Ardsley, New York*

Part I 560 pages, illustrated. 1967

Part II 432 pages, illustrated. 1969

"The book is a good introduction to the series. It has provided a sound basis for subsequent volumes and should serve as an important reference text for students and researchers in polymer chemistry."—B. D. Gesner, Bell Telephone Labs., *SPE Journal*

"The book is highly recommended."—Arthur Tobolsky, *The American Scientist*

CONTENTS:

Part I: General aspects of free-radical polymerization, *G. E. Ham.* The mechanism of cyclopolymerization of nonconjugated diolefins, *W. E. Gibbs and J. M. Barton.* Styrene, *M. H. George.* Mechanism of vinyl acetate polymerization, *M. K. Lindemann.* Polymerization of vinyl chloride and vinylidene chloride, *G. Talamini and E. Peggion.* Occlusion phenomena in the polymerization of acrylonitrile and other monomers, *A. D. Jenkins.* Polymerization of acrolein, *R. C. Schulz.* Heats of polymerization and their structural and mechanistic implications, *R. M. Joshi and B. J. Zwolinski.*

Part II: Mechanism of emulsion polymerization, *J. W. Vanderhoff.* Elucidation of emulsion polymerization mechanism based upon copolymer studies, *W. F. Fowler, Jr.* Mechanism of the emulsion polymerization of ethylene, *H. K. Stryker, G. J. Mantell, and A. F. Helin.* Mechanism of stereospecific polymerization of propylene, *W. "E" Smith.* Anionic polymerization,

(continued)

HAM *(continued)*

M. Morton. Mechanisms of cationic polymerization, *Z. Zlámal.* Radiation-induced polymerization, *Y. Tabata.*

HENCH and DOVE *Physics of Electronic Ceramics*

In 2 Parts

(Ceramics and Glass: Science and Technology Series, Volume 2)

edited by LARRY L. HENCH and DEREK B. DOVE, *College of Engineering, University of Florida, Gainesville.*

Part A 584 pages, illustrated. 1971

Part B 576 pages, illustrated. 1972

A highly useful treatise which deals with the physical basis for the behavior of electronic ceramics.

Fundamental physical theories describing each type of electronic ceramics are presented, with discussions included that relate the theories to the applications of the materials. Of special value to graduate students who have had a course in modern physics and also of interest to materials scientists and engineers in electronics, communications, and ceramics.

CONTENTS:

Part A: Quantum mechanics and ceramics, *J. C. Slater.* Band structure and electronic properties of ceramic crystals, *D. Adler.* Electrical conduction in low mobility materials, *I. Bransky and N. M. Tallan.* Defect structure and electronic properties of ceramics, *R. W. Vest.* Conduction domains in solid mixed conductors and electrolytic domain of calcia stabilized zirconia, *J. Patterson.* Semiconducting glasses, *J. D. Mackenzie.* Electronic processes in amorphous semiconductors, *E. A. Davis.* Heterogeneous semiconducting glasses, *H. F. Schaake.* The determination of local order in amorphous semiconducting films, *D. B. Dove.* Negative capacitance effects in amorphous semiconductors, *M. Allen, P. Walsh, and W. Doremus.* Some conduction phenomena in amorphous materials, *K. L. Chopra.* Applications of thin film dielectrics in microelectronics, *N. N. Axelrod.* Substructure and electrical conduction in amorphous thin films, *N. Fuschillo and A. D. McMaster.* Structure of surface defects, *D. L. Stoltz and J. J. Hren.* Electronic surface properties, *P. Mark.* Electron spin resonance and defects in solids, *W. S. Brey, R. B. Gammage, and Y. P. Virmani.* Theory of linear dielectrics, *A. D. Franklin.* Polycrystalline insulators, *H. C. Graham and N. M. Tallan.* Electrical conduction in glass and glass-ceramics, *D. L. Kinser.* Dielectric breakdown of ceramics, *G. C. Walther and L. L. Hench.*

Part B: Some structural mechanisms in ferroelectricity, *R. Pepinsky.* Thermodynamic phenomenology of ferroelectricity in single crystal and ceramic systems, *L. E. Cross.* Dynamical effects in solid state phase transformations, *J. D. Axe.* Theory of antiferromagnetism and ferrimagnetism, *J. B. Goodenough,* Microstructure and processing of ferrites, *F. J. Schnettler.* Microwave garnet compounds, *G. R. Harrison and L. R. Hodges, Jr.* The optical absorption of glasses, *N. J. Kreidl.* Light scattering from glass, *J. J. Hammel.* Electro-optical and magneto-optical effects, *Y. R. Shen.* The influence of the composition of the gain of Nd-doped glasses, *C. F. Rapp.* Solid state reactions in the preparation of zircon stains, *R. A. Eppler.* Computer color control for ceramic wall tile, *W. K. Culbreth, Jr.* Ceramics and glasses – some uses in the communications industry, *D. G. Thomas.*

HENCH and GOULD *Characterization of Ceramics*

(Ceramics and Glass: Science and Technology Series, Volume 3)

edited by LARRY L. HENCH and ROBERT W. GOULD, *University of Florida, Gainesville*

672 pages, illustrated. 1971

Focuses on the two major directions which comprise the distinct discipline of ceramic characterization: the exploration of the factors that control the properties of the final product and the rapid development of high resolution analytical technisues used for ceramic materials. A particularly timely textbook for an advanced undergraduate or graduate materials science curriculum. Valuable for all materials scientists and ceramic and materials engineers working on the development of an effective and economical materials characterization program.

CONTENTS: Introduction to the characterization of ceramics, *L. L. Hench.* **Part 1: Chemical Analysis:** General analytical chemistry, *P. Rankin.* X-ray spectroscopy, *R. W. Gould.* Atomic absorption flame spectrometry, *J. D. Winefordner.* **Part 2: Phase State and Structure:** X-ray diffraction, *R. W. Gould.* Transmission electron microscopy and electron diffraction, *C. F. Tufts.* Analysis of microstructural defects, *R. W. Newman.* Petrographic analysis, *V. D. Fréchette.* Thermal analysis, *R. K. Ware.* Point defect analysis, *W. J. James and G. Lewis.* **Part 3: Size, Shape, Strain, and Surface of Powders:** Physical characterization, *D. R. Lankard and D. E. Niesz.* Small angle x-ray scattering, *R. W. Gould.* X-ray line profile analysis, *R. W. Gould.* Scanning electron microscopy, *S. R. Bates.* Light scattering, *J. H. Boughton.* Characterization of powder surfaces, *L. L. Hench.* **Part 4: Microstructure:** Electron microprobe, *G. Lewis.* Quantitative stereology, *R. T. DeHoff.* Applied stereology, *S. W. Freiman.* **Part 5: Surfaces:** Characterization of ceramic surfaces, *L. Berrin and R. C. Sundahl.*

KATON *Organic Semiconducting Polymers*

(Monographs in Macromolecular Chemistry Series)

edited by J. E. KATON, *Miami University, Oxford, Ohio*

328 pages, illustrated. 1968

CONTENTS: Basic physics of semiconductors, *D. E. Hill*. Theoretical aspects of the electronic behavior of organic macromolecular solids, *H. A. Pohl*. Recent experimental aspects of the electronic behavior of organic macromolecular solids, *S. Kanda and H. A. Pohl*. Semiconducting organic polymers containing metal groups, *B. A. Bolto*. Semiconducting biological polymers, *D. D. Eley*.

KETLEY *The Stereochemistry of Macromolecules*

In 3 Volumes

edited by A. D. KETLEY, *W. R. Grace & Co., Clarksville, Maryland*

Vol. 1 424 pages, illustrated. 1967

Vol. 2 400 pages, illustrated. 1967

Vol. 3 476 pages, illustrated. 1968

CONTENTS:

Volume 1: Ziegler-Natta polymerization: Catalysts, monomers, and polymerization procedures, *D. O. Jordan*. The mechanism of Ziegler-Natta catalysis. I. Experimental foundations, *D. F. Hoeg*. Mechanism of Ziegler-Natta polymerization. II. Quantum-chemical and crystal-chemical aspects, *P. Cossee*. Copolymerization of olefins by Ziegler-Natta catalysts, *I. Pasquon, A. Valvassori, and G. Sartori*. Polymerization of dienes by Ziegler-Natta catalysts, *W. Marconi*. Manufacture and commercial applications of stereoregular polymers, *M. Compostella*.

Volume 2: Stereospecific polymerization of vinyl-type monomers and dienes by alkali-metal-based catalysts, *D. Braun*. Stereospecific polymerization of vinyl ethers, *A. D. Ketley*. Ionic polymerization of aldehydes, ketones, and ketenes, *G. F. Pregaglia and M. Binaghi*. Stereospecific polymerization of epoxides, *T. Tsuruta*. Stereochemistry of free-radical polymerizations, *W. Cooper*. Conformational effects induced in polymers by rigid matrices, *N. Marans*. Simple stereoregular polymers in biological systems, *J. N. Baptist*.

Volume 3: Chain conformation and crystallinity, *P. Corradini*. High-resolution nuclear magnetic resonance of synthetic polymers, *J. C. Woodbrey*. Vibrational analyses of the infrared spectra of stereoregular polypropylene, *T. Miyazawa*. Optically active stereoregular polymers, *M. Farina and G. Bressan*. Physical properties of stereoregular polymers in solid state, *J. F. Johnson and R. S. Porter*. Properties of synthetic linear stereoregular polymers in solution, *V. Crescenzi*. Macromolecules as information storage systems, *A. M. Liquori*. Automata theories of hereditary tactic copolymerization, *H. H. Pattee*. Effect of microtacticity on reactions of polymers, *M. M. van Beylen*. Degradation of stereoregular polymers, *H. H. G. Jellinek*.

KURYLA and PAPA *Flame Retardancy of Polymeric Materials*

a series edited by WILLIAM C. KURYLA and ANTHONY J. PAPA, *Union Carbide Corporation, South Charleston, West Virginia*

Vol. 1 352 pages, illustrated. 1973

Vol. 2 296 pages, illustrated. 1973

A series concerned with the various modes of rendering polymeric materials fire resistant, which emphasizes specific reagents and techniques in use today. Treats each class of polymer separately to aid the fabricator in gaining an understanding of the specific problems associated with its flammability characteristics, and to review the science and practical solution to its flame retardancy. Of special benefit to industrial polymer chemists, plastics engineers, and fabricators of polymeric materials.

CONTENTS:

Volume 1: Available flame retardants, *W. Kuryla*. Inorganic flame retardants and their mode of action, *J. Pitts*. Fire retardation of polyvinyl chloride and related polymers, *M. O'Mara, W. Ward, D. Knechtges, and R. Meyer*. Fire retardation of wool, nylon, and other natural and synthetic polyamides, *G. Crawshaw, A. Delman, and P. Mehta*.

Volume 2: Fire retardation of polystyrene and related thermoplastics, *R. Lindemann*. Fire retardation of polyethylene and polypropylene, *R. Schwarz*. Flame retardation of natural and synthetic rubbers, *H. Fabris and J. Sommer*. Flame retardancy of phenolic resins and urea- and melamine-formaldehyde resins, *N. Sunshine*.

LEFEVER *Aspects of Crystal Growth*

(Preparation and Properties of Solid State Materials Series, Volume 1)

edited by R. A. LEFEVER, *Sandia Laboratories, Albuquerque, New Mexico*

Vol. 1 296 pages, illustrated. 1971

Concerns certain aspects of the growth and properties of single crystals. Directed to both beginning and experienced crystal growers, material scientists, and solid state physicists.

CONTENTS: A review of the preparation of single crystals by fused melt electrolysis and some general properties, *W. Kunnmann*. The role of mass transfer in crystallization processes, *W. Wilcox*. Exploratory flux crystal growth, *A. Chase*.

LOWRY *Markov Chains and Monte Carlo Calculations in Polymer Sciences*

(Monographs in Macromolecular Chemistry Series)

edited by GEORGE G. LOWRY, *Western Michigan University, Kalamazoo*

344 pages, illustrated. 1970

Written for the polymer chemist who, although not primarily concerned with mathematical theories, desires a working knowledge of the topics treated. Begins with an introduction to the principles involved and later exemplifies some significant applications of Markov chain theory and Monte Carlo methods.

CONTENTS: Introduction: Deterministic and stochastic approaches, *G. G. Lowry*. Markov chains, *J. Myhre*. Monte Carlo methods, *M. Fluendy*. Polymer conformation as a Markov chain problem, *J. Kinsinger*. Polymer conformation and the excluded-volume problem, *S. Windwer*. Higher order Markov chains and statistical thermodynamics of linear polymers, *J. Mazur*. Copolymer composition and tacticity, *F. P. Price*. Molecular-weight distributions, *G. G. Lowry*.

McCULLOUGH *Concepts of Fiber-Resin Composites*

(Monographs and Textbooks in Material Science Series, Volume 2)

by R. L. MCCULLOUGH, *Boeing Scientific Research Laboratories, Seattle, Washington*

128 pages, illustrated. 1971

Presents basic concepts of composite material systems. Introduces the study of composite systems by discussing where and how composite materials are used and how their components are selected. A valuable reference for materials scientists, and students and research management engaged in the exploration of composite materials.

CONTENTS: Materials • Composite structures • Composite properties • The interphase region • Synopsis.

MAY and TANAKA *Epoxy Resins: Chemistry and Technology*

edited by CLAYTON MAY, *Shell Development Company, Emeryville, California* and YOSHIO TANAKA, *Research Institute for Polymers and Textiles, Yokohama, Japan*

704 pages, illustrated. 1973

Brings together the contributions of a number of outstanding researchers in the field of epoxy resins. Not only emphasizes the chemistry and technology of epoxy resins, but also deals with many industrial applications. Of great value for polymer chemists and technicians, and a wide variety of engineers.

CONTENTS: Introduction to epoxy resins, *C. May*. Synthesis and characteristics of epoxides, *Y. Tanaka, A. Okada, and I. Tomizuka*. Epoxide-curing reactions, *Y. Tanaka and T. Mika*. Curing agents and modifiers, *T. Mika*. Properties of cured resins, *D. Kaelble*. Epoxy-resin adhesives, *A. Lewis and R. Saxon*. Epoxy-resin coatings, *G. Somerville and I. Smith*. Electrical and electronic applications, *A. Breslau*. Epoxy laminates, *J. DelMonte*. Polymer stabilizers and plasticizers, *W. Port*. Analysis of epoxides and epoxy resins, *H. Jahn and P. Goetzky*. Toxicity, hazards, and safe handling, *H. Borgstedt and C. Hine*.

MILLICH and CARREHER *Interfacial Synthesis*

edited by FRANK MILLICH, *Department of Chemistry, University of Missouri, Kansas City,* and CHARLES E. CARREHER, JR., *Department of Chemistry, University of South Dakota, Vermillion*

in preparation. 1973

Summarizes the accomplishments in interfacial synthesis to date. Speculates on mechanism, discusses the complex matter of synthetic control, and points out the beneficial aspects of interfacial synthesis in comparison to alternative methods of synthesis. Of fundamental concern to graduate students and teachers of organic chemistry, physical chemists, macromolecular biochemists, and polymer chemists who engage in interfacial synthesis.

CONTENTS: Stirring in organic chemical synthesis, *J. Rushton*. High-speed stirring in interfacial synthesis, *J. Rushton*. Problems and solutions in kinetics and mechanisms, *J. Bradbury and P. Crawford*. Interface effects on chemical reaction rate, *F. MacRitchie*. Liquid-vapor interfacial polycondensations, *L. Sokolov*. Copolycondensation and macroscopic kinetics, *L. Sokolov and V. Nikonov*. The role of the particle-water interface in polymerization, *J. Gardon*. Biochemical reactions at an interface, *R. Baier and D. Cadenhead*. Commercial application of interfacial synthesis, *E. Oliver and Y. Yen*. Polycarboxylic esters, *S. Temin*. Polycarbonates, *H. Vernaleken*. Polycondensations with carbon suboxide, *I. Daniewska*. Polyamides, *V. Nikonov and V. Savinov*. Polyesteramides, *I. Panayotov*. Polyurethanes, *T. Tanaka and T. Yokoyama*. Polyureas, *K. Stueben and A. Barnabeo*. Polyphosphonates, polyphosphates, and polyphosphites, *F. Millich, J. Teague, L. Lambing, and D. Hackathorn*. Other phosphorus containing polymers, *C. Carreher, Jr*. Organometallic polymers, *C. Carreher, Jr.* Modification of natural polymers by interfacial methods, *M. Horio.* Interfacial modifications of poly(vinyl alcohol) and related polymers, *M. Tsuda.* High temperature resistant polymers made by interfacial polymerization, *H. Mark and S. Atlas.*

MYERS and LONG *Characterization of Coatings:* *Physical Techniques*

In 2 Parts

(Treatise on Coatings Series, Volume 2)

edited by RAYMOND R. MYERS, *Paint Research Institute, Kent State University, Ohio,* and J. S. LONG, *Department of Chemistry, University of Southern Mississippi, Hattiesburg*

Part I 696 pages, illustrated. 1969

Part II in preparation. 1973

Explores the scientific frontier that has developed since the appearance of Mattiello's treatise on coatings. Emphasizes the urgent need of the paint industry to master new technological concepts and instrumental techniques to match the rapid pace of development of its products. Written for the working paint scientist, the laboratory assistant, technician, and superintendent. Also a valuable reference for the formulator and personnel engaged in the production of raw materials for the paint industry.

CONTENTS:

Part I: The intrinsic properties of polymers, *A. Tawn.* Surface areas, *D. Gans.* Adhesion of coatings, *A. Lewis and L. Forrestal.* Mechanical properties of coatings, *P. Pierce.* The ultimate tensile properties of paint films, *R. Evans.* Gas chromatography, *J. Haken.* Thermoanalytical techniques, *P. Garn.* Microscopy in coatings and coating ingredients, *W. Lind.* Radioactive isotopes, *G. Coe.* Infrared spectroscopy *C. Smith.* Ultraviolet and visible spectroscopy, *F. Spagnolo and E. Scheffer.* Color of polymers and pigmented systems, *G. Ingle.* Photoelastic coatings, *A. Blumstein.*

Part II: Dielectric properties, *S. Negami.* Gel permeation chromatography, *K. Boni.* Infrared Fourier transform spectroscopy, *M. Low.* Interfacial energetics, *D. Gans.* Nuclear magnetic resonance, *M. Levy.* Particle sizing, *B. DeWitt.* Scanning electron microscopy, *L. Princen.* Solubility, *J. Gordon and J. Teas.* Transport properties, *G. Park.* Viscometry, *K. Oesterle.* X-ray analysis, diffraction, and emission, *R. Scott.*

MYERS and LONG *Film-Forming Compositions*

In 3 Parts

(Treatise on Coatings Series, Volume 1)

edited by RAYMOND R. MYERS, *Paint Research Institute, Kent State University, Ohio,* and J. S. LONG, *Department of Chemistry, University of Southern Mississippi, Hattiesburg*

Part I 584 pages, illustrated. 1967

Part II 448 pages, illustrated. 1968

Part III 608 pages, illustrated. 1972

Devoted to materials which form, or aid the formation of, continuous films. Discusses vehicles and resins, placing considerable emphasis on procedures for developing a suitable vehicle for conveying dissolved or suspended solids to a substrate and imparting to the surface those protective and decorative properties for which the coating was designed. Of inestimable value to chemists, formulators, laboratory assistants, technicians, and production superintendents of the coatings industry. Also valuable for laboratory personnel of raw material suppliers, chemists in the plastics and similar industries, and as an excellent reference treatise for libraries.

Part I: Acrylic ester emulsions and water-soluble resins, *G. Allyn.* Acrylic ester resins, *G. Allyn.* Alkyd resins, *W. M. Kraft, E. G. Janusz, and D. Sughrue.* Asphalt and asphalt coatings, *S. H. Alexander.* Chlorinated rubber, *H. E. Parker.* Driers, *W. J. Stewart.* Epoxy resin coatings, *G. R. Somerville.* Hydrocarbon resins and polymers, *D. F. Koenecke.* Hydrocarbon solvents, *W. W. Reynolds.* Natural resins, *C. L. Mantell.* Polyethers and polyesters, *A. C. Filson.* Urethane coatings, *A. Damusis and K. C. Frisch.* Vehicle manufacturing equipment, *A. F. Steioff.*

Part II: Styrene-butadiene latexes in protective and decorative coatings, *F. A. Miller.* Starch polymers and their use in paper coating, *T. F. Protzman and R. M. Powers.* Cellulose esters and ethers, *J. B. G. Lewin.* Drying oils – modifications and use, *A. E. Rheineck and R. O. Austin.* Paint and painting in art, *S. Rees Jones.* Rosin and modified rosins and resins, *C. L. Mantell.* Urea and melamine resins, *H. P. Wohnsiedler and W. L. Hensley.* Vinyl resins *W. H. McKnight and G. S. Peacock.* Vinyl emulsions, *H. D. Cogan and A. L. Mantz.*

Part III: Dimer acids in surface coatings, *J. Boylan.* Emulsion technology, *L. Princen.* Phenolic resins for coatings, *S. Richardson and W. Wertz.* Plasticization of coatings, *F. Ball.* Fatty polyamides and their applications in protective coatings, *D. Wheller and D. Peerman.* Polycarbonate resins, *D. Fox and K. Goldblum.* Reactive polyesters, *F. Ball.* Varnishes, *L. Montague.* Reactive silanes as adhesion promoters to hydrophilic surfaces, *E. Plueddemann.* Surface–active agents, *T. Ginsberg.* Shellac, *J. Martin.* Tall oil in surface coatings, *R. Perez.* Silicones in protective coatings, *L. Brown.*

MYERS and LONG *Formulations*

(Treatise on Coatings Series, Volume 4)

edited by RAYMOND R. MYERS, *Department of Chemistry, Kent State University, Ohio,* and J. S. LONG, *University of Southern Mississippi, Hattiesburg*

Part I in preparation. 1973

NEUSE and ROSENBERG *Metallocene Polymers*

(Reviews in Macromolecular Chemistry Series, Volume 5, Part I)

by EBERHARD NEUSE, *F. J. Weck Company, City of Industry, California,* and HAROLD ROSENBERG, *Air Force Materials Laboratory, Wright-Patterson Air Force Base, Ohio*

170 pages, illustrated. 1970

Presents a comprehensive and critical account of the progress made in the synthesis and characterization of metallocene polymers.

CONTENTS: Introduction • Macromolecular compounds with pendent metallocenyl groups • Macromolecular compounds with intrachain metallocenylene groups • Conclusions.

O'CONNOR *Instrumental Analysis of Cotton Cellulose and Modified Cotton Cellulose*

(Fiber Science Series, Volume 3)

edited by ROBERT T. O'CONNOR, *Agricultural Research Service, U.S.D.A., New Orleans, Louisiana*

512 pages, illustrated. 1972

Describes the applications of instrumental procedures specifically developed by the textile chemist to meet today's demands. Particularly geared to textile chemists and others in the textile industry, and to paper and wood manufacturers. Also of interest to polymer chemists, analytical chemists, and other researchers involved with the processes by which fibers are blended and modified.

CONTENTS: Elemental analysis: Detection, identification, and quantitative determination of metals and nonmetallic elements, *R. T. O'Connor.* Infrared spectroscopy and physical properties of cellulose, *C. Y. Liang.* Light microscopy in the study of cellulose, *M. L. Rollins and I. V. de Gruy.* Electron microscopy of cellulose and cellulose derivatives, *M. L. Rollins, A. M. Cannizzaro, and W. R. Goynes.* Instrumental methods in the study of oxidation, degradation, and pyrolysis of cellulose, *P. K. Chatterjee and R. F. Schwenker, Jr.* X-ray diffraction, *V. W. Tripp and C. M. Conrad.* Wide-line nuclear magnetic resonance spectroscopy, *R. A. Pittman and V. W. Tripp.* The infrared spectra of chemically modified cotton cellulose, *R. T. O'Connor.*

PEARL *The Chemistry of Lignin*

by IRWIN A. PEARL, *The Institute of Paper Chemistry, Appleton, Wisconsin*

360 pages, illustrated. 1967

CONTENTS: Nebulous concept of lignin • Isolation of lignin • Chemical structure of lignin • Biosynthesis and formation of lignin • Reactions of lignin in major pulping and bleaching processes • Chemical reactions of lignin • Physical properties of lignin and its preparations • Biological decomposition of lignin • Thermal decomposition of lignin • Linkage of lignin in the plant • Utilization of lignin and its preparations.

PETERLIN *Plastic Deformation of Polymers*

edited by A. PETERLIN, *Research Triangle Institute, Research Triangle Park, North Carolina*

318 pages, illustrated. 1971

Explores the actual mechanism of deformation that occurs during the formation of fibers and films. Focuses primarily on three topics: what happens on a molecular, crystalline, and supercrystalline level; how the original structure influences the deformation; and what determines the useful mechanical properties of fibers and films. A valuable tool for the staff of research and development laboratories working with plastics, rubbers, fibers, and films, and for students and faculty involved in material science and polymer chemistry.

CONTENTS: Infrared studies of the role of monoclinic structure in the deformation of polyethylene, *Y. Kikuchi and S. Krimm.* Infrared studies of drawn polyethylene. Part I. Changes in orientation and conformation of highly drawn linear polyethylene, *W. Glenz and A. Peterlin.* Structure of oriented polyacrylonitrile films, *J. L. Koenig, L. E. Wolfram, and J. G. Grasselli.* Morphology and deformation behavior of "row-nucleated" polyoxymethylene film, *C. A. Garber and E. S. Clark.* Plastic deformation of polypropylene. VI. Mechanism and properties, *F. J. Baltá-Calleja and A. Peterlin.* Polyethylene crystallized under the orientation and pressure of a pressure capillary viscometer. Part I., *J. H. Southern and R. S. Porter.* Heat relaxation of drawn polyoxymethylene, *A. Siegmann and P. H. Geil.*

Retraction of cold-drawn polyethylene and polypropylene, *D. Hansen, W. F. Kracke, and J. R. Falender*. Electron spin resonance studies of free radicals in mechanically loaded nylon 66, *G. S. P. Verma and A. Peterlin*. Transition from linear to nonlinear viscoelastic behavior. Part I. Creep of polycarbonate, *I. V. Yannas and A. C. Lunn*. Yielding behavior of glassy polymers. III. Relative influences of free volume and kinetic energy, *K. C. Rusch and R. H. Beck, Jr*. Yielding of quenched and annealed polymethyl methacrylate, *D. H. Ender*. Yield phenomenon in oriented polyethylene terephthalate, *M. Parrish and N. Brown*. Electron paramagnetic resonance investigation of molecular bond rupture due to ozone in deformed rubber, *K. I. DeVries, E. R. Simonson, and M. L. Williams*. Factors affecting the depth of draw in a cold-forming operation, *H. L. Li, P. J. Koch, D. C. Prevorsek, and H. J. Oswald*. Quantitative structural characterization of the mechanical properties of isotactic polypropylene, *R. J. Samuels*.

RAVVE *Organic Chemistry of Macromolecules:* An Introductory Textbook

by A. RAVVE, *Continental Can Company, Chicago, Illinois*

512 pages, illustrated. 1967

CONTENTS: **Part I: Introduction:** Historical introduction and definitions • Physical properties of macromolecules • Molecular weights of polymers • **Part II: Polymerization Reactions-Mechanisms:** Addition polymerization: Mechanism of free-radical polymerization • Ionic polymerization • Polymerization with the aid of complex catalysts • Stereospecific polymerization • Bulk, solution, suspension, and emulsion polymerization. **Part III: Common Addition Polymers:** Macroalkanes • Polymers and copolymers from dienes and polyenes • Styrene and styrene-like polymers and polyacrylics • Halogen-bearing addition polymers and vinyl esters and ethers. **Part IV: Condensation Polymers:** Mechanism of polycondensation reactions • polyesters • Polyamides • Polycarbamates, polyureas, and polycarbodiimides • Phenoplasts • Aminoplasts • Ladder and semiladder polymers. **Part V: Naturally Occurring Polymers:** Polysaccharides • Proteins • Polynucleotides. **Part VI: Reactions of Polymers:** Graft and block copolymers • Reactions of polymers • Degradation of polymers.

REICH and STIVALA *Autoxidation of Hydrocarbons and Polyolefins: Kinetics and Mechanisms*

by LEO REICH, *Picatinny Arsenal, Dover, New Jersey*, and SALVATORE S. STIVALA, *Department of Chemistry, Stevens Institute of Technology, Hoboken, New Jersey*

544 pages, illustrated. 1969

CONTENTS: Introduction • Oxidation of simple hydrocarbons in absence of inhibitors and accelerators • Oxidation of simple hydrocarbons in presence of antioxidants • Oxidation of simple hydrocarbons in presence of metal catalysts • Weak chemiluminescence during hydrocarbon autoxidation • Qualitative aspects of autoxidation of saturated polyolefins • Quantitative aspects of autoxidation of saturated polyolefins • Investigation of polyolefin oxidation by various techniques.

REMBAUM and SHEN *Biomedical Polymers*

edited by ALAN REMBAUM, *California Institute of Technology, Pasadena* and MITCHEL SHEN, *University of California, Berkeley*

304 pages, illustrated. 1971

A collection of papers given at the Symposium on Biomedical Polymers held in Pasadena in 1969. Of interest to scientists in polymer chemistry, polymer physics, biochemistry, bioengineering, materials science, pharmacology, and surgery.

CONTENTS: Problems in blood-tissue reactions to polymeric materials, *B. Zweifach*. Past, present, and future of artificial kidney treatment, *B. Barbour*. The chemistry and properties of the medical-grade silicones, *S. Braley*. Correlation of the surface charge characteristics of polymers with their antithrombogenic characteristics, *S. Srinivasan and P. Sawyer*. Persistent polarization in polymers and blood compatibility, *P. Murphy, A. Lacroix, S. Merchant, and W. Bernhard*. Selection, characterization, and biodegradation of surgical epoxies, *A. Cupples and R. Schubert*. Foreign body reactions to plastic implants, *D. Ocumpaugh and H. Lee*. Rapid *in vitro* screening of polymers for bio-compatibility, *C. Homsy, K. Ansevin, W. O'Bannon, S. Thompson, R. Hodge, and M. Estrella*. Improved membranes for hemodialysis, *F. Martin, H. Shuey, and C. Saltonstall, Jr*. Control of polymer morphology for biomedical applications, 1. Hydrophilic polycarbonate membranes for dialysis, *R. Kesting*. Surgical adhesives in ophthalmology, *M. Refojo*. Medical uses for polyelectrolyte complexes, *M. Vogel, R. Cross, H. Bixler, and R. Guzman*. Potentialities of a new class of anticlotting and antihemorrhagic polymers, *T. Yen, M. Daver, and A. Rembaum*. Synthesis and properties of a new class of potential biomedical polymers, *A. Rembaum, S. Yen, R. Landel, and M. Shen*. Recognition polymers, *D. Bradley*. The challenge for high polymers in medicine, surgery, and artificial internal organs, *H. Lee and K. Neville*.

RICHARDSON *Optical Microscopy for the Materials Sciences*

(Monographs and Textbooks in Material Science Series, Volume 3)

by JAMES H. RICHARDSON, *Aerospace Corporation, El Segundo, California*

(continued)

RICHARDSON *(continued)*

704 pages, illustrated. 1971

Provides in one volume a comprehensive survey of the techniques for preparation and optical examination of specimens in the broad area of materials sciences. A highly useful text for the university or vocational school student studying the microstructure of materials or metallography and also a valuable practical reference for all researchers using microscopy, as well as for engineers and industrial metallographers.

CONTENTS: The Brightfield optical microscope • The microscopy of phase structures • Photomicrography • Photomacrography • Specimen preparation • Reagents and techniques for specimen preparation • Laboratory safety • Examination of the specimen • Accessories • Laboratory design.

ROGERS *Permselective Membranes*

edited by CHARLES E. ROGERS, *Case Western Reserve University, Cleveland, Ohio*

224 pages, illustrated. 1971

Encompasses a broad range of topics pertaining to the expanding field of permselective membranes, including theoretical aspects of transport behavior, new and unusual methods for the preparation or modification of membrane materials, and the effects of experimental conditions on the permselectivity of membranes to both ionic and nonionic penetrants. Of value to biophysicists, polymer chemists, physicists, biochemists, and chemical engineers, as well as biologists and physiologists.

CONTENTS: Transport of dissolved oxygen through silicone rubber membrane, *S. Hwang, T. Tang, and K. Kammermeyer*. Gas transport in segmented block copolymers, *K. Ziegel*. Transport of noble gases in poly(methyl acrylate), *W. Burgess, H. Hopfenberg, and V. Stannett*. Permeation of gases at high pressures, *S. Stern, S. Fang, and R. Jobbins*. Permeation of gases through modified polymer films III. Gas permeability and separation characteristics of gamma-irradiated Teflon FEP copolymer films, *R. Huang and P. Kanitz*. Theoretical interpretation of the effect of mixture composition on separation of liquids in polymers, *M. Fels and R. Huang*. Permselectivity of solutes in homogeneous water-swollen polymer membranes, *H. Yasuda and C. Lamaze*. Ion-exchange selectivity coefficients in the exchange of calcium, strontium, cobalt, nickel, zinc, and cadmium ions with hydrogen ion in variously cross-linked polystyrene sulfonate cation exchangers at 25°C, *M. Reddy and J. Marinsky*. Ion and water transport through permselective membranes, *N. Lakshminarayanaiah*. Permeability of cellulose acetate membranes to selected solutes, *H. Lonsdale, B. Cross, F. Graber, and C. Milstead*. Transport through permselective membranes, *C. Rogers and S. Sternberg*.

SCHEY *Metal Deformation Processes: Friction and Lubrication*

(Monographs and Textbooks in Material Science Series, Volume 1)

edited by JOHN A. SCHEY, *University of Illinois at Chicago Circle*

822 pages, illustrated. 1970

A comprehensive treatment of all aspects of friction and lubrication in metal deformation processes. Of aid to metallurgists, mechanical engineers, chemists, and physicists.

CONTENTS: Background and system of approach, *J. Schey*. Friction effects in metalworking processes, *J. Schey*. Friction, lubrication, and wear mechanisms, *C. Riesz*. Lubricants, *C. Riesz*. Lubricant properties and their measurements, *J. Schey*. Rolling lubrication, *J. Schey*. Wire drawing lubrication, *J. Newnham*. Hot extrusion lubrication, *S. Kalpakjian*. Forging lubrication, *S. Kalpakjian*. Cold forging and cold extrusion lubrication, *J. Newnham*. Sheet metal working lubrication, *J. Newnham*.

SEGAL *High-Temperature Polymers*

edited by CHARLES L. SEGAL, *North American Aviation, Inc., Canoga Park, California*

208 pages, illustrated. 1967

CONTENTS: Introduction, *C. L. Segal*. Thermally stable polymers with aromatic recurring units, *C. S. Marvel*. Inorganic polymer chemistry, *J. R. Van Wazer*. Kinetics and gaseous products of thermal decomposition of polymers, *H. L. Friedman*. Studies of stability of condensation polymers in oxygen-containing atmospheres, *R. T. Conley*. Thermal degradation of polymers. III: Mass spectrometric thermal analysis of condensation polymers, *G. P. Shulman*. Viscoelastic relaxation mechanism of inorganic polymers. V: Counterion effects in bulk polyelectrolytes, *A. Eisenberg, S. Saito, and T. Sasada*. Thermal stability of carborane-containing polymers, *J. Green and N. Mayes*. Synthesis and thermal stability of structurally related aromatic Schiff bases and acid amides, *A. D. Delman, A. A. Stein, and B. B. Simms*. New high-temperature polymers. II: Ordered aromatic copolyamides containing fused and multiple ring systems, *F. Dobinson and J. Preston*. Synthesis of fusible branched polyphenylenes, *N. Bilow and L. J. Miller*.

SEGAL, SHEN, and KELLEY
Polymers in Space Research

edited by CHARLES L. SEGAL, *Whittaker Corporation, San Diego,* MITCHEL SHEN, *University of California, Berkeley,* and FRANK N. KELLEY, *Air Force Rocket Propulsion Laboratory, Edwards, California.* Associate Editors: GEORGE F. PEZDIRTZ, *NASA Langley Research Center, Hampton, Virginia,* and W. DAVID ENGLISH, *Astropower Laboratory, McDonnell Douglas Aeronautics Company, Huntington Beach, California*

480 pages, illustrated. 1970

CONTENTS:

Part I: Recent Developments in the Synthesis, Characterization, and Evaluation of Thermally Stable Polymers

Introduction, *C. Segal and G. Pezdirtz.* Aromatic polymers: Single- and double-stranded chains, *J. Stille.* Thermally stable spiropolymers, *J. Hodgkin and J. Heller.* Isomeric and substituent effects in some dibenzoylbenzene-diamine polymers, *A. Volpe, L. Kaufman, and R. Dondero.* Arylsulfimide polymers. III. The syntheses of some monomeric aryl-1,2-disulfonic acids and derivatives, *G. D'Alelio, Y. Giza, and D. Feigl.* Properties of heterocyclic condensation polymers, *G. Berry and T. Fox.* Relative thermophysical properties of some polyimidazopyrrolones, *R. Jewell.* Thermal decomposition of polyimides in vacuum, *T. Johnston and C. Gaulin.* Thermomechanical behavior of an aromatic polysulfone, *J. Gillham, G. Pezdirtz, and L. Epps.* Panel discussion on thermally stable polymers, *C. Segal, J. Stille, G. Pezdirtz, G. D'Alelio, H. Levine, and W. Gibbs.*

Part II: Properties of Polymers at Low Temperatures

Introduction, *M. Shen and W. English.* Relaxation behavior of polymers at low temperatures, *J. Sauer and R. Saba.* Thermal properties of polymers at low temperatures, *W. Reese.* Multiple transitions in polyvinyl alkyl ethers at low temperatures, *W. Schell, R. Simha, and J. Aklonis.* Internal friction study of diluent effect on polymers at cryogenic temperatures, *M. Shen, J. Strong, and H. Schlein.* Stress-strain behavior of adhesives in a lap joint configuration at ambient and cryogenic temperatures, *G. Tiezzi and H. Doyle.* Some properties of nitroso rubbers in fluorine at ambient and cryogenic temperatures, *S. Toy, W. English, W. Crane, and M. Toy.* Cryogenic properties of a polyurethane adhesive, *R. Robbins.* Some effects of structure on a polymer's performance as a cryogenic adhesive, *R. Gosnall and H. Levine.*

Part III: Solid Propellants

Introduction, *F. Kelley.* Recent developments in solid-propellent binders, *H. Marsh, Jr.* Saturated hydrocarbon polymers for solid rocket propellents, *A. Di Milo and D. Johnson.* Preparation and curing of poly (perfluoroalkylene oxides), *J. Zollinger, J. Throckmorton, S. Ting, R. Mitsch, and D. Elrick.* Functionality and functionality distribution measurements of binder prepolymers, *A. Muenker and B. Hudson, Jr.*

SERAFINI and KOENIG
Cryogenic Properties of Polymers

edited by TITO T. SERAFINI, *NASA-Lewis Research Center, Cleveland,* and JACK L. KOENIG, *Case Western Reserve University, Cleveland, Ohio*

312 pages, illustrated. 1968

CONTENTS: Cryogenic positive expulsion bladders, *R. F. Lark.* Adhesives for cryogenic applications, *L. M. Roseland.* Glass-, boron-, and graphite-filament-wound resin composites and liners for cryogenic pressure vessels, *M. P. Hanson.* Mechanical behavior of poly(ethylene terephthalate), *I. M. Ward.* Effect of film processing on cryogenic properties of poly(ethylene terephthalate), *R. E. Eckert and T. T. Serafini.* Mechanical properties of epoxy resins and glass/epoxy composites at cryogenic temperatures, *L. M. Soffer and R. Molho.* Mechanical relaxation of poly-4-methyl-pentene-1 at cryogenic temperatures, *M. Takayanagi and N. Kawasaki.* Transitions in glasses at low temperatures, *R. A. Haldon, W. J. Schell, and R. Simha.* Mechanical behavior of poly(ethylene terephthalate) at cryogenic temperatures, *C. D. Armeniades, I. Kuriyama, J. M. Roe, and E. Baer.* Infrared studies of chain folding in polymers II. Poly(ethylene terephthalate), *J. L. Koenig and M. J. Hannon.* Crystallization of poly(ethylene terephthalate) from the glassy amorphous state, *G. S. Y. Yeh and P. H. Geil.* Strain-induced crystallization of poly(ethylene terephthalate), *G. S. Y. Yeh and P. H. Geil.* Molecular motion in polytetrafluoroethylene at cryogenic temperatures, *E. S. Clark.* Synthesis of ultrahigh molecular weight poly(ethylene terephthalate), *L.-C. Hsu.* Development of vulcanizable elastomers suitable for use in contact with liquid oxygen, *P. D. Schuman, E. C. Stump, and G. Westmoreland.* Synthesis of fluorinated polyurethanes, *R. Gosnell and J. Hollander.*

SKEIST *Reviews in Polymer Technology*

edited by IRVING SKEIST, *Skeist Laboratories, Inc., Livingston, New Jersey*

260 pages, illustrated. 1972

Consists of intensive, up-to-date reviews on various aspects of polymer technology Directed to chemists, engineers, technicians, commercial planners, and others working in the polymers and plastics fields.

CONTENTS: Coupling agents as adhesion promoters, *P. Cassidy and W. Yager.* Processing powdered polyethylene, *A. Zimmerman.* Plastics and other polymers in building, *I. Skeist and J. Miron.* Fire retardance of polymeric ma-

(continued)

SKEIST *(continued)*

terials, *I. Einhorn*. Recent advances in photo-cross-linkable polymers, *G. Delzenne*. Organic colorants for polymers, *T. Reeve*.

SOLOMON *Step-Growth Polymerizations*

(Kinetics and Mechanisms of Polymerization Series, Volume 3)

edited by DAVID H. SOLOMON, *C.S.I.R.O., Melbourne, Australia*

416 pages, illustrated. 1972

Presents a critical and constructive assessment of developments in step-growth polymerization and considers the application of theoretical concepts to commercial systems. Highly recommended to students of polymer science and researchers working in the area of step-growth polymerization, including polymer chemists and other scientists in the paint, coatings, and plastics industries.

CONTENTS: Polyesterification, *D. H. Solomon*. Polyamides, *D. C. Jones and T. R. White*. Polyurethanes: The chemistry of the diisocyanate-diol reaction, *D. J. Lyman*. Cyclopolycondensation, *P. M. Hergenrother*. The reactions of formaldehyde with phenols, melamine, aniline, and urea, *M. F. Drum and J. R. LeBlanc*. Diels-Alder polymerization, *W. J. Bailey*. Inorganic polymers, *J. R. MacCallum*.

STEWART *Infrared Spectroscopy: Experimental Methods and Techniques*

by JAMES E. STEWART, *Durrum Instrument Corporation, Palo Alto, California*

656 pages, illustrated. 1970

A guide to instrumentation and experimental methods and techniques for the infrared spectroscopist. Primarily for those involved with spectroscopy research and for graduate students in chemistry and physics interested in spectroscopy.

CONTENTS: Infrared spectroscopy • The infrared spectrophotometer • Elements of geometric optics • Elements of physical optics • Optical components of infrared spectrophotometers • Optical systems of infrared spectrophotometers • Slit functions and spectral modulation transfer functions of monochromators • Interference spectroscopy • Mechanics of infrared spectrophotometers • Elements of electronics • Infrared detectors • Electronic systems of infrared spectrophotometers • Electromechanical transfer functions of infrared spectrophotometers • Photometric accuracy of infrared spectrophotometers • Experimental methods of infrared transmission spectroscopy • Experimental methods of infrared reflection spectroscopy • Experimental methods of infrared emission spectroscopy • Advanced methods of infrared spectroscopy.

SZEKELY *Blast Furnace Technology: Science and Practice*

edited by JULIAN SZEKELY, *State University of New York, Buffalo*

414 pages, illustrated. 1972

Represents the efforts of academic and industrial researchers, metallurgists, plant operators, and designers concerned with the most up-to-date aspects of ironmaking technology. Valuable reading for all production engineers, plant operators, and designers concerned with ironmaking technology, and also of importance to metallurgical engineers, research scientists—chemists, physicists, engineers—and students in this field.

CONTENTS: Single particle studies applied to direct reduction and blast furnace operations, *R. Bleifuss*. Structural effects in gas–solid reactions, *J. Szekely and J. Evans*. The use of catalysts to enhance the rate of Boudouard's reaction in direct reduction metallurgical processes, *Y. Rao and B. Jalan*. Thirty psi high top–gas pressure operation at NSC Nagoya works, *T. Yatsuzuka, Y. Yamada, and A. Tayama*. Practical application of mathematical models in ironmaking, *D. Christie, C. Kearton, and R. Thomas*. A mass–transport model of erosion of the carbon hearth of the iron blast furnace, *J. Elliott and J. Popper*. Contribution to the study of the reaction mechanism occurring in high temperature zone of the blast furnace, *R. Vidal and A. Poos*. The place of direct reduction in a modern blast furnace–BOF plant, *J. Peart and D. George*. The blast furnace control problem, *J. A. Laslo*. Modern blast furnace design in Germany, *F. Lenger*. The blast furnace–a transition, *F. Berczynski*. Projected performance of a blast furnace with prereduced burdens, *J. Agarwal*.

SZEKELY *The Steel Industry and the Environment*

edited by JULIAN SZEKELY, *Center for Process Metallurgy, State University of New York, Buffalo*

312 pages, illustrated. 1973

Contains the proceedings of the Second C. C. Furnas Memorial Conference on The Steel Industry and The Environment held at the State University of New York at Buffalo in November, 1971. Brings together authors representing different viewpoints on the questions raised in examining the interaction of the steel industry and the environment. Of utmost importance to plant operators, designers, metallurgists, environmental scientists, and all others concerned with the impact of the steel industry on the environment.

CONTENTS: The role of the federal government in environmental pollution control, *K. Johnson*. Control of air pollution in the British iron and steel industry, *F. Ireland*. Health and the steel industry environment, *K. Spring*. The economic impact of the installation and operation of pollution abatement devices, *J. Barker*. Desulfurization of coke oven gas: Technology, economics, and regulatory activity, *R. Dunlap, W. Gorr, and M. Massey*. Treatment of cold-mill wastewaters by ultrahigh-rate filtration, *C. Symons*. Experience with pollution abatement, *C. Black and W. Sebesta*. The interaction of the socioeconomic and ecological environment in American steel mill towns, *L. Thaxton and R. Genton*. Plant availability versus clean air: An economic dilemma that can be solved, *R. Heller*. A survey of wastewater treatment techniques for steel mill effluents, *T. Centi*. Emission of sulfurous gases from blast-furnace slags, *R. Kaplan and G. Ringstorff*. Treatment of waste gases from the basic oxygen furnace in West Germany, *E. Weber*. On the oxidation of cyanides in the stack region of the blast furnace, *H. Sohn and J. Szekely*. Reclaimed scrap and solid metallics for steelmaking, *J. Elliott*.

TALLAN *Electrical Conductivity in Ceramics and Glass*

(Ceramics and Glass: Science and Technology Series, Volume 4)

edited by NORMAN M. TALLAN, *Aerospace Research Laboratories, Wright-Patterson Air Force Base, Ohio*

in preparation. 1973

A text which thoroughly describes several aspects of the electrical conductivity of ceramic materials, and discusses their conductivity, physical dependence on their electronic and ionic defect structures, and the transport mechanisms by which charge and mass move through ceramic materials. Additionally stressed is the use of conductivity measurements to characterize the defect structure and transport properties of ceramics. A great aid to advanced students in materials science, ceramics, and glass.

CONTENTS: General concepts of electrical transport, *D. Adler*. Experimental techniques, *R. Blumenthal and M. Seitz*. Defect structure of ceramic materials, *R. Brook*. Electronic conduction mechanisms, *I. Bransky and J. Wimmer*. Controlled valency effects in electronic conductors, *J. Wagner*. Highly conducting ceramics and the conductor-insulator transition, *J. Honig and R. Vest*. Ionic conductivity and electrochemistry of crystalline ceramics, *J. Patterson*. Conductivity of glass and other amorphous materials, *J. Mackenzie*. Microstructural and polyphase effects, *J. Wimmer and H. Graham*.

TSURUTA and O'DRISCOLL *Structure and Mechanism in Vinyl Polymerization*

edited by TEIJI TSURUTA, *Department of Synthetic Chemistry, Faculty of Engineering, University of Tokyo,* and KENNETH F. O'DRISCOLL, *Department of Chemical Engineering, State University of New York, Buffalo*

552 pages, illustrated. 1969

Presents a general survey of studies on this subject in terms of physical organic chemistry. Topics are organized to focus on the most important chemical features of vinyl compounds and their response to variations in chemical and physical circumstances.

CONTENTS: Historical development of the theory of the reactivity of vinyl monomers, *M. Imoto*. Structure and reactivity of vinyl monomers, *T. Tsuruta*. Initiation in free radical polymerization, *K. F. O'Driscoll and P. Ghosh*. Termination mechanism in radical polymerization, *A. M. North and D. Postlethwaite*. Organometallic compounds as radical-type initiators for vinyl polymerization, *S. Inoue*. Heterogeneous metal peroxides, *T. Otsu*. Polymerization of α, β-disubstituted olefins, *Y. Minoura*. Polymerization of α, β-unsaturated carbonyl compounds, *D. M. Wiles*. Cationic polymerization of vinyl monomers by metal alkyl catalysts, *T. Saegusa*. Rate constants of elementary reactions in cationic polymerization, *T. Higashimura*. Elementary steps in anionic vinyl polymerization, *J. Smid*. Molecular rearrangements in polymerization of vinyl monomers, *A. D. Ketley and L. P. Fisher*.

VOGL *Polyaldehydes*

edited by OTTO VOGL, *Central Research Division, E. I. du Pont de Nemours & Company, Wilmington, Delaware*

152 pages, illustrated. 1967

CONTENTS: Preface, *O. Vogl*. Polyaldehydes: introduction and brief history, *O. Vogl*. Polymerization of formaldehyde, *N. Brown*. Polymerization and copolymerization of trioxane, *M. B. Price and F. B. McAndrew*. Polymerization of aliphatic aldehydes, *O. Vogl*. Polymers of haloaldehydes, *I. Rosen*. NMR studies of polyaldehydes, *E. G. Brame, Jr. and O. Vogl*. Polymerization of fluorothiocarbonyl compounds, *W. H. Sharkey*. Crystal structure of polyaldehydes, *P. Corradini*. Morphology of polyoxymethylene, *P. H. Geil*.

VOGL and FURUKAWA *Polymerization of Heterocyclics*

edited by OTTO VOGL, *Department of Polymer Science and Engineering, Univer-*

(continued)

VOGL and FURUKAWA *(continued)*

sity of Massachusetts, Amherst, and JUNJI FURUKAWA, *Kyoto University, Japan*

216 pages, illustrated. 1973

Reviews the polymerization of cyclic ethers and thio ethers, lactones, and lactams. Also covers preparation, polymerization, and properties of perfluoro epoxides, kinetics of cyclic ether polymerization, and the influence of ring strain on the rate of polymerization and living polymers based on cationic ring opening polymerization. Valuable to scientists interested in polymer science, heterocyclic chemistry, polymer engineering, and materials science.

CONTENTS: Introduction, *J. Furukawa.* Polymerization of cyclic ethers, *T. Saegusa.* Polymerization of perfluoro epoxides, *H. Eleuterio.* Specific nature of the polymerization of heterocyclics, *N. Enikolpoyan.* New trioxane copolymers, *H. Cherdron.* Alkylene sulfide polymerizations, *F. Lautenschlaeger.* Lactone polymerization and polymer properties, *G. Brode and J. Koleske.* Lactam polymerization, *J. Sebenda.*

WALKER and THROWER ***Chemistry and Physics of Carbon:*** *A Series of Advances*

a series edited by PHILIP L. WALKER and PETER A. THROWER, *Department of Material Sciences, Pennsylvania State University, University Park*

Vol. 1 400 pages, illustrated. 1965
Vol. 2 400 pages, illustrated. 1966
Vol. 3 464 pages, illustrated. 1968
Vol. 4 416 pages, illustrated. 1968
Vol. 5 400 pages, illustrated. 1969
Vol. 6 368 pages, illustrated. 1970
Vol. 7 424 pages, illustrated. 1970
Vol. 8 480 pages, illustrated. 1973
Vol. 9 272 pages, illustrated. 1973
Vol. 10 288 pages, illustrated. 1973
Vol. 11 in preparation. 1974

CONTENTS:

Volume 1: Dislocations and stacking faults in graphite, *S. Amelinckx, P. Delavignette, and M. Heerschap.* Gaseous mass transport within graphite, *G. F. Hewitt.* Microscopic studies of graphite oxidation, *J. M. Thomas.* Reactions of carbon with carbon dioxide and steam, *S. Ergun and M. Mentser.* Formation of carbon from gases, *H. B. Palmer and C. F. Cullis.* Oxygen chemisorption effects on graphite thermoelectric power, *P. L. Walker, Jr., L. G. Austin, and J. J. Tietjen.*

Volume 2: Electron microscopy of reactivity changes near lattice defects in graphite, *G. R. Hennig.* Porous structure and adsorption properties of active carbons, *M. M. Dubinin.* Radiation damage in graphite, *W. N. Reynolds.* Adsorption from solution by graphite surfaces, *A. C. Zettlemoyer and K. S. Narayan.* Electronic transport in pyrolytic graphite and boron alloys of pyrolytic graphite, *C. A. Klein.* Activated diffusion of gases in molecular-sieve materials, *P. L. Walker, Jr., L. G. Austin and S. P. Nandi.*

Volume 3: Nonbasal dislocations in graphite, *J. M. Thomas and C. Roscoe.* Optical studies of carbon, *S. Ergun.* Action of oxygen and carbon dioxide above 100 millibars on "pure" carbon, *F. M. Lang and P. Magnier.* X-ray studies of carbon, *S. Ergun.* Carbon transport studies for helium-cooled high-temperature nuclear reactors, *M. R. Everett, D. V. Kinsey, and E. Römberg.*

Volume 4: X-ray diffraction studies on carbon and graphite, *W. Ruland.* Vaporization of carbon, *H. B. Palmer and M. Shelef.* Growth of graphite crystals from solution, *S. B. Austerman.* Internal friction studies on graphite, *T. Tsuzuku and M. H. Saito.* Formation of some graphitizing carbons, *J. D. Brooks and G. H. Taylor.* Catalysis of carbon gasification, *P. L. Walker, Jr., M. Shelef, and R. A. Anderson.*

Volume 5: Deposition, structure and properties of pyrolytic carbon, *J. C. Bokros.* The thermal conductivity of graphite, *B. T. Kelly.* The study of defects in graphite by transmission electron microscopy, *P. A. Thrower.* Intercalation isotherms on natural and pyrolytic graphite, *J. G. Hooley.*

Volume 6: Physical adsorption of gases and vapors of graphitized carbon blacks, *N. N. Avgul and A. V. Kiseleyv.* Graphitization of soft carbons, *J. Maire and J. Méring.* Surface complexes on carbons, *B. R. Puri.* Effects of reactor irradiation on the dynamic mechanical behavior of graphites and carbons, *R. E. Taylor and D. E. Kline.*

Volume 7: The kinetics and mechanism of graphitization, *D. B. Fischbach.* The kinetics of graphitization, *A. Pacault.* Electronic properties of doped carbons, *A. M. Marchand.* Positive and negative magnetoresistances in carbons, *P. Delhaes.* The chemistry of the pyrolytic conversion of organic compounds to carbon, *E. Fitzer, K. Mueller and W. Schaefer.*

Volume 8: The electronic properties of graphite, *I. Spain.* Surface properties of carbon fibers, *D. McKee and V. Mimeault.* The behavior of fission products captured in graphite by nuclear recoil, *S. Yajima.*

Volume 9: Carbon fibers from rayon presursors, *R. Bacon.* Control of structure of carbon for use in bioengineering, *J. Bokros, L. LaGrange, and F. Schoen.* Deposition of pyrolytic carbon in porous solids, *W. Kotlensky.*

Volume 10: The thermal properties of graphite, *B. Kelly and R. Taylor.* Lamellar reactions in graphitizable carbons, *M. Robert, M. Oberlin, and J. Mering.* Methods and mechanisms of growth of synthetic diamond, *F. Bundy, H. Strong, and R. Wentorff, Jr.*

† *Volume edited by Philip L. Walker*

Volume 11: Structure and physical properties of carbon fibers, *W. Reynolds.* Highly oriented pyrolytic graphite, *A. Moore.* Evaporated carbon films, *I. McLintock and J. Orr.* Deformation mechanisms in carbons, *G. Jenkins.*

WARD *Chemical Modification of Papermaking Fibers*

(Fiber Science Series, Volume 4)

by KYLE WARD, JR., *Institute of Paper Chemistry, Appleton, Wisconsin*

256 pages, illustrated. 1973

Bridges the gap between research and industrial applications in the field of chemical modification of papermaking fibers. Deals with the chemical changes which produce new or improved properties in paper products. Of particular importance to researchers and technologists in the paper, textile, and related industries, and students of polymer and organic chemistry.

CONTENTS: Introduction • Esterification • Etherification • Oxidation • Crosslinking • Graft polymerization onto cellulose.

WASLEY *Stress Wave Propagation in Solids: An Introduction*

(Monographs and Textbooks in Material Science Series, Volume 5)

by RICHARD J. WASLEY, *Department of Chemistry, University of California, Livermore*

344 pages, illustrated. 1973

Provides the fundamentals necessary for the study of the propagation of short duration, high-intensity, nonelastic, mechanical stress disturbances in solids. The first part of the book treats some of the dynamic analyses of elastic solid media which obey Hooke's law. The last section discusses some of the theoretical and experimental aspects of one-dimensional stress waves and shock loading and response. For scientists and engineers desiring further knowledge in the field of stress wave propagation. Also of interest to advanced undergraduate and graduate students of engineering and physics.

CONTENTS: Elasticity: Quasistatic and dynamic response • Wave propagation in extended media • Wave propagation in semi-extended media: Reflection and refraction • Wave propagation in circular cylindrical rods • Selected applications of concepts of elasticity • Nonelastic material behavior • One-dimensional stress wave investigations • Nonelastic (shock) one-dimensional strain wave investigations.

WEINBERG *Tools and Techniques in Physical Metallurgy*

In 2 Volumes

edited by FRED WEINBERG, *University of British Columbia, Vancouver*

Vol. 1 416 pages, illustrated. 1970

Vol. 2 376 pages, illustrated. 1970

Aids the non-specialist in understanding and making use of the new instruments and techniques of physical metallurgy.

CONTENTS:

Volume 1: Temperature measurement, *R. Bedford, T. Dauphinee, and H. Preston-Thomas.* X-ray diffraction, *C. M. Mitchell.* Crystal growth and alloy preparation, *F. Weinberg and J. T. Jubb.* Quantitative metallography, *J. R. Blank and T. Gladman.* Metallography, *H. E. Knechtel, W. F. Kindle, J. L. McCall, and R. D. Buchheit.*

Volume 2: Electron microscopy, *E. Smith.* Scanning electron microscopy, *O. Schaaber.* Field-ion microscopy, *B. Ralph.* Thermionic-emission microscopy, *W. L. Grube and S. R. Rouze.* Electron-probe microanalysis, *L. C. Brown and H. Thresh.* Emission spectrography and atomic absorption spectrophotometry, *G. L. Mason.*

WILSON *Radiation Chemistry of Monomers-Polymers-Plastics*

by JOSEPH E. WILSON, *Department of Chemistry, Bishop College, Dallas, Texas*

in preparation. 1974

Provides an up-to-date survey of the radiation chemistry of monomers, polymers, and plastics. Gives essential information on radiation properties, measurement, and detection, and the primary chemical results of the interaction of radiation with matter. Of particular interest to polymer and radiation chemists.

CONTENTS: (tentative): Types and sources of radiation • Fundamental effects of the irradiation of matter • Short-term chemical effects of radiation absorption • Radiation chemistry of small molecules • Radiolytic polymerization in homogeneous systems • Radiolytic polymerization in the solid state • Radiation-induced polymerization in thermosetting, polyester, and emulsion systems • Irradiation of polymers: Crosslinking versus scission • Radiolytic grafting of monomers on polymeric films • Radiolytic grafting on fibers.

YOCUM and NYQUIST *Functional Monomers: Their Preparation, Polymerization, and Applications*

In 2 Volumes

edited by RONALD H. YOCUM, *The Dow Chemical Company, Freeport, Texas,* and

(continued)

YOCUM and NYQUIST *(continued)*

EDWIN B. NYQUIST, *The Dow Chemical Company, Midland, Michigan*

Vol. 1 712 pages, illustrated. 1973

Vol. 2 321 pages, illustrated. 1973

A practical reference work which deals with functional monomers. Presents a broad technical background on the preparation and polymerization of individual functional monomers and their applications to various areas of industry. Of special interest to both academic and industrial chemists, particularly those working in the paint, coatings, and textile industry.

CONTENTS:

Volume 1: Acrylamide and other alpha, beta, and unsaturated acids, *D. C. MacWilliams.* Reactive halogenated monomers, *C. F. Raley and R. J. Dólinski.* Hydroxy monomers, *E. B. Nyquist.* Sulfonic acids and sulfonate monomers, *D. A. Kangas.*

Volume 2: Reactive heterocyclic monomers, *D. Tomalia.* Acidic monomers, *L. Luskin.* Basic monomers, vinylpyridines and aminoalkyl (METH) acrylates, *L. Luskin.*

ZIEF *Purification of Inorganic and Organic Materials: Techniques of Fractional Solidification*

edited by MORRIS ZIEF, *J. T. Baker Chemical Co., Phillipsburg, New Jersey*

340 pages, illustrated. 1969

Of interest to the chemist, chemical engineer, and metallurgist.

CONTENTS: Analysis of ultrapure materials, *C. L. Grant.* Optical-emission spectrochemical analysis—arc, spark, and flame, *C. L. Grant.* Spark-source mass spectrography, *P. R. Kennicott.* Atomic-absorption spectroscopy, *J. W. Robinson.* Infrared spectrophotometry, *K. E. Stine and W. F. Ulrich.* Gas-liquid chromatography, *R. A. Keller.* Differential thermal analysis and differential scanning calorimetry, *E. M. Barrall, II, and J. F. Johnson.* Electrical resistance-ratio measurement, *G. T. Murray.* Reduction of cyclohexane content of benzene under steady flow conditions, *J. D. Henry, Jr., M. D. Danyi, and J. E. Powers.* Purification of aromatic amines, *B. Pouyet.* The freezing staircase method, *C. P. Saylor.* Purification of aluminum, *J. L. Dewey.* Concentration of humic acids in natural waters, *J. Shapiro.* Fractionation of polystyrene, *J. D. Loconti.* Purification and growth of large anthracene crystals, *J. N. Sherwood.* Purification of indium antimonide, *A. R. Murray.* Purification of alkaline iodides (KI, RbI, CsI), *D. Ecklin.* Zone melting of metal chelate systems, *K. Ueno, H. Kobayashi, and H. Kaneko.* Purification of dienes, *R. Kieffer.* Purification of kilogram quantities of an organic compound, *J. C. Maire and M. Delmas.* Rapid purification of organic substances, *M. J. van Essen, P. F. J. van der Most, and W. M. Smit.* Investigation of zone-melting purification of gallium trichloride by a radiotracer method, *W. Kern.* Purification of potassium chloride by radio-frequency heating, *R. Warren.* Purification of a metal by electron-beam heating, *R. E. Reed and J. C. Wilson.* Heating by hollow-cathode gas discharge, *W. Class.* Continuous zone refining of benzoic acid, *J. K. Kennedy and G. H. Moates.* Purification of naphthalene in a centrifugal field, *E. L. Anderson.* Zone-melting chromatography of organic mixtures, *H. Plancher, T. E. Cogswell and D. R. Latham.* The concentration of flavors at low temperature, *M. T. Huckle.* Containers for pure substances, *E. C. Kuehner and D. H. Freeman.*

ZIEF and SPEIGHTS *Ultrapurity: Methods and Techniques*

edited by MORRIS ZIEF, *J. T. Baker Chemical Co., Philipsburg, New Jersey,* and ROBERT M. SPEIGHTS, *American Metal Climax, Inc., Golden, Colorado*

720 pages, illustrated. 1972

Brings together for the first time the four essential and interrelated parameters of ultrapurity: preparation, handling, containment, and analysis. Reflects the continuing progress in the preparation of ultrapure chemicals, the explosive growth in developments pertaining to the handling and containment of these materials, as well as the necessity for complete analysis.

Directed to all those working in research, development, or analysis of ultrapure products.

CONTENTS: Purification of alkali halides, *F. Rosenberger.* Purification of organic solvents by frontal-analysis chromatography, *H. Engelhardt.* The preparation of pure sodium and potassium, *R. L. McKisson.* Sublimation of phosphorus pentoxide, *R. D. Mounts.* Purification of proteins by membrane ultrafiltration, *G. J. Fallick.* The purification of p-xylene by partial freezing, *J. R. Gruden and M. Zief.* Purification of isopropylbenzene by preparative gas-liquid chromatography, *J. R. Gruden and M. Zief.* The preparation of ultrapure chemicals by fractional distillation, *H. Plancher and W. E. Haines.* Purification by dry-column chromatography, *F. M. Rabel.* Preparation of ultrapure water, *V. C. Smith.* Preparation and characterization of cholesterol, *I. L. Shapiro.* Contamination problems in trace-element analysis and ultrapurification, *D. E. Robertson.* Airborne contamination, *J. A. Paulhamus.* Glass containers for ultrapure solutions, *P. B. Adams.* Vitreous silica, *G. Hetherington and L. W. Bell.* Ceramics, *C. Garnsworthy.* High-purity chemicals—a challenge to practical analysis, *A. J. Barnard, Jr.* Emission spectroscopy, *E. C. Snooks.* Flame spectrophotometric trace analysis, *D. C. Burrell.* Neutron-activation analysis, *J. J. Kelly.* Visible spectrophotometry, *R. H.*

Weiss. Coulometric titration, *G. W. Higgins*. Information sources for ultrapurification and characterization, *T. E. Connolly*.

ZIEF and WILCOX *Fractional Solidification*

edited by MORRIS ZIEF, *J. T. Baker Chemical Company, Phillipsburg, New Jersey,* and WILLIAM R. WILCOX, *Aerospace Corporation, Los Angeles*

736 pages, illustrated. 1967

CONTENTS: Introduction, *W. R. Wilcox*. **Part I: Basic Principles:** Phase diagrams, *G. M. Wolten and W. R. Wilcox*. Mass transfer in fractional solidification, *W. R. Wilcox*. Constitutional supercooling and microsegregation, *G. A. Chadwick*. Polyphase solidification, *G. A. Chadwick*. Heat transfer in fractional solidification, *W. R. Wilcox*. **Part II: Laboratory Scale Apparatus:** Laboratory scale apparatus, *E. A. Wynne and M. Zief*. Batch zone melting, *E. A. Wynne*. Progressive freezing, *D. Richman, E. A. Wynne, and F. D. Rosi*. Continuous-zone melting, *J. K. Kennedy and G. H. Moates*. Column crystallization, *R. Albertins, W. C. Gates, and J. E. Powers*. Zone precipitation and allied techniques, *I. A. Eldib*. **Part III: Industrial Scale Equipment:** Proabd refiner, *J. G. D. Molinari*. Newton Chambers' process, *J. G. D. Molinari*. Rotary-drum techniques, *J. C. Chaty*. Phillips fractional-solidification process, *D. L. McKay*. Desalination by freezing, *J. C. Orcutt*. **Part IV: Applications:** Ultrapurification, *P. Jannke, J. K. Kennedy, and G. H. Moates*. Ultrapurity in pharamaceuticals, *P. Jannke*. Ultrapurity in electronic materials, *J. K. Kennedy and G. H. Moates*. Ultrapurity in materials research, *J. K. Kennedy, G. H. Moates, and W. R. Wilcox*. Ultrapurity in crystal growth, *G. H. Moates and J. K. Kennedy*. Bulk purification, *J. D. Loconti*. Analytical applications of fractional solidification, *A. S. Yue*. Materials preparation, *D. Richman and F. D. Rosi*. **Part V: Economics:** Economics of fractional solidification, *J. C. Chaty and W. R. Wilcox*. **Part VI: Appendix:** Introduction, *M. Zief and C. E. Shoemaker*. Survey of inorganic materials, *C. E. Shoemaker and R. L. Smith*. Survey of organic materials, *M. Zief*.

Solvent Properties of . Solutions

(...tant Science Series. Volume 2)

...d by KOZO SHINODA, *Department of ...emistry, Yokohama National Univer... ...ity, Japan*

376 pages, illustrated. 1967

SLADE and JENKINS
Thermal Analysis

(Techniques and Methods of Polymer Evaluation Series, Volume 1)

edited by PHILIP E. SLADE, JR., *Monsanto Company, Pensacola, Florida,* and LLOYD T. JENKINS, *Chemstrand Research Center, Durham, North Carolina*

264 pages, illustrated. 1966

SLADE and JENKINS
Thermal Characterization Techniques

(Techniques and Methods of Polymer Evaluation Series, Volume 2)

edited by PHILIP E. SLADE, JR., *Monsanto Company, Pensacola, Florida* and LLOYD T. JENKINS, *Chemstrand Research Center, Durham, North Carolina*

384 pages, illustrated. 1970

STEVENS *Characterization and Analysis of Polymers by Gas Chromatography*

(Techniques and Methods of Polymer Evaluation Series, Volume 3)

by MALCOLM P. STEVENS, *American University of Beirut, Lebanon*

216 pages, illustrated. 1969

SWISHER *Surfactant Biodegradation*

(Surfactant Science Series, Volume 3)

by R. D. SWISHER, *Monsanto Company, St. Louis, Missouri*

520 pages, illustrated. 1970

WALTON *Radome Engineering Handbook:* *Design and Principles*

(Ceramics and Glass: Science and Technology Series, Volume 1)

edited by JESSE D. WALTON, JR., *Georgia Institute of Technology, Atlanta*

616 pages, illustrated. 1970

JOURNALS OF INTEREST

BIOMATERIALS, MEDICAL DEVICES, AND ARTIFICIAL ORGANS
An International Journal

editor: T. F. YEN, *University of Southern California, Los Angeles*

The aim of this new international journal is to bridge the gap between the theoretical aspects and practical applications of artificial organs and other medical devices, and implantation materials. The basic principles responsible for the success of artificial organs are stressed in order to encourage new research in this field.

4 issues per volume

JOURNAL OF MACROMOLECULAR SCIENCE—*Chemistry*

editor: GEORGE E. HAM, *White Plains, New York*

This international journal provides scientists with a cross-section of the outstanding contributions from laboratories around the world—published four and one-half months from publisher's receipt of last manuscript. The fields covered include anionic, cationic, and free-radical addition polymerization and copolymerization, the manifold forms of condensation polymerization, polymer reactions, molecular weight studies, temperature-dependent properties, rheology, effects of radiation of all forms, polymer degradation, and many others.

8 issues per volume

JOURNAL OF MACROMOLECULAR SCIENCE—*Physics*

editor: Phillip H. Geil, *Case Western Reserve University*

A periodical devoted to the publication of significant fundamental contributions concerning the physics of macromolecular solids and liquids. Papers deal with research in transition mechanisms and structure property relationships, the physics of polymer solutions and melts, glassy and rubbery amorphous solids, and individual polymer molecules and natural polymers, as well as all the areas generally contained in polymer state physics.

4 issues per volume

JOURNAL OF MACROMOLECULAR SCIENCE—*Reviews in Macromolecular Chemistry*

editors: George B. Butler, *University of Florida, Gainesville,* and Kenneth F. O'Driscoll, *State University of New York, Buffalo,* and Mitchell Shen, *University of California, Berkeley*

Topics in this journal are reviews of certain recent chronological periods, and also have the advantage of reflecting the authors' knowledge, interpretation, and concise summary of the state of knowledge in the given area. Because of the nature of the journal, and the short time between completion of a manuscript and its publication, reviews of this nature more closely approximate a current review than could otherwise be accomplished.

2 issues per volume

POLYMER-PLASTICS TECHNOLOGY AND ENGINEERING

editor: Louis Naturman, *Stamford, Connecticut*

This journal reflects the increasing importance that polymer applications, processing developments, and mass production of new polymer products will have in the coming years. The emphasis of the articles that comprise the journal will also consider plastics technology and engineering as an important new feature of this publication.

2 issues per volume

JOURNAL SUBSCRIPTION INFORMATION

Subscriptions are entered on a calendar year basis. When a subscription is entered, it entitles the subscriber to all issues in the particular volume.

All journal subscriptions are processed after payment has been received. Only prepaid orders will receive service.

Cancellations requested for other than publisher's error will be accepted only prior to the publication of the first issue of each current volume, and will be subject to a handling charge.

Please add **foreign postage** for delivery to all countries outside the U.S. and Canada.

Air mail postage is available upon request.

Indexes to journals are bound into the last issue of each volume with the exception of review journals which are not indexed.

Back volume information and prices are available upon request.

A complimentary copy of any journal is available upon request.

Date____________________

MARCEL DEKKER, INC.
95 Madison Avenue
New York, New York 10016

Please send me a complimentary copy of the following journal(s):

____

____

____

Name____________________________________

Position__________________________________

Company__________________________________

Address__________________________________

City________________State____________Zip________